An open letter

"TO SELECTED ACADEMICS"

3

Written By Peet (P.S.J.) Schutte

ISBN--13: 978-1499354485

ISBN- ISBN-10: 1499354487

Locating Singularity

This book uses the flowing work as reference :

© KOSMOLOGIESE EN ASTRONOMIESE TEGNIKA

An open letter

TO SELECTED ACADEMICS

ISBN 0-9584410-9-X

An open letter TO SELECTED ACADEMICS is THE ACADEMIC PROLOGUE AND AN ACADEMIC INTRODUCING LETTER TO ACADEMICS PRESENTING A NEW COSMIC THEORY AS IS STATED IN MATTER'S TIME IN SPACE THE THESES, ISBN 0 – 6 2 0 – 2 7 0 4 1 – 1 **which consists of the following seven books in title:**

1) **A Cosmic Birth...Dismissing Nothing** I.S.B.N. 0-620-31609

2) AN OPEN LETTER ON: **XEPTED ASTRONOMICAL MISTAKES** ISBN 0-9584410-1-4

3) AN OPEN LETTER ON: **INTER GALACTICA SPACE TRAVEL** ISBN 0-9584410-2-2

4) AN OPEN LETTER ON: **CORRECTING COSMOLOGY** ISBN 0-9584410-5-7

5) MATTER'S TIME IN SPACE: **THE HYPOTHESIS** ISBN 0-9584410 –6-5

6) AN OPEN LETTER ON: **" STARSSTUFFN'** ISBN 0-9584410-3-0

7) **" SEVEN DAYS OF CREATION"** ISBN 0-9584410-4-9

© KOSMOLOGIESE EN ASTRONOMIESE TEGNIKA

An open letter

TO SELECTED ACADEMICS

ISBN 0-9584410-9-X

THIS LETTER IS THE ANNOUNCING OF THE BOOK IN SEVEN VOLUMES CALLED MATTER'S TIME IN SPACE: THE THESIS ISBN 0-9584410-8-1 VOLUMES 1-7

This letter was the letter that was sent to near enough eighty Universities through out the world in regard of announcing a new cosmic theory. It now is turned into a separate and individual commercial book.

TO WHOM IT MAY CONCERN,

An open letter TO SELECTED ACADEMICS ISBN 0-9584410-9-X Is THE ACADEMIC NOTIFYING OF

MATTER'S TIME IN SPACE: THE THESIS ISBN 0-9584410-8-1 Written by PEET SCHUTTE

Dear Professor,

I am Petrus Stephanus Jacobus Schutte going by the name of Peet and who is the author of the above-mentioned book(s). I hope you find your reading of this book presented as an open letter a most fruitful experience. I feel I need to warn you, the person reading this letter, that the work contained herein strays widely from mainstream science and for that there is a very good reason. However, in the least, the content is thought provoking. I researched the work of a man that is most exceptional and therefore should be placed much more prominent in the allocated position his work has in the history of mankind. His contribution in the gathering of information that furthered the entire human species in their accumulation of knowledge as well as the human understanding in cosmic affairs stands second to none in comparison to most others whilst most people are not even aware of the full implication of his work. Whilst recognising the work of Johannes Kepler, Mainstream science bluntly ignores the impact of his work, and in that, they miss the full vastness of the wide influence of his work. Newton shrouded Kepler under a blanket and every one since, kept Kepler there. It is therefore almost absolutely realistic to say that what you are about to read in this open letter sent to you for your attention was never yet printed in the near or the far past although the work has been with us for about four hundred years during which time it went unnoticed. It seems to me that any research predating Newton never came into use or into practise. My investigation of Kepler's work brought about a conclusion that no one yet arrived at concerning the findings of Kepler because no one scrutinised Kepler's formula. Kepler found planets rotating around a centre but Newton saw a circle and added what is mathematically required to indicate such a circle. Newton added a mathematical $4\Pi^2$ to the formula of Kepler and removed the distance symbolising measure that Kepler introduced using **k**. On the other side, Newton changed the symbol of **k** by using the mathematical equated symbols G $(m + m_p)$. This is just a longer and probably a more detailed manner of indicating **k** and better defining of **k** but it symbolises precisely to the point what **k** stands for nonetheless. I wish to draw your attention to the matter of Johannes Kepler's findings that Mainstream science considers as resolved and closed for many a century while it is not. My investigating Kepler helped me to resolve other unresolved matters but it was only possible by using Kepler's work.

I too am well aware that at first glance you will immediately arrive at the opinion that the theme of the letter has to be considerably below the standard of an intellectual Master such as you must be, due to the position you hold, and because of that, the normal research work you do. Nevertheless, I hope that this writing may spark interest even at such a low academic level and grade in scientific sophistication and development because I am about to prove that I discovered:

1) The location, the position and the value of **singularity** as a factor forming space-time

2) Finding **space-time** by dissecting Kepler's formula in relation to valuing singularity

3) Finding space-time, **proving space-time** and **aligning space-time** with gravity

4) The **working principals** behind and manifesting **of gravity** as a cosmic occurrence.

5) The **Roche limit**, and explaining the resulting of a law coming about from singularity.

6) The **Lagrangian system**, how and why that becomes the building form of the Universe.

7) The **Titius Bode law** and I show mathematically how gravity comes about from that

8) The **Coanda effect** and the producing of gravity through reproducing space-time.

9) The **sound barrier** by proving it **is gravity** generated **by motion** in space becoming independent where motion creates independence. Breaking the sound barrier is the motion in space duplicating space by crossing over gravity borders. It is $a^3 = kT^2$ where $(k > T^2)$ or $(k > T^2)$

I first started my studies in the field of Cosmology as a spontaneous development of my natural curiosity spawned from childhood interests in the field of cosmology, which I developed even before I went to school. The studies were a reaction (I would imagine) that was part of my personal childhood development in how I was forming a personal concept of a lifelong interest that followed me into my future. At first I conducted all my earlier studying mostly on the basis that inspired me to find out more about what made the Universe tick, with no intention ever on my part to reach a point where I would be writing books on the subject. At first I was investigating cosmology on a part time basis. This went on, on and off, or the best part of twenty odd years (*as* time and *when* time would permit). Then in later life with my health deteriorating I committed myself to more intense investigation and my effort developed onto involving a study using time that is only permitted by a person when that person is involved in such a quest on a full time basis. That quest has now been going on for the last seven years in full devotion and if one includes all the years invested on my part including the twenty odd years before, part time, then the time I have spent in completing my theory when adding all in comes down to almost twenty eight years. This is to say that I did not come to realise what I am about to introduce on a light-hearted conclusion. I mention this because I wish to ensure the reader that he should have no doubt about my most sincere commitment in producing a cosmic theory on matters concerning the start and the working of the Universe during and before the Planck era. At first I began by arguing that there is something that is blocking our progress. There is some barrier preventing humans passing a threshold whereby our understanding will pass such an obstacle. If there were any way that any one may break through that barrier there is that is preventing normal research to go pre-Big Band, it would be accomplished by finding the barrier whereby then our vision we use to focus would pass such a limit. If we wished on progress in our pursuit of the very first cosmic moment then we have to find and cross the barrier that blocks our view. We have to look deeper and in another direction should the desire driving us be strong enough to commit us to reach into the very birth of the cosmos. We have to rethink the strategy that we use. Max Planck was one of the most brilliant men of all times and even he, notwithstanding all his personal brilliance, accomplished little. There are parts missing in what we have and that which we have at our disposal to use, because if there was no such an obvious barrier then the Wise-Men involved in science would by now have found the way to break through the seal that is locking us out of the critical past which will uncover the origin of the Universe's infancy stage. I went about trying to find what everyone since Adam, (meaning all of the rest of mankind and myself) were missing throughout the ages of speculating and interpreting while philosophising about whatever we find inspirational. The obvious we saw; that was clear. Therefore I had to find a route that would lead into the not so obvious that all of us were missing, notwithstanding the best efforts of the best qualified to accomplish such a breakthrough. My effort involved trying to accommodate that which was in the cosmos available to use by the cosmos in all phases of developing. If I had any hope of finding the answer, such an answer had to be simple because I am not very inclined to unravel what is deemed as complicated. The simplicity had to be locked in what was not yet understood about that which was in the cosmos as it formed part of the process used in forming the cosmos. My realising this brought me to focus not on that which we understand. There is not a lot we actually understand because even gravity is very poorly understood. In fact gravity is so poorly understood that there is not one person alive that can claim the prestige of understanding gravity and among the dead there is even less that can make such a claim. There are several phenomena that are presented in nature and acknowledged by science but also discounted by science and therefore not presented as accepted science. By admitting that that what we have available to us to use concerning our research of cosmology in an attempt to better our understanding of cosmology, is useless to use, then one realises that not having what there might be makes what we already have useless. It then is useless to use what there is as part of the big picture we are trying to paint because what we use is not really part of the picture. This leads one to believe that the picture of the cosmos Mainstream science is painting, is being painted without painting a full picture.

In my first attempt to understand the full picture of what science was painting I found so many colours missing there was no picture painted that anyone could appreciate. This is what made me decide to go on researching the 'unknown' in the hope it might clarify the 'known' and as the book unfolds. You as the reader may agree that I was correct in pursuing the misunderstood and rejected phenomena. Finding the missing phenomena helped me to place the phenomena mentioned above in a theory where the principles also mentioned above form a part of the overall gravity used in binding the Universe. I believe what is in the Universe is not able to be coincidental because of too many influences contributing to what there is - notwithstanding the fact that that is the manner which science uses when they refer to the Bode law. What is in the Universe has a role as it had a role, which is the same role that phenomena has had and in future will have. This is establishing a very new idea about the working relationship between particles and in explaining it by using Kepler's studies. Redefining the work of Kepler's views brings a new Universe to light involving new concepts that are based on old principles but principles in updating man's view about cosmology are very new in that capacity. Through that new vision I was able to come to realise what the reasons might be why Kepler never saw it fitting to include the measure of Π in his formula. I do not suggest his neglect thereof was intentional, nevertheless the formula he devised without using Π proved that there was no need for the inclusion of Π since his figures brought about a correct answer in the final end result leaving a well concluded fitting answer. The numbers he produced brought about a specific space a^3 contained in a circle T^2 at the distance of **k** from a defining centre thus the calculations did not require the use of Π to find a meaning. In that Kepler did not see a need to include Π. I would not go as far as declaring with absolute certainty on his behalf that he did it deliberately, however there never arrived such a necessity. It is prudent to agree on whether or not such a need is necessary, because if one is agreeing about such changing not being required a new Universe emerges. The circle that Kepler discovered came about without ever forcing Π into the frame because it is clear that the circle formation came about as a natural consequence and came spontaneously delivering an equation while he was working. In this book I prove that the reason for adding Π to the rest of Kepler's formula is unnecessary. This unnecessary addition is because when going one step further in the investigation one will find that **k** and **a** and **T** are symbolising the same value with the only difference being that each one represents a different dimension to our six dimensional or six sided Universe we enjoy. In fact I shall show that Π replaces "**a**" and "**k**" and "**T**" and that Π is the true value that should be replacing each factor as to indicate the correct value to the sides nominating Π. We humans work on a numerical base using ten as a basis where we count to nine and re-establish a new decimal numbering line by adding a nought behind the number in value. This is using the numerical basis of ten, which I suspect we took from ancient knowledge about cosmology and not from using our fingers and toes as the earliest calculating processors. In this letter there is unfortunately no room to explain my suspicion but another fact I do prove is that the cosmos uses Π in the cosmic numerical basis as a means to measure and quantify. Therefore in fact the Kepler formula should read instead of $a^3 = T^2 k$ as it does it must be $\Pi^3 = \Pi^2 \Pi$ where I shall show that Π represents singularity wherefrom the entire Universe sprang from Π and by forming as $\Pi^3 = \Pi^2 \Pi$ it is confirming that space is equal to the motion thereof. Kepler's greatest achievement was showing that the cosmos is space –time $a^3 = T^2 k$ while time is the motion of space in space. The value of Π is the primeval and most basic of measures applying as an accepted cosmic legal value that the cosmos used exclusively in the very beginning and as it does today. The measure of Π in the Universe, values particle development that brought about all development ever conducted in the Universe. Only after this stage did the rest come including mathematics and went on to freeze spilled singularity into frozen material. Reading this statement may sound suspiciously senseless but as the book unfolds the sensibility will become apparent. The full implication of such a statement will become clear when one dissects different facts coming from studying Kepler. My discovery of this fundamental basis of legal valuing ensured me again that there was no need for someone the likes of Newton to add Π in any form to the work of Kepler because Kepler discovered the ultimate Π in the Universe, the Π giving the Universe form and gravity. The concept of Π that is the only single form of all other forms available that can by duplication of Πs assemble the value of gravity. When replacing the symbols with Π the facts of the Universe become self-explanatory because the most basic form that forms the cosmos has a definitive and uncompromising value.

But getting this far took me down roads overgrown by ignorance and which I had to uncover myself as if hacking away miles of overgrowth with a machete chopper. All of the disbelief science showed to my work in the past and their refusal to see past Newton made any and all attempts on my part as bad as they could be, strangling and smothering my attempts to announce my uncovering of the newly found insight on my part.

For decades I tried to come to terms with the inability there is in science to explain the cosmos in real terms, when using the science of official reputation. That which there is makes a mockery of science because the undisputable clues left in the cosmos makes what little correct explaining there is available, seem like a comedy of errors, when it is mixed in with all the other near Dark Age errors we still use after so many centuries that provided countless opportunities to revise the old muck. By applying current accepted Astronomy as such the phenomenon found all over the cosmos is still beyond the explaining ability of Mainstream science. This is true and it is a shame because it also is an undeniable fact in spite of the vast knowledge and progress in other forms of science taken in the manner science uses when it approaches cosmology. Cosmology truly lagged behind while the understanding and advancing of physics, mathematics and chemistry as subjects were flourishing. By comparison I saw how little there was available in explaining cosmic phenomenon and how much improvement in understanding the other departments such as chemistry, electronics, medicine etc. could offer as results were coming about from research. Even where there is a little explaining available in cosmology it turns out that such explaining is confusing to say the least and at best it highlights the manner in which science is applying double standards. For decades photographs were the only progress forthcoming as an addition to improve the meagre field in cosmology and that improvement was artificially stimulating cosmology. By providing a false impression of advancement, everyone missed what and how much was missing...To the connoisseur desperately looking for more than the obvious stirred in with some out-dated misinformation dating back to the Middle Ages, it all seemed as if it was a picture portraying the ridiculous to make the sublime look good. The pictures only proved the opposite of what progress in cosmology will represent. In truth and as such in cosmology the cover up that was hiding the lack of progress about the science of true cosmology was only forthcoming in the improving of electronic optical telescopic advances and spectroscopic progress. There were only photographs carrying beautiful pictures which pleased the less informed except the photographs did not bring progress to cosmology at any intellectual level by promoting insight. The explaining that the photos demanded about the subject had the opposite effect of installing hope because what it did do was underline what lack in any notable progress there truly is in our understanding of cosmology and laws in the cosmos.

While such Hubble telescopic images might seem to be as clear as daylight it was more than clear there was little academic value to them. To the person in need of more stimulation than being impressed with pictures of God's marvellous Creation and the sightseeing that always accompanies such pictures, such persons always felt very disappointed. The pictures did give satisfaction to those more easily impressed, but the rest of us seeking knowledge accompanied by understanding the images left us despondent. Although they leave the vast majority in total amazement there are those less impressed about not knowing the 'why' and the 'how' in such amazing pictures. I know the group I fall into may be the greater minority and the majority may only demand the portraying of the images, which is what that easily satisfied group demand. The rest of us rouse with anguish at the lack of information about what is known and what lies behind what those pretty pictures are conveying. Nevertheless there can be no real progress in scientific understanding about the images portrayed by the Hubble telescope, and others, if no one is able to show the slightest clue of a deeper understanding of what is going on in the Universe. Everyone is almost breathless waiting the commentating by the most informed which accompanies the magnificent cosmic portraying of God's Creation. When we are portraying the new images, we should also be investigating that what we see that the cosmos is at the moment portraying. The lack of actual believable explanation coming from investigating by means of telescopic imaging should impress one and all, but the impressing must not be based on the colours in the images but the sensible information attached to the image investigated. It is *that* that we wish to see. What we wish to see must at least be accompanied by scientifically backed information, which provides the proven understanding coming from science. When science is employing new explanations with such photos it should also be discarding senseless baggage carried over from the past. Most images contradicted Newton and for saying that, every Academic I ever came across in the past ostracized me. That bothers me little! I know I cannot possibly be the only person absolutely discontented with what Mainstream science accepts as science. Here I refer to the out of date theorising Mainstream science still accepts amongst many others as how they suggest stars and planets are forming. One cannot promote cosmology in honesty and advocate scientific fact whilst dishing up such fairy-tale nonsense to students. Moreover I hold the opinion that amongst Academics in particular there must be many if not most that share my personal serious doubts or have an inclination to share some of them. This I say when considering the overall doubtful picture painted about what there is and what one believes there should be. I just cannot believe those forming the most intellectual group of mankind are unaware of the mismatching facts seen over the broader picture because the contradiction and lack of a plan, makes what there is so very doubt provoking. Newton dismissed the formula Kepler presented as all factors forming motion. That is where the apple cart derailed.

In honesty we have to realise that we cannot dismiss the whole formula that Kepler produced as being motion. It is so much more than just motion. It is $a^3 = T^2 k$:
That is what Kepler brought into civilization for all time to come. He saw space a^3 being in isolation due to the time it uses to move T^2 claiming such space forming independence according to the lines k indicate.
Let us look at the factors in more detail before we proceed with the rest of the book.

a^3 symbolises a mathematical interpretation of implicating the three-dimensional space.
T^2 is representing the period or time that Kepler suggested we should use to calculate time that holds the orbiting planet in direct contact with the space in relation to a very specific centre.

k is the space taken from the centre to the end of the line from which the planets must have grown if one accepts the Big Bang growth of particles and the affect of the Hubble constant on all cosmos material. The specific value about the centre is most important because from the specific centre gravity always applies the strongest influence.

One cannot justify Newton's dismissing of Kepler's formula as that all factors only contribute to the motion indicated because that is misleading. We all accept that the true cosmic form *would be* and most probably *is* a sphere. Everyone accepts the Universe as a whole as a sphere...but why would the sphere form? What would be the reason why the original form that we devote to the Universe would take on a sphere as a natural form? Apparently our imagination grabs the sphere as form. In all natural events the gravity in that space which stands apart and independent from all other space takes on by cosmic pre-casting the sphere as form of shape ... **it is because gravity chooses the smallest space to hold the strongest force**.

I am of the opinion that gravity is about dismissing space to the advance of heat increasing in such a specific and concentrated space using the concentration as measure for the heat as well as the space holding the heat in space. According to Kepler that is what he found to be true. Space a^3 will always be circling space around as T^2 in any position from the centre **k**. That is what Kepler said when he said $a^3 = T^2 k$. Kepler indicated space a^3 will forever fight for independence and show separate individuality in remaining apart as identifiable cosmic components by means of motion. Every space will cling to independence indicated by **k** through fighting off the integrating of another coverall unifying unit by applying the motion of T^2! The problem we have to solve in this letter is what will the cosmos use to secure such independence between all particles? What sets space apart from the rest of space? First we have to admit that Kepler was the one that introduced the following.

Kepler gave us the answer to the following but no one ever took notice!
Kepler was the one that discovered **space / time** as $k = a^3/T^2$
Kepler was the one that discovered **singularity** as $k^0 = a^3/T^2 k$
Kepler was the one that discovered **gravity** is holding **space-time** relative by the measure of distancing **k** as $k = a^3/T^2$ and $k^{-1} = T^2/a^3$
Everyone able to read mathematics has to realise that Newton suggested collisions between cosmic structures must eventually come about as gravity erodes the distance separating the cosmic structures multiplied by the product of the mass of both structures from both ends. Newton said the multiplying mass of both structures destroys the distance between the structures by using the eroding force of gravity in the square. The cosmos then must end in a Big Crunch with all material joining together but that joining is not forthcoming at all...and that only indicates how much insufficient understanding there is on offer in cosmology by the educated–to-be-wise-about-these-matters. There is precious little available to explain about their field of cosmology amongst the ranks of Astronomers. So...let's us return to the beginning of cosmology before every one became oh so wise and see what there is to see.

The cosmos informed Kepler of another gravity, which the cosmos applies much more widely and is used by nature all over the Universe.

While we are in gravity the manner in which gravity applies in our use of gravity makes us part of the Earth by mass forcing us onto the Earth as a semi unit with all other Earth belongings. Is that which we have truly gravity?

By using mathematics, the cosmos spoke to Kepler personally and by the use of mathematics as the medium, it provided Kepler with information about the cosmos coming directly from the cosmos.

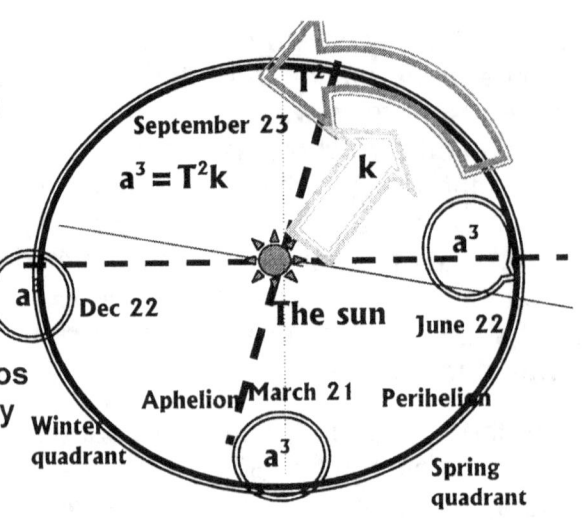

The picture we see coming from the Hubble telescope shows why, in the perfect Universe...but can the Universe be perfect when... we see a radius between the Sun and individual planets is not using a regular distance as one would expect of gravity in being a force driven by the mass and in that sense the mass is producing the gravity that always remains even because the mass doesn't alternate. As the mass is never changing on either side, that steady mass has to keep the gravity steady. But in our imperfect understanding of the Universe we find that the radius that should be constant varies considerably proving either that mass somehow adds by measure unnoticed while the structure is in orbit and later allows the same amount of mass to escape undetected; or it's the seasons adding and removing mass at will. This is an absolute contradiction to reality if mass was the factor determining the radius we find between the Sun and the planets. This suggests strongly that we'd better be getting very suspicious about the idea of mass contributing to gravity. But in contrast to this, science is unshaken about their confidence in the perfection about facts they use in terms of correctness. It is well known amongst all persons that science only uses dependable and ultra reliable facts coming from sources beyond doubt. Referring to any work done by any scientist will find a remark about science only accepting facts they use to work with. It is accepted overall by all communities that in science those in science use one hundred percent accurate facts or they use no facts at all. If our view was as perfect as science would lead us to believe it then must be the Universe that is imperfect as it otherwise would not behave so mystifyingly. The unshaken confidence science uses has us believing at first consideration that the drawing of gravity should produce an even diameter positioned between the Sun and the planets because of ever dependable evenly distributed gravity... but I believe there is a perfect Universe and our understanding carries the doubtful suspicions. Delving deeper uncovers even more contradictions and the level of accuracy contained by our scientific understanding then arouses more suspicion about the correctness of science. Remember Newton changed what the cosmos told Kepler leaving much suspicion as to how far the misdirection takes science. We have to correct the facts we doubt because when correcting the facts they use in science concerning our view about science, such correcting brings along a better understanding and then the Universe has to become ever more perfect as one learns to understand the perfect Universe even better. But it does require an open and clear mind and it needs no culture driven preconception that should confirm interpretations about facts surmised even before they are carefully studied. It becomes obvious that Newton never gave careful attention to Kepler's findings because if he did he would have seen what gravity was. Kepler described gravity without using the name that later was given as 'gravity'. Kepler did not give the name gravity, but Kepler's studies gave Kepler the insight to coin the concept of gravity. Nevertheless it was a name and not the concept that was later named by Newton. The naming was the contribution of the Englishman. The concept that Newton later introduced is totally incompatible to the concept that Kepler introduced. What he (Newton) introduced as the force of gravity, he connected to mass, which diverts totally from Kepler's findings. With giving a name, the Englishman also changed the concept that Kepler introduced. Kepler made no mention of size or mass as part of the phenominon that later was named as gravity, yet it must be gravity that holds the Universe together. The concept the Englishman changed when he introduced what he introduced with the name he introduced. That what he introduced, he corrupted beyond recognition. The concept that accompanied his new name strayed completely from what Kepler introduced. Newton brought in something that was mismatching what Kepler saw in Kepler's view of the phenominon that holds the Universe true to form. The name was dominant but even more dominant and totally inaccurate was the other concept Newton introduced. In truth Newton only gave the world a name of an idea, which he then corrupted as far as cosmic physics are concerned. It is important to admit that as far as cosmology is concerned Newton gave the concept the name but *only* the name and not the concept of gravity. Newton's persuasion on matters of gravity as gravity functions between cosmic structures orbiting one another as we find in outer space is inaccurate. What Kepler saw, Newton saw differently and used the opportunity that Kepler left by not giving any name to the process he (Kepler) and Tycho Brahe worked on for two life spans. Newton did seize the opportunity to name what he, Newton, saw but that what Newton saw did not include that which Kepler uncovered. In Kepler's era the name or title was lacking but Kepler established the concept of gravity and the formulation thereof. The concept came from Kepler even before the name gravity was used by Newton to describe in the concept of whatever we today (after Newton) became accustomed to believing what the concept of gravity is about. With the help of Newton everyone since Newton confused Kepler and Newton on the issue of gravity and this confusion even begins with Newton. Gravity might not have been named but became a proven concept and factor after Kepler formulised it, which is before Newton named it. The concept of gravity that Kepler saw is about the manner in which the structures orbit because there is a space that circles around a centre and this process has kept planets secured, connected and rotating around the Sun which is the same concept that is keeping the Universe secure and comes about with a process Newton later named as 'gravity'. What Kepler saw is not the same as what Newton saw when he saw two objects drawing closer by pulling on each others mass. Then later on Newton named, what he thought he saw as the force that Kepler saw but introduced another completely different concept. Kepler saw cyclic formations keeping the Universe together and never approaching each other. Newton ignored what he wished not to see but he changed as he saw fit and what he thought that should be. His experience as a young man drove him to establish a process he formulated as the process that is keeping the Universe together. In that act he corrupted as much as ignored the work of Kepler, which he also named as the same gravity that he saw as a young man. Why he chose to ignore Kepler's findings on gravity we shall never know but why the world still chooses to ignore Kepler's findings about gravity almost four hundred years after the fact I shall never know. My saying this has literally made Academics ignore me as they would avoid the plague. I am not pretending nor do I exaggerate when I say there were those in Academic institutions that questioned my mental development. Some went as far as seeing me as a joker of sorts and I have correspondence to show evidence to that fact. I know by now while Newtonians are reading this letter I have aroused the tempers of every Academic reading this far, therefore let's see what is being ignored by the Academics which I blame to do just that. .

We live through seasons which comes from being that at one point, (a^3_1) the distance between the sun and the earth is less than at another point we call a^3_3

Let us put a value of a^3_1 = one and a^3_3 = three. This means that each year, for the past 4 500 000 000 years the effect of the common gravity between the earth and the sun has a greater effect than at another point six months later. That means at one point the earth should be drawn or pulled closer to the sun   and after another six months interval the earth should stand less effected by the sun's gravity, therefore it   should move away from the sun.

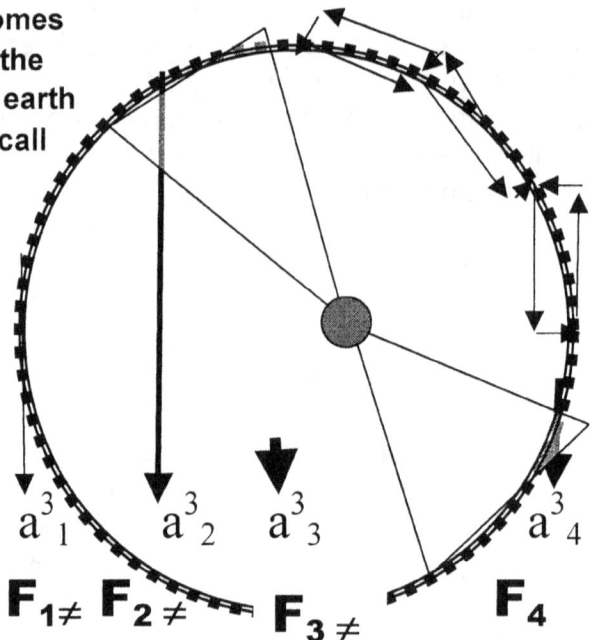

$$a^3_1 \qquad a^3_2 \qquad a^3_3 \qquad\qquad\qquad a^3_4$$

$$F_1 \neq \; F_2 \neq \;\; F_3 \neq \qquad\qquad F_4$$

May I remind you THAT NEWTON'S OWN LAWS ARE IMPLIED, and again the planets disobey these laws completely!! In the modern age all evidence point away from Newton's vision

Kepler said gravity in space is about the area a^3 that would always keep equilibrium with the time T^2 it takes to travel the distance of the full circle position placed by the indicator **k**, therefore adjusting **k** as the need arrives. With **k** shifting in length a^3 will have to readjust and therefore T^2 will find a new relating value each time. This was the finding of Kepler and came after his intense study of orbiting planets.

Before I attempt any investigation into this matter there must be coherence in our agreeing about what gravity is. If you the reader insist that the falling of objects is the only gravity found, your further reading will convince you little. Anything we do decide upon must support the fact that it is gravity that prevents planets from dislodging from the grip the Sun has on them. Gravity is not about the Sun trying to catch the Earth by attracting the Earth...no, there is so much more to gravity. We must be under no illusions about what gravity is and that being the focus of our discussion and where that gravity is because we have to identify and not confuse the gravity we are looking at. We are now discussing the gravity, which is keeping planets circling around the sun, and stars around specific galactica centre. In that we do not find one example to use as proof in connection to stars coming tumbling down on galactica centres and crushing into galactica centres. If that is gravity keeping structures in orbit around specific centres we must look at the behaviour of the structures in gravity. We have to find a reason why the planets do not reduce the radius between them as Newton suggested but we must trace the reason why it is gravity, which is keeping them apart because if anything, they are departing as they extend the radius connecting them to the sun. That is gravity because it applies throughout the Universe. The gravity Kepler found is the general gravity that is keeping structures from colliding and in that the principles are avoiding collision or on the other hand avoiding abandoning each other. It is about confirming respect for one another's independence and clearly staying at a predetermined distance while at the same time both are sharing a common space unit. That then must be the defining of gravity we have to study to find the Universal enticing gravity holding the Universe together. By close investigation one will find three factors in urgent need of investigation. There is firstly a centre that draws the object closer. This gravity is clearly a synonym to what Newton saw as gravity. If it were not drawing the object closer the object would not be orbiting around the centre and applying motion. It will draw and absorb all rotating things in its field of gravity.

The fact it does not draw the object into its ranks is because there is another gravity standing alongside this first mentioned gravity. Our recognising the first gravity forces us to accept the presence of another part of gravity. This forces us to recognise the second gravity. When saying this we are not using Newton's cosmic formula concept $F = G (M.m)/r^2$ because that can barely be what is out there happening. What Newton saw was falling. If that what Newton saw is the only gravity then whatever Kepler saw including all other parts of everything out there that are spinning around some centre must come closer to one another and connect in collisions. While that is not happening we must start to look past Newton to new grounds we can investigate. We have to go beyond Newton and admit there is more than that what Newton led us to believe because it is clear that what Newton had us believe...is not happening. That confirms the presence of the second gravity. The fact proves that everything is departing and not arriving. Even the moon is drifting away from the Earth and this information comes about from the most advanced investigation up to date, including a moon visit and the placing of measuring devices there.

Looking at the gravity intensely we find the roving structure travels in a straight line, which repeats another circle around another centre but because of the influence of a centre keeping the roving structure attached to such a centre the motion allows a circle to form by reforming motion from the original straight line to that of a partial circle. There is a centre; a connecting line travelling between what the two points establishes the specifics of a centre within a circle and the end of the circle. According to Newtonians the centre supposedly draws the rotating object closer. That is half the story.

I suggest we do some deliberation and in deliberating may I remind you THAT NEWTON'S OWN LAWS ARE IMPLIED, and again the planets disobey these laws completely!! In the modern age all evidence points away from contracting and favours eternal expanding.

The latest news confirms that the lot is apparently not coming any closer!

Kepler said gravity in space is about

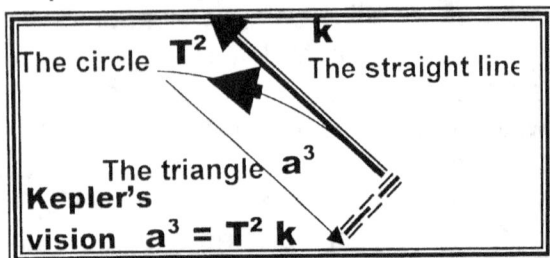

the area a^3 that would always keep equilibrium with the time T^2 it takes to travel the distance of the full circle position placed by the indicator k, therefore adjusting k as the need arrives. With k shifting in length a^3 will have to readjust and therefore T^2 will find a new relating value each time. This was the finding of Kepler and came after his intense study of orbiting planets. The line formed in rotation is never straight but always part of a larger circle. The line is eternally connected to a specific domineering centre that always insists on a circle by rotation.

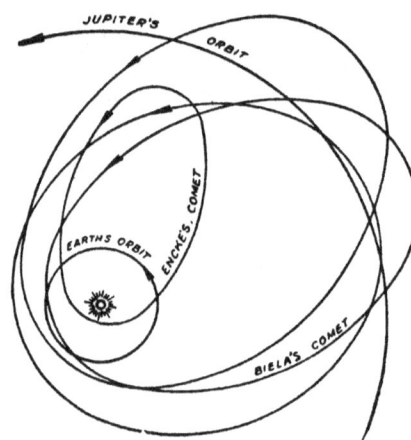

Every object in outer space holds as much turning as it commences its run in a straight line. It turns as much as it goes straight. The comets towards are equal to the comets passing. There is a space defining between the sun and the comet and the comet uses the space defining to motion around the sun on a cyclic basis. Every time the comet passes the comet is not on a death-defying route onto disaster that can or cannot be avoided. The comet is on a set path and the path has circled around the sun more times than a person has hair on his or her head. In the circle the line circling the Sun is hidden as much as there is the straight line of the circle. That is not what Newton says. Newton says the two are heading for disaster. However Kepler says quite another story than what Newton says.

> Gravity is what prevents planets escaping from the grip the sun has on them and gravity is what is forcing the planets to orbit in circles around the sun. That is what Kepler said gravity is when he investigated the solar system. Kepler said gravity is $a^3 = T^2k$ by witch the sun prevents the planets escaping out of the gravitational strangle hold the sun has on the planets. That is gravity in all possible forms gravity can have...gravity can only be motion.

In Kepler's formula $k = a^3/T^2$ the smaller k becomes the smaller a^3 becomes and the bigger T^2 then gets. T^2 represents the gravity that positions the space a^3 at distance k from the centre capturing the structure through gravity applying T^2. We all are very aware which star is the mighty gravity producer. So has mass the least say when gravity is generated? It seems most likely to be true.
$k=a^3/T^2$; The distance depends on the position that the orbiter develop space-time

> **$a^3 = T^2\, k$ then $k^3 = k^2\, k$ and this is showing that the space k^3 is equal = to the motion $k^2\, k$ of the space k^3 seen form one specific point.**

$a^3 = k\, T^2$;The space depends on the distance the space develop from the centre and the speed the space moves around the centre.

$T^2 = a^3 / k$; The speed the space orbits around the centre depends on the distance of development and the size into which the space developed.

Gravity has two factors influencing space, which is a straight line **k** and a circle going around the centre T^2. $a^3 = k\, T^2$

Translating Kepler's mathematical expression $a^3 = T^2 k$ correctly to the verbal statement in English Kepler said that there is a space a^3 which is equal = to the motion in the time duration T^2 thereof between two specific points which holds a relation to a centre wherefrom there forms a straight line **k** and is located on the spot where space begins the circle. Therefore that spot has the least space.

The value of Kepler's space he indicated as a third dimension a^3 does not depend on indicating a structure a^3 that is in rotation T^2 but only needs one position having a constant of some sorts. Any point where **k** may indicate a position one will find a value matching a^3 and the matching location will fit T^2 at that point. That is the relation there is in the solar system between all planets and the sun. The sun always indicates the centre and the planets always indicate the rotation. But $a^3 = T^2\, k$ is only producing a relevancy of three dimensions that is equal to two plus one dimension.

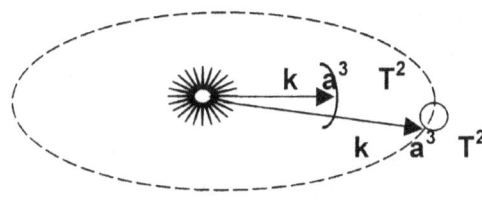

From the sun there are three points moving between two points from one point to two other points giving the six dimensions we find in space. It is space in time or space converting space through the movement of time. It is a location of a point in the third dimension a^3 that will move according to the second dimension T^2 that will implicate **k** as a reference in the first dimension. It is about dimensions in reference to one another.

Let us take it from a point where the sun provides a centre as one starting edge of **k** then that centre **k** will provide a line from the centre and the line **k** will provide three spots in a formation that produces a structure by the square T^2 of the dimension. Not once did Kepler indicate size as a contributing factor to a^3. That means every single point that **k** indicates there are three positions a^3 implicating sides of a double dimension. In the same manner is **k** not limited to distance or is T^2 lesser by size. $k = a^3 / T^2$

a^3 the fact that any symbol uses a value to the third power indicates space or a volumetric established and separate unit. It is space because it is volume using the third dimension.

T^2 is an indication of motion, moving from one point to another point or following a flat distance between two points. It is motion that is taking time in the second dimension.

k^1 is the symbol used to indicate a straight line between two points with a definite beginning and a specific end position.

Newton did not see this just because Newton declared that he (Newton) did not know what gravity is and... contrary to this... Kepler proved long before Einstein that gravity is strongest where space collapse...and the method Kepler used are so simple to read! What is there mathematically not correct in my interpretation of Kepler's manner of translating mathematics to English and why is any changing thereof by Newton or any other person necessary in any way? Test the following symbolic values in the mathematical expression and also test the principal behind the expression in which Kepler stated them. Find where the translation thereof is incorrectly presented and in such a case needed alterations to secure the correct reporting thereof. Kepler translated the cosmic information as follows:

$$a^3 = T^2 k$$

What formed the grounds for any individual to develop any need to change Kepler's translations from the cosmic given to mathematics and then from mathematics to English?
What did I not translate correctly from the mathematical expressed to English?
By my translating Kepler's work correctly I came upon answers not yet uncovered by Mainstream Science
Kepler gave the World mathematic translated cosmic answers that Kepler uncovered long before Newton, Einstein and others got wise about cosmology...
Such is the advantage of recollecting Kepler facts that it does answer many questions, which went unnoticed and therefore not spoken about up to now and some were previously never even thought about.

$a^3 = 4 / 3 \Pi r^3$ is

Newton's circle

$a^3 = T^2 k$ is

Kepler's circle

$a^3 =$ 4Π $r^3/3$

$a^3 =$ T^2 k

Newtonians should have realised centuries ago that Newton and Kepler did not have the same mathematics in mind. In the manner the two measured the circle Kepler used a different measure about the circle than Newton did.
But Kepler formulated his ideas according to cosmic information while Newton saw this formulating as mathematically being incorrect.

That is what Kepler said. There are three dimensions a^3 between any two points T^2 flowing as time from the centre of the sun, which is indicated by the line k.

The implication of the relevancy produced by the use of the formula $k = a^3 / T^2$ brings about that when dividing T^2 into a^3 there is k left.

The fact is that a^3 is a three dimension (^{3}) of single k (^{1}) showing one or T^2 is two dimensions of k being the one dimension. It means that k is a part of space a^3 or T^2 which is time. It is the same thing in a double dimension or space being a triple of k then k is one factor and k cannot show a position of zero.

If $k = 0$ then there is no possibility of $k = a^3 / T^2$ because $k = 0$ then $0^3 / 0^2 = 0$. That does not make sense. Mathematically space cannot be zero because those being of the opinion of space being zero or nothing must first prove mathematically that space is zero.

Moreover they then must prove mathematically how does zero grow through the Hubble constant. By translating Newton's vision of the circle in completing a cycle would become zero through rotation…well that does not count the use of the formula a^3. If k cannot be zero then k could not start from zero.

With $k = a^3 / T^2$ no point can be zero because k shows space $a^3 = k T^2$ is no reference to the volumetric mathematical formula used to calculate $a^3 = 4/3 \, \Pi \, r^3$.

Nor does it show the use of the circle in the second dimension being $a^2 = \Pi \, r^2$.

In the case of the Newton formula the circle factor becomes the square as indicated by the duration of the time T^2. The factor standing in for the line which normally would be r and then be the square value is in the case of Kepler not the value indicating the square. That means Kepler never indicated a circle of mathematical procedure but said mathematically the distance of the planet from the sun k holds space a^3 in relation to time T^2

Newton's mathematical vision was the way to calculate the space by using a mathematical formula used as

$$a^3 = 4\pi \, X \, (r^3/3)$$

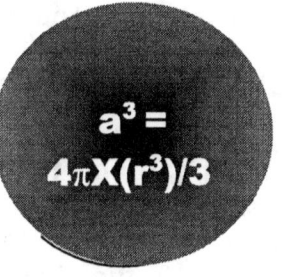

Kepler's cosmic vision was that in the formula $a^3 = k \, T^2$ the space a^3 is equal to the movement $T^2 k$ of the space, which comes about as time T^2 in relation to a distance k

The lot is more likely moving away from the sun. The lot is not coming closer!

$F = \dfrac{M_1 M}{r^2} G$ This is the suggested formula confirming the behaviour of planets used by Newtonian scholars underlining the argument that contraction is coming about between all cosmic objects.

Later Newton introduced gravity to the vision he had about gravity in his youth when he saw the apple fall from a tree and landed close to him. This event impressed him at first and later on the entire world. What he witnessed was an apple falling from a tree where both the apple and the tree was part of the Earth and not a cosmic event. In the mathematical sense it does not make sense when Newton's argument is taken out to apply in outer space.

The lot is more evidently moving further apart

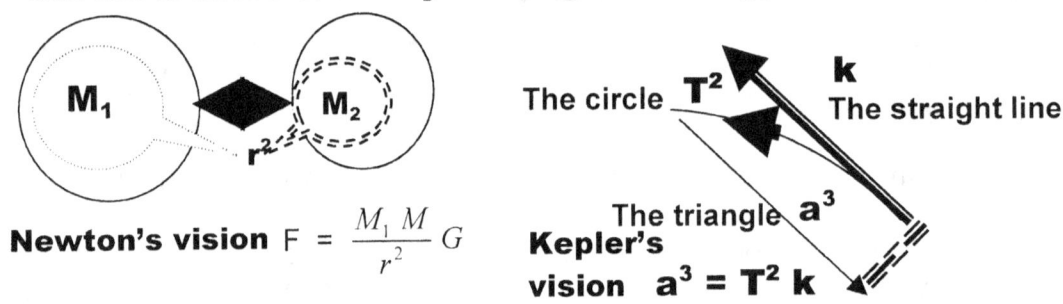

Newton's vision $F = \dfrac{M_1 M}{r^2} G$

Kepler's vision $a^3 = T^2 k$

$F = \dfrac{M_1 M}{r^2} G$ this is referring to one "force" but actually there are three:

▲ The pulling away of the smaller space. a^3

► The double counter-acting referee. T^2

▼ The pulling towards within the larger space k

Newton saw just the opposite...Newton saw both compromising their individual as well as each other's position. However, since the mass in both cases is unchanged and the mass is the factor that is establishing the force that is used by the circle to hold the radius steady and in place, these facts point to a balance that formed bringing about the above-mentioned steadiness. In the view of science, however it is the mass that draws the orbiting either objects closer or is keeping them apart. The mass does not change and since that mass of both produces the radius between both, the logic is that there has to be an even and steady radius that develops. The radius has to be equal all the time since the mass never changed throughout the rotation. The radius must be the same from any and all given points that form the rotating circle which must keep the radius equal from every angle...yet we know that Kepler proved this not to be the case even before Newton's naming and changing of Kepler's work came about.

$F = \dfrac{M_1 M}{r^2} G$ This is the suggested formula confirming the behaviour of planets used by Newtonian scholars underlining the argument that contraction is coming about between all cosmic objects. What Newton witnessed, if my memory serves me correctly was an apple falling from a tree where both the apple and the tree were part of the Earth and this did not constitute - or lead to - or come as a result of - a catastrophic cosmic event happening. In the mathematical sense it does not make sense when Newton's argument is taken out and used in outer space. What Newton saw with his falling apple was a mass influencing another mass to reduce the distance as the influencing involved motion that came about. In outer space there is another gravity where in the case of those cosmic structures in outer space there is no mass pulling each other about or pulling one another onto each other. In the case where there is particles falling from space onto the Earth, that falling also results from gravity, as much as it varies from the cosmic gravity. There is another type or form of gravity different to the concept Newton introduced. The concept Newton introduced is not the cosmic gravity Kepler formulised. What Kepler introduced is a duel where both objects are clearly in an eternal compromise therefore neither party relents its position. Newton saw just the opposite...Newton saw both compromising their individual as well as each other's position. But since the mass in both cases is unchanged and the mass is the factor that is establishing the force that is used by the circle to hold the radius steady and in place, these facts point to a balance that formed bringing about the above-mentioned steadiness. In the view of science however it is the mass that either draws the orbiting objects closer or is keeping them apart. The mass does not change and since that mass of both produces the radius between both, the logic is that there has to be an even and steady radius that develops. The radius has to be equal all the time since the mass never changed throughout the rotation. The radius must be the same from any and all given points that form the rotating circle which must keep the radius equal from every angle...yet we know that Kepler proved this not to be the case even before Newton's naming and changing of Kepler's work came about. What we see is that there is one factor that is trying to run away being a lesser space within the pulling powers of a larger space (the second factor) trying to capture and control and a referee (the third factor) is seeing to it that the even-handedness is at all times applying in the fight. That gravity which I am familiar with and know is there). In some part but not in all out representing all the gravity there might be because I cannot see the jerking, as much as I do not feel it. That is then most probably another gravity I can see and which is Kepler's gravity which **a³=T²k** represents. We have a motion of pulling...yes and that is what Newton saw...but then there is another motion of establishing a motion trying to depart, leaving the centre by tearing away from the centre and thirdly there is a motion that sees to it that the balance evolves as rotation. That is what Kepler said when he saw all three factors whereas Newton saw but one of the three. The one space is filling the next space as the space duplicates the position it had in the next moving moment that brings about the next position through motion. This eventually will have confined the next point by using a circle motion, which at first was intended to be a straight line, which is stopped by another straight line. The quest in this book is to find out why the other two factors apply in outer space as only one of the factors comes about on Earth under normal applying conditions.

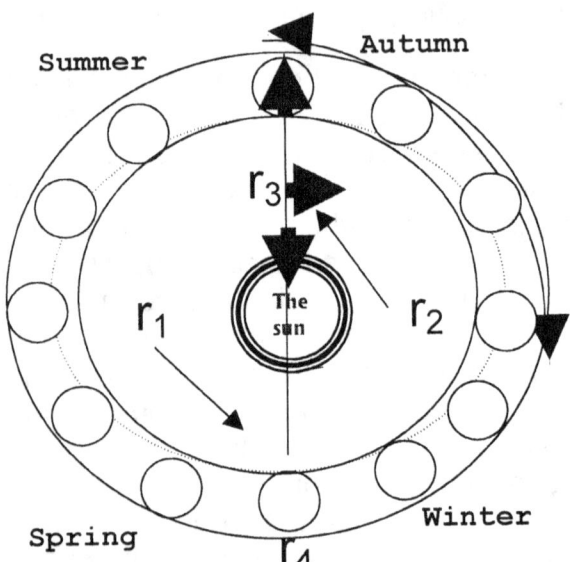

Kepler's investigation indicates to the fact that the orbiting structure is in a motion that is going on where one strength is in a fight with a second strength and the two are pretty much matching in strength because not one of the two is very much winning the dual so no one is winning or losing the fight.

As the two factors are in a motion directional dispute there is obviously one of the two factors or strengths fighting to cut loose from the other one's grip and run off. If there were not such a force trying to escape, the first force would have a quick and decisive victory by reeling in the loser just as Newton predicted. The fleeing object and its matching fighting partner has a third party referee that allows the fight to go in a specific direction as long as there is no decisive victor.

This book, which I produced in the form of an open letter, is on a quest to find the missing two factors and I can declare with some delight and with even more certainty that I found the missing factors. By Newton's introducing gravity as a force with the formula $F = G(M_1.M_2)/r^2$ a precedent was set of gravity being a contracting force forcing distances supposedly to grow smaller. Apply Newton's view to comet behaviour. Newton insists that the Sun has gravity reducing distance between the objects and while lecturers are teaching this during the day, at night they all witness how the comet follows this principle in detail showing Newton as a prophet. No sooner does the final conclusion draw near by orchestrating the final demise of the distance separating the two cosmic components when the opposite changes all concepts taught by institutions of science, the next minute out of the blue with no pre warning of the comet changing its mind, the comet defies all logic in scientific circles that apparently even included defying Newton and his logic. Because at the very point you'd think there is no chance of any return where gravity supposedly should peak because the comet is so close to the Sun and due to that fact makes the collision unavoidable...then the comet chooses that very point to dart away into the blackness of outer space, missing the definite collision by miles. By the time the collision is truly unavoidable with the radius between the Sun and the comet being as small as it realistically can be the comet starts gaining on the radius distance in spite of Newtonian denial of any possibility that such an event can in fact take place. The radius that should be shrinking further is instead enlarging. The radius that now begins to stretch proves Newton incorrect and it even depicts Newton as possibly being a fraud. The gravity applied that focussed on the comet reducing the radius between it and the Sun was not acting predictably by maintaining the reducing of the distance until collisions come about as Newton insisted on. In our reading the Newton formula in English it says that $F = G (M_1.M_2)/r^2$ which when one translates that which is said in mathematics to a verbally spoken linguistic dialect, the translation then suggests that a force is committing the material that forms the factors involved, and forcing the material into a path that is leading to a collision. It says that the two will eventually collide because of the non-retractable mass inside each one that enforces the pulling which by the mass in each case is creating the force. The unchangeable ability of the mass and the unavoidable pulling each mass creates would bring about such a collision. The mass contributes a force making a collision imminently unavoidable. The collision is beyond any attempts of diverting any oncoming objects away from the inevitable possibility of contact. The force that mass contributes is ruling out all possible evading each other or avoiding the destruction. (By enforcing a mass it created force that removes all chances from diverting away from the collision that is about to occur). Such a force then removes all possibilities of avoiding the oncoming collision. The force will not allow any attempt to try and bring into the equation other possibilities in as much as rerouting the approaching object and changing the course in the imminent collision that is due and in due course will come about between the comet and the sun. That which I explained is what Newton mathematically suggested with the formula. That is not what Kepler said notwithstanding so many arguments with Academics that I had in the past who tried to prove to me that the two visionaries views were equal and the same. Well…it's not the same because when we go onto translate Kepler to the verbal English the letters that come out do not even spell the same words.

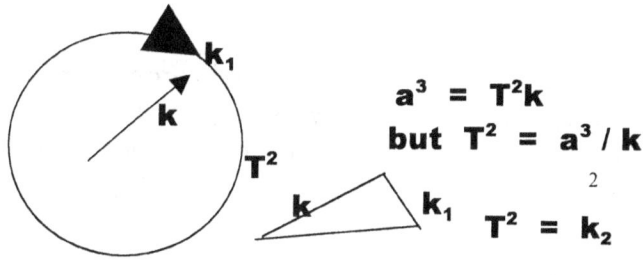

$$a^3 = T^2k$$
$$\text{but } T^2 = a^3 / k$$
$$T^2 = k_2{}^2$$

It is conducive to remember that there is another part of the two relevancies applying where one is a^3 that is relevant to k but also there is the point where k has a duty to place a relation to a^3

Translating Kepler's mathematical expression $a^3 = T^2k$ correctly to the verbal statement in English, Kepler said that there is a **space a^3** which is **equal =** to the motion in the **time duration T^2** thereof between two specific points which is a straight line **k** that holds a relation from a centre to an end where the two ends run from the beginning of **k** to connect at the end of **k.** I might not be the smartest boy on the block but I'm not that stupid either. I know how to translate… and I translate as follows:

a^3 must have a volumetric interpretation because the third dimension is sure evidence of multiple conjunctions of dimensions put together in three sides opposing three sides having the third dimension in place. The fact that any symbol uses a value to the **third power a^3** indicates **space** or a volumetric established and separate unit. Using a cube by three dimensions symbolises a cube, a room, a space to be filled, a unit able to hold other ingredients on the inside when empty or partly filled. It is space because it is volume using the third dimension.

T^2 is an indication of something having a cubic nature other than the square forming motion that is provided by the motion the square indicates, which is where the moving object is representing a third dimensional object that is moving from point to point and it is this point to point that multiplies into the square. The space is moving as a unit from one point to another point and the moving between the points are represented by a flat square or following a flat distance between two points. The cubic space was in one instant in one place and then the second instant in the other and because time can never stand still or become single dimensional (this I am about to prove as the letter unfolds) insisting that time must always support the motion it consists of or time cannot be. It is motion that is taking time, which is motion in the second dimension moving the space in the cube.

k^1 is the symbol used to indicate a straight line between two points with a definite beginning and a specific end position. It is the location where the cube is holding space and where the space was and where the cube in space is going to be in very the next split instant that follows. That will then in multiplying form the square that indicates the time the journey took to move the cube of space from one point where **k** is indicating the location of the space to where the next indicating of **k** will shift the space being the cube pointing at the end of **k.** Since time represents the square and with **k** being the distance that proves that the **k** represents the distance the space representing the cube went to take the time represented by the square through the motion. It is the distance moving space in the cube to complete time in duration in the square of motion; therefore **k** is permitted to be in the single dimension.

There are infinitely more implications in the statement Kepler delivered than what is merely a contribution to motion and only motion as Newton was of the opinion. What is there mathematically not correct in my interpretation of Kepler's manner of translating mathematics to English and why is any changing thereof by Newton or any other person necessary in any way?

We can test any of the following symbolic values in the mathematical expression and also test the principals behind the expression in which Kepler stated them. By such testing we will find that time after time there were never any corrections in the translations required since the translation thereof was never incorrectly presented and in that a case asked for no alterations to secure the correct reporting of the cosmic information being translated. By taking the formula on face value it can change as follows: $a^3 = T^2 k$ can become $k = a^3 / T^2$

That proves that the establishing of distance **k** will produce space a^3 and set space a^3 in motion T^2 where such motion is in opposition to singularity, which means gravity or contraction is the deliberate opposite of expanding $a^3 /k = T^2$. In the beginning the expanding then also involved three more points all just outside the border of singularity but within the atom exclusivity. It extends k while it introduce a returning relevancy back to singularity k^0 by creating motion in spin and duplicating space by reducing space.

With this mathematical reality what then later formed the grounds for any individual to develop any need to change Kepler's translations from the cosmic given to mathematics and then from mathematics to English while the guilty party is renowned for his superior skills in mathematics?

Kepler translated what he found to be the cosmic given to mathematics which we humans are able to interpret from the mathematical expressed to the verbally pronounced and written but Newton still saw a need to change what the cosmos said about how the cosmos is presented and by no one less than by its own interpretation of its self structured composition.

When viewing my interpreting of what Kepler said I might have asked myself countless times what did I not translate correctly from the mathematical expressed to English after encountering a battery of Academic onslaught and resentment on my Newtonian views because after all it is directly diverting strongly from the teachings presented by Mainstream science and the diverting is not coming in a small way.

In truth from my diverting I came across very new ideas I am able to prove. By my translating Kepler's work correctly I came upon answers not yet uncovered by Mainstream Science

Kepler gave the World mathematically translated cosmic answers he received from the cosmos that Kepler uncovered long before Newton, Einstein and others got wise about cosmology...and later the wise came up with old news (old views as far as Kepler expressed their views before they, the wise were born with the purpose of coming to the conclusion that those wise men eventually did) and where the conclusions that the wise concluded brought much surprise to the world with the originality of the later Masters' initiative while Kepler said the same thing ages before...!)

Such is the advantage of recollecting Kepler facts that it does answer many questions, which went unnoticed and therefore not spoken about up to now and some were previously never even thought about.

Newton said a sphere is $a^3 = 4/3 \, \Pi \, r^3$, which is mathematically correct, however

Kepler said the cosmos told him a cosmic sphere is $a^3 = k \, T^2$ There are the two distinct possibilities which Newton saw and which Kepler saw and both are most valid. Between the two concepts there is literally one Universal difference and the two can never be mistaken as promoting the same principles. 'Ever try to answer facts about the Universe in as much as...what brings about the expanding? Kepler said the Universe plus its entire content is expanding centuries before Edwin Hubble realised what he was seeing through his telescope.

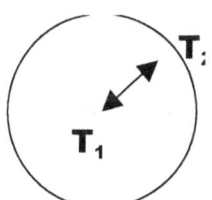

Kepler was the very first person to mathematically introduce **space a^3 centre k** and **time T^2**. Not only did he introduce **space-time a^3 / T^2** but he also placed **space a^3** and **time T^2** in a relevancy long before Einstein did and placed **gravity in space-time a^3 / T^2** even before Newton named gravity. He showed that space **k** is growing in the measure of what means the Universe attend to by promoting space-time as a^3 / T^2 = k^1. Kepler was the person who placed gravity as the ingredient in the Universe that determines **space a^3** and **time T^2** and much more. Kepler was the first one that said that gravity comprises of two factors being **k** or linear gravity and **circular gravity or T^2** as gravity keeps space in form while all is staying together.

Although not one Academic has ever openly admitted to me that they as members and part of Mainstream science are more aware than I am of all the facts and doubts I point out to them, such evidence then becomes clear whenever I mention the matter to them I get more than the impression it does not come as a surprise to them and hit them like a brick between the eyes. The lack of surprise and initial doubt they should show at first when they discover the incorrectness of evidence in their theory is a tell-tale sign confirming my suspicions about their evidently knowing all this information all along. They clearly seem very agitated about every detail I show when I bring the mistakes and double talk to their attention in the hope that they may confirm my doubts.

 Never is there a whisper of a surprise or a hint of a suggestion that would initiate an argument carried on by the bewilderment or the astonishing surprise they should feel confirming my arguments because there is a mild complacency in their voices. My jumping them total unexpectedly about matters they never contemplated in the least leaves them unturned. The rush in blood pressure that should be a factor on their part and part of the instant where total surprise will bring about some confusing thoughts that will inspire the unleashing of an argument in defending their holy grail should at least carry a surprise in an attempt to save what they believe as being the Gospel in science and with that defending their honour. They lack embarrassment, which they should have in their disputing of my claim as they fight off my allegations with a countering of denial claiming foul on my part as they are in shock when finding out about any doubts. A lack of true emotion on their part is a telling sign that they also may have some serious thoughts on the quiet about any inclination presenting a flawed view about what they always thought they knew to be true.

There is only that eerie dismissing of the seriousness and the lack they show in excitement that would deny or support my credibility as I present my findings. If they know about the inconsequential facts in science why is it not generally acknowledged and pronounced as a matter of fact? Why is there the covering up and hiding facts that we associate with some professional criminals such as politicians. The fact that Academics are aware of this evidence in general terms about the misinformation and doubting evidence about Newton's cosmic vision but moreover underlying this is their total denial of knowing about it and that is what is so seriously unforgivable. The fact that all Academics are aware of my evidence even before my presenting them with such evidence is beyond doubt. If that is the case then why are they forever trying to kill my viewpoint and forever try to silence me where I am only the messenger because I bring the solution and the answer? Please note that the answer and the solution are unbelievably simple and unsophisticated. It lacks all the splendour and grandeur expected by all Academics concerned. It is because it is so simple that it went amiss for four hundred years. It is because it is so simple that it misses the grandeur that will entice them. Instead every academic accuses me of not understanding Newton while they can't show me what part it is that I can't understand and I on the other hand can't see what there is not to understand..

Newton said that it is the reducing of the distance between the objects that would bring about the un-reversible reducing that will end in a total demolishing of the radius that is between the cosmos structures, but instead we find the gravity applying in outer space is one of the instances where gravity provides an orbit circle that gravity seems never to completed as the orbiting objects follow from closing any circle that is leading into a following circle up to where the circle is completed in cyclic precision. That is not the gravity that Newton identified although Newton admitted that there is a presence of a centre forming a point in the middle between the two objects. He was unable to know what caused or even the presence of the Coanda principle, which forms so critical a part of my theory. The formula concerning cosmic balanced gravity however leaves no room for the admitting of such a point and by not leaving a possible inclusion of such a point in his formula Newton did by such gesture in principle repeal his admission of such a centre. This had me cast doubt on what is taught at institutions of learning. It motivated me to venture back to an era before Newton came to influence science. **I came to acknowledge Kepler as I came to understand Kepler. The accepting of that what I understand in Kepler involves much more reading into what Kepler said by finding what Kepler did not say in the way that he did say what he said than the reading about what Kepler said as it is written in the precise detail and to the letter used in his statements. He never directly stated what he said. Again I must stress this point: when I refer to what Kepler said it most likely means reading into the part that is being a part of the part that he did not say when he was saying what he said but I accept that he meant to say what I am reading and translating from Kepler as part of what he did not say but meant to say.** I have to read more with my mind than with my eyes. This comes as a result of interpreting Mathematics to the verbally expressed. I had to learn to read with my mind and not my eyes and I found that that is the manner in which one has to approach cosmology. From the first time I discovered what manner one should use if one wished to read into Kepler's findings I saw Kepler was all about uncovering the unknown. Realising that, the conclusions I drew by reading in such a way cemented my better understanding of Kepler's work, which then helped me improve my insight into Kepler's work as it increased my understanding about cosmology several fold. This helped me to realise what implications were to be found underneath Kepler's discoveries. From my realising what approach I should use, it helped me to improve my cosmic realising by using the method of reading Kepler and from that I could come to appreciate what Kepler introduced.

Only then did it bring insight and proof to me as a student of Kepler and this proof I found by dissecting what Kepler **did not** say instead of what he **did** say, which I now present to you with this letter, you being a superior intellectual person. Kepler said $a^3 = T^2 k$ and that correctly translates to a mathematical expression $k^0 = a^3 / T^2 k$ which in the verbal statement in English translates that Kepler said that there is a **space a^3** which is **equal =** to the motion in **the time duration T^2** thereof between two specific points which holds a relation onto a centre k^0 where from there forms **a straight line k** that is centred on the spot where space begins from k^0 **that produces k** as well as producing the circle therefore that spot $k^0 = a^3 / T^2 k$ has hold k^0 at a value of having the least space. The line **k** is centred onto a spot where space begins specifically at k^0. This point not only produces the line k^0 but represents also the space that forms the eventual circle T^2. Therefore from the centre holding k^0, k^0 leads to **k** that forms the roving space a^3, which is rotating at a distance **k** where T^2 forms the outer limit of k^0. Mathematically $a^3 = T^2 k$ will be $k^0 = a^3 / (T^2 k)$ because $k^0 = 1$. But $k^0 = 1$ also present the single dimension where all factors are a product of one. If one can locate k^0 one will find singularity. That is where gravity is because gravity is strongest where space is least. Then that suggests that gravity is strongest at k^0 because space is least. That is gravity because that is what keeps the orbiting object in orbit but also that is what Newton completely missed when he changed Kepler's work. Newton failed to recognise gravity as the only ingredient in Kepler's formula. He admitted he missed this because he admitted he did not know what gravity is while Kepler explicitly showed what gravity is. Gravity is what keeps the orbiting object orbiting. $k = a^3 / T^2$ is **distance^1** = **space** ^{3}/ **time**^{2} forming from a pivoting centre k^0. That is a cycle and moreover it is a cycle formed **by space/time**. What Kepler said is that space is a^3 **in motion T^2 k.**

That says **space^3 (a^3)** relates directly to **time^2** that uses the symbol T^2. This is also what I refer to when I say one has to read what Kepler did **not** say when one wishes to see what he **meant** to say. Kepler introduced space^3 –time^2 long before Einstein's date of birth appeared on any calendar although Einstein is credited with the formulating of the concept of space-time and giving it a name. Going even further Kepler stated that the space a^3 is on the move T^2 around in a circle at a distance **k.** That is what that comet we are discussing is doing. The space^3 (Comet) is circling the Sun using a radius **k** to establish the cyclic time^2 as a period of continuous motion and continuous motion is gravity. That reads much more correctly and closer to the truth than what Newton predicted what according to him (Newton) was happening in space. Remember in this statement I am separating cosmic principles applying from the way that gravitational principles apply on Earth. I distinguish that which is the rule in the cosmos from what we find ourselves trapped in on Earth. The two just don't mix. I am removing cosmic physics from normally accepted physics because the gravity concerned is not the same.

The proof I bring is real however simple it may seem. It has none of the mind-blowing complexities normally associated in the presenting of investigative analyses of Astronomy. I realise the information in this book carries the arguments in a childlike manner which are very simple to follow, and for that in the past I have been blamed over and over again as being unprofessional. In my answer to that I can only reply by using another question: Are only professionals adequately equipped with minds that make them (the professionals) the only ones able to think? We being part of the human race are all thinkers. Everyone as a human being can think. Every person on Earth is a thinking thinker that uses his brainpower by exploring thoughts mainly and normally to his or her personal benefit. It is what we think about that produces the results of our efforts by which we accomplish what ever we are thinking about. I have met professional Academics that I found foolish as much as there are other cases where the so-called amateurs can credit themselves with much more wisdom and insight. Albert Einstein as a patent clerk was that much but to name one. Please understand that I do not compare my achievements or myself in any way, shape or form with the likes of a Master such as Einstein although I speak my mind when not being totally in agreement with some of his or other views. My unsophisticated retracing of Mainstream physics concerning the Big Bang in detail helps to reinvestigate established principles and moreover investigate proof in the light of modern evidence. In principle I distinguish between Kepler and Newton in that Newton is one hundred percent correct concerning gravity on Earth but as far as outer space forms gravity the conclusions of Kepler and Newton do not match and they had totally different ideas about what they saw in gravity. I am in disagreement with some basic principles that science acknowledges and I divert strongly from all accepted roads Mainstream physics follow. My doing that promted those who are considered and accepted as self-proclaimed members of Mainstream Physics have categorised my views in the past as incoherent. That I do not accept. I admit that my line of thought is extraordinary and controversial but only to Mainstream science and not to the standards lay down by nature. Since the concepts I follow start at the beginning, and I take Kepler at the point where modern cosmology began and in that mindset I re-evaluate Kepler's work. I start by tracing a new approach as to what I see Kepler found. The main condition of my investigation is to establish a divorce between what Kepler said and what Newton thought to add to what Kepler said. It is this divorce I create that Mainstream science finds repugnant or even in some persons' opinion repulsive. I believe the repugnancy does not come from or is not manifested in any part of my work to the letter as such, but rather what my work suggests and who is doing the suggesting. To my view in cosmology such adding to Kepler by Newton was unnecessary and it diverts Kepler's work away from cosmology. But as the generations moved on Newton became religiosity in the mind of science wherever science was taught. To students there is little or no choice in the matter since the only choice left to them is one of understanding by forcefully accepting or die an academic death since Newton is academically accepted without asking questions or raising an opinion. For the second choice, the less accepting students are greeted with a Dear John good-bye letter sending them off into the unknown sunset that such a future outside physics will bring them. That is brain washing.

From studying Kepler I saw that we have to gauge what we find in the Universe. What we find is not that what we realise with our eyes but that what we observe by using our minds to translate from visions coming from our eyes to our minds. We have to test the part that we are seeing much more than merely accept what there is to see on face value. We have to not only see what other life beings blessed with much less insight most probably also should see. We must stop using our eyes in the same manner as animals do and start seeing with our mind, as humans should do. By being the superior evolved species that we are, it gives us the ability to read into that which only we can see and that we only can see by using our intellectual mindset. By seeing with an intellectual understanding what there is to see when we see what we can observe, we should therefore have the ability to be in understanding by looking at what we can see but moreover understand that which we cannot see. It is the same as playing chess. See what there should be moved instead of noticing an object not having an ability to move by own initiative. This I first found to be true about Kepler's work and when I started projecting this method of observing what the Universe is, as it scattered most previous perceptions I found that using the new method brought along answers so fast I could sometimes hardly keep up with the interpreting thereof. But as is the case with Kepler so is the case with the entire study of cosmology: One should see what there is about the cosmos, which is unseen to us and then we may find so much more in the cosmos unseen to us representing that which we cannot see and that which we cannot read because we have to learn to read what is not written in light. Armed with this realising I then proceeded from that point by further arguing and debating the full implication of Kepler's contribution. Kepler placed cosmic structures in relevance to one another and so does the Big Bang Theory. The backbone of the Big Bang is that relevancies apply in dynamics and such dynamics are placing all structures without any reservations independent from each other. As the Big Bang progresses all filling the Universe to the inside is in the same Universe that was then at the time of the Big bang just as much as it will always be and the lot remain the same, however the relations that the elements comply to bring across new relevancies with new positions to fill. The father of the Big Bang concept is a person by the name of Father LE MAÎTRE, GEORGE ÉDOUARD (1894-1966) who was a Belgian priest and cosmologist. He was the first person to embrace the fact that the Universe expanded from an infant stage. His model of an expanding Universe (1927) was superior to that of W. de Sitter in that it took into account mass, gravitation and the curvature of space. Similar models were proposed in the early 1920s by the Russian mathematician Alexander Alexandrovich Friedmann (1888-1925) but Friedman compiled various such possibilities. Lemaître argued further (1931) that the quantum theory supported an origin in the explosion of a 'primeval atom' or 'cosmic egg' into which was originally concentrated all mass and energy. As modified by A.S. Eddington, Lemaître's model provided the springboard for G. Gamow's Big Bang theory. In the wider picture of science in general a lot changed to just allow such turnabout in thought since the day of Isaac Newton. From Newton's attraction and contraction many things came into place that allowed change in the most hardened minds. Accepting facts about the Big Bang concept is quite radical. By promoting expansion the Big Bang theory contradicts gravity and our accepting of the Big Bang has to change all other concepts. By accepting the Big Bang other changes are also involved.

KEPLER, JOHANNES (1571-1630)
The German mathematician and astronomer KEPLER, JOHANNES (1571-1630) became Tycho Brahe's assistant in Prague in 1600 A. D. where he undertook to complete the tables of planetary motion Tycho had begun. Kepler first calculated the orbit of Mars. He spent much time trying to reconcile Tycho' s accurate observations of the planet with a circular orbit, but concluded (in Astronomia nova,

published in 1609) that Mars moved instead in an elliptical orbit. Thus, he established the first of his laws of planetary motion. A theory that the Sun controlled the planets by a magnetic force led him to the second and third of his laws, which were published as part of his treatise on theoretical astronomy, Epitome astronomiae Coernicanae (1618-21). The Rudolphine Tables (named after Tycho's patron, the Holy Roman Emperor Rudolph II) of planetary motion appeared in 1627 and were still in use in the 18th century. Kepler also wrote De Stella nova, on the supernova of 1604 and Diptirce on optics and the theory of the telescope. The overall view followed in this book **an open letter To Selected Academics ISBN 0-9584410-9-X** places the true significance of his work in true contents. In KEPLER'S EQUATION is the equation that relates the eccentric anomaly of a body in an elliptical orbit to its mean anomaly. The equation is $E - e \sin E = M$., where E is the eccentric anomaly, M the mean anomaly, and e the eccentricity of the orbit. It is important as one of the mathematical relations enabling the position of a planet about the Sun, or a satellite about is planet, to be calculated from the orbital elements for any time. However this only relates to the solar system, and KEPLER'S LAWS only apply in the contents of the solar system. The three laws governing the orbital motions of the planets, discovered by J. Kepler is as follows: The first law states that the orbit of a planet is an ellipse with the Sun at one focus of the ellipse. The second law states that the radius vector joining planets to the Sun sweeps out equal areas in equal times. The third law states that the square of the orbital period of each planet in years is proportional to the cube of the semi major axis of the planet's orbit. The first law gives the shape of the planet's orbit; the second describes how the planet must continuously vary its speed as it follows its orbit, moving fastest at perihelion and slowest at aphelion. The third law gives the relationship between the planets' average distances from the Sun and their periods of revolution.

Instead of studying the true value and contribution of to Kepler's laws an Englishman going by the name of I. Newton placed his own interpretation to Kepler's laws, and in doing this, he wilfully destroyed the principle working of the Creation. Saying this I hear the alarming hooters announce Newtonian dismay. In the past my experience was that all the revered Academics lost their appetite for any further investigation of my work. That is sad as much as it is regrettable. Through Newton's tunnel vision, he applied his own misinterpretations to the correct presumptions of Kepler and through the Newtonian tunnel vision Academics did not move an inch away from repeating the same procedure. In the past it was this that had Academics shying away from me because at the point where I raise criticism of the Newtonian viewpoint I am rejected. The point where I declare my suspicions concerning they're accuracy and the correctness about their theories, which is where I should then be raising their doubts about their way of thinking is the point where in stead I raise their suspicions about my way of thinking. That is what caused the rejection of my criticism about Academic Newtonian science and evoked their criticism in the past about my views instead of them following the logic by investigating what I said. Their rejection of self-investigating got me and my work rejected to a point where the applecart lost its wheels on every occasion. It is where Academics read my remarks and what brings (seemingly in an instant) wrath to Academics. I say this because I realise that reading my remarks or hearing me remarking about this notion brought much resentment on their part and if the reader at the present moment is a Newtonian, boiling his/her blood. It is blood boiling because I believe they see my remarks as belittling that which they feel they have accomplished. This is not the case but still my remarks have the same effect on the Academic as pouring icy cold water down the back of his shirt. I mention this because I know it has happened many times before and if possible I wish to avoid this response. Therefore I ask you kindly to please be warned about the negativity you must feel towards me where you are the Newtonian and I am not. Before you lose interest in reading this letter any further please allow me to finish. In the past Academics thought me to be presumptuous and that normally became the point where all the Academics find their interest vanishes. That should not be because if Newton's work is as utterly accurate as those with faith in his work believe it is, then every aspect about Newton should stand above any and all reprimanding or any form of doubt causing a notion to reprimand. The testing of Newton's work should withstand all testing notwithstanding the person or the prominence of such a person's social or academic standing in the Academic society or even the prominence that such testing will deliver. From what I see about Kepler's work it is a flow of circumstances that lead to Academics neglecting Kepler's work and the realising of the theory I suggest is not forthcoming due to my personal brilliance. I do not consider myself to be the brilliant in any way as to be the one that can remove the verbal splinter from the eye of the Academic. Yet…if there is a splinter what else should I then do…Newton reduced the implication that Kepler findings hold by introducing to the law of gravitation. He then went about and changed it to three laws of motion. It is clear that while he formulated the laws on motion he missed the way Kepler introduced gravity as space a^3 coming about through motion T^2 and that gravity is space a^3 within space k within motion T^2. Newton also missed the fact that gravity is at its strongest where motion and space cease to be. This is most important to recognise about gravity in one of the two forms it has. I. Newton generalized Kepler's first law, verified the second law, and showed that the third law should be amended to the form; $4 \pi^2 a^3 / T^2 = G (m + m_p)$. In this, the value of "T" and "a" are the period of revolution and semi major axis of the orbit of a planet of mass m_p about the Sun of mass m, and G is the gravitational constant.

It should be clear to any person investigating Johannes Kepler and his work that Isaac Newton hijacked Kepler's work and any time there is the slightest referring to Kepler about the research Tycho Brahe and Johannes Kepler did such referring to Kepler always lead to and always include the mentioning of Isaac Newton changing the work of Johannes Kepler. It is as if the World never could acknowledge Johannes Kepler because the work of Johannes Kepler would be completely wrong and misleading if it were not for the intervention of Isaac Newton saving the skin of the less admirable Johannes Kepler. This comes in the midst of every one realising that Kepler used the information he received directly from the cosmos. I do stress this on many occasions throughout the letter because the embarrassing part is that Newton changed the work of The Universe and not of the man called Kepler. Should you reading the letter entertain the opinion of Newton and feel any urge to defend Newton you should ask the question as to who is standing corrected, is it Kepler or is it the cosmos that gave Kepler the information he concluded? The cosmos supplied all the information by using mathematics, which Kepler then had to translate. But Newton destroyed the accuracy by altering what the cosmos said and directly by adding to that what he (Newton by name) thought that the cosmos left out. This set a precedent by

Newton in cosmology and also set a trend, which was retained in all future cosmological development and it lasted in cosmology for three hundred and fifty years. In this book you are reading, I am about to show that such practise should no longer be accepted in cosmology. In the process the world of Mathematics developed by the world of cosmology stood still for almost four hundred years. Faculties contributing to cosmology and feeding off cosmology improved as much as they developed, but when cosmologists see the Roche limit in action in the lens of the Hubble telescope and refer to the event as "stars blowing bubbles" being the ultimate response coming from those persons who are supposedly the Masters of cosmology affairs, then the truth of what I just said comes down on you like a ton of bricks. Everyone having any remote interest in cosmology will find they are being very disillusioned by such "official" testimony about the evidence the Ultra Wise report about. This book is about showing how great Johannes Kepler was and how enormous his work was. It will show he preceded all ideas of everyone that came later and officially introduced the novelty of such ideas. Back during the time Kepler was introducing his work the stature and the magnitude of his work was beyond any person's understanding (including Isaac Newton) and this prevailed for most of half a millennium. I do not say I am the brilliant one to uncover Kepler in the face of everyone failing that came before me, but as I am not a Newtonian such bias was not part of my repertoire and denying me the fortune of being a Newtonian added to my fortune of realising Kepler. Yet as you will notice, the work I contribute is much below the sophisticated norm of modern investigative research and the levels that modern research accomplishment demands to better the effort of the understanding ability in the splendour that investigative research work should deliver in view of our modern times. It is only pure neglect in science circles that moved science past Kepler. Not seeing and therefore not investigating through almost half a millennium has paved a road past the inferior levels that the researching of Kepler's work holds because it was rocket science four centuries ago but the brilliance of it has faded since then. My contribution holds no astonishing flair that may add to science in general. Only failure to notice what I see on the part of those truly brilliant can explain my being able to present my contribution about my work in investigating Kepler. Only by their passing such degrading levels of the Academic establishment in the past and the present can bring the blame for such an obvious discrepancy because any involvement in the work at such an inferior level as that which I bring cannot interest and excite a salted Academic and when thinking about it, the idea is totally unthinkable. This letter, although it is on this inferior level is about correcting this tendency and has in mind the effort to put in writing what would place Kepler in the greatness and glory he deserves. As I already said, if Kepler was wrong then the cosmos was wrong about facts and applying relevancies and tendencies in the cosmos. I yet again wish to reiterate we should never for one moment forget that Kepler received his information directly from studying the cosmos so how could the cosmos stand corrected? In spite of all the brilliance attributed to Newton nonetheless if Newton had the mind to change Kepler's work and my saying this includes all persons agreeing with such changing by Newton of the work of Kepler those persons admit that he or she or Newton never took any time to really and truly investigate what the cosmos told Kepler. Through understanding the work of Kepler I prove gravity, the Titius Bode law, singularity, space-time, space-time relevancy, the Lagrangian system, the Coanda effect and the Roche principle, the sound barrier, the principle behind the

Black hole. The precondition for my ability in doing so is that I have to remove Newton's opinion about Kepler's work from Kepler's work. Whenever cosmology comes into question and all the phenomena, which I mentioned just now remains unexplained and by that token alone it shows to what degree did cosmology remain undeveloped. Whenever there is any mention of Newton, Kepler is never mentioned. But the reverse is always applying. Mainstream physics holds the opinion that Kepler may only have an opinion if Newton can change the opinion. Kepler gave space-time, gave gravity, gave singularity, gave the Plank theory, gave the theory on relativity but no one ever found Kepler's work deserving enough to launch any investigation such as I did. I belabour this because of what revulsion my rejection of Newton unleashed. That is one barrier much unnecessary but it has been an insurmountable barrier this far.

NEWTON, ISAAC (1642-1727) and NEWTON'S LAWS OF MOTION
An English physicist and mathematician who developed his principal theories about gravitation, optics and mathematics between 1665 and 1666. In 1668, he made the first working reflecting telescope. Most of his work remained unpublished for long periods, partly because of criticisms by c. Huygens and the English scientist Robert Hooke (1635-1703) of his early work on the corpuscular theory of light. However, in 1684 E. Halley persuaded him to organize his work on the celestial mechanics of the Solar System, which was published as the Principia. Newton's other major work, Opticks, was not published until 1704. It contains his corpuscular theory of light, and the theory of the telescope. His greatest mathematical achievement was his invention of calculus, independently of the German mathematician Gottfried Wilhelm Leibniz (1646-1716). His profound influence on physics and astronomy is reflected in the phrase 'Newtonian revolution'. Three laws published in 1687 by I. Newton concerning the motion of bodies.

1. A body continues in a state of uniform rest of motion unless acted upon by an external force.
2. The acceleration produced when a force acts is directly proportional to the force and takes place in the direction in which the force acts.
3. To every action there is an equal and opposite reaction.
4. However there is one more law on motion that went undetected by Newton…This book is not about trying to disprove Newton…it is about adding too science more than there now is available without removing any that science already accumulated.

In this book I use Kepler's formula to either prove or to disprove the following accepted principals in cosmology and if any person in the past gave only the slightest attention to Kepler's work, many statements would have come much sooner delivered by someone else or may never have come at all. By applying Kepler's formula correctly in this letter I can either agree with or in other cases deny the following principles.

It began with NICOLAUS COPERNICUS who changed the status quo. COPERNICUS, NICOLAUS (1473-1543) was, according to the Anglo Americans, a Polish churchman and astronomer although this is just more politically inspired propaganda because his parents were both German (in Polish, Mikolaj Kopernigk). While he was completing his studies, he had realized that the Earth revolves around the Sun and not vice versa. Such a view was in that time, held to be heretical. As I pointed out in the first few articles, the Church regarded the geocentric world-view of Ptolemy as consistent with its doctrines. Copernicus set down his basic ideas around 1510 in the Commentariolus, which he circulated anonymously, because of the Islam link. In 1512-- 29 he conducted his study and concluded the observations that he needed to support his theory, while carrying out ecclesiastic and local administrative duties. In this time, he had to defend his mother in court on charges of witchcraft. In 1539, the Austrian astronomer and mathematician Georg Joachim von Lauchen (1514-74), known as Rheticus, became a pupil of Copernicus and began to spread his ideas. The published work was openly spread as the Copernican system, in spite of the life-threatening dangers connected with such a "crime", in 1543 in the book De revolutionibus orbium coelestium. However, the reality of a heliocentric Solar System was only commonly accepted, after the work of Galileo and J. Kepler. The ideas introduced developed along and proved to be correct until such a time it met a solid wall with the investigation of Max Planck.

PLANCK CONSTANT
(Symbol h) A constant that relates the energy of a photon to its frequency. It has the value 6.62076×10^{-34} Js. It is named after the German physicist Max Karl Ernst Ludwig Planck (1858 – 1947). PLANCK ERA. In the Big Bang theory, the fleeting period between the Big Bang itself and the so-called Planck time when the Universe was 10^{-43} s old and the temperature were 10^{34}K. In this period, quantum gravitational effects are thought to have dominated. Theoretical understanding of this phase is virtually non-existent. It is named after Max Planck (1858-1947).

PLANCK'S LAW
A mathematical description of the energy radiated at different wavelengths by a black body: $E = hf$, where E is the energy of a photon and f its frequency. It was formulated in 1900 by Max Planck (1858-1947), who realized that energy is radiated in discrete packets, which he called quanta, and it formed the basis of quantum theory. The quantum of light is a photon, the energy of which depends on its wavelength.

There is one rule which is well established and which Mainstream science agrees about. It is one aspect, which forms the very principle that holds the theory about the cosmic starting together under the covering of a verbal blanket. All in science agree that it all started with singularity but I manage to go one step further where I prove that it is also where it ends, as singularity reunites space-time, which is from where Creation split in the very beginning.

Singularity is as follows: Singularity: a mathematical point at which certain physical quantities reach infinite values, for example, according to the general relativity, the curvature of space-time becomes infinite in a black hole. In the big bang theory the Universe was born from singularity in which the density and temperature of matter were infinite. From singularity flows space-time.

Space-time is as follows: Space-time is a four dimensional position of the Universe where the position of an object is specified by three coordinates in space and one position in time. According to the theory of special relativity there is no absolute time, which can be measured independently of the observer, so events that are simultaneous as seen from one observer occur at different times when seen from a different place. Time must therefore be measured in a relative manner as are positions in three-dimensional Euclidean space, and this is achieved through the concept of space-time. The trajectory of an object in space-time is called world line. General relativity relates to curvature of space-time to the positions and motions of particles of matter.

SPECIAL THEORY ON RELATIVITY
A theory proposed by A. Einstein in 1905, based on the proposition that the speed of light in a vacuum is constant throughout the Universe, and is independent of the motion of the observer and the emitting body. A consequence of this proposition is that three things happen as an object's velocity approaches the speed of light: its mass goes up, its length shortens in the direction of motion, and time slows down. Hence, according to special relativity, no object can ever reach the speed of light because its mass would then become infinite, its length would become zero, and time would stand still. In addition, Einstein concluded that the mass of a body is a measure of its energy content, according to the famous equation $E = MC^2$, where c is the speed of light. This equation describes the conversion of mass into energy in nuclear reactions within stars.

GRAVITATIONAL COLLAPSE
The collapse of a body that is unable to support itself against its own gravity. Gaseous bodies undergo such collapse if they are not hot enough for their gas pressure to balance gravity. This can happen in the early stages of star formation, or when nuclear burning ceases in a star's core. The time taken for such collapse decreases rapidly with increasing density, varying from about 100 000 years for the birth of a new star to less than a second for the formation of a neutron star. Star clusters may undergo a similar collapse if the random motions of their constituent stars are insufficient to offset gravitational effects, either during their formation or at an advanced stage of their evolution.

GRAVITON
A hypothetical particle or quantum of gravitational energy, predicted by the general theory of relativity. gravity - motions have not been observed but are predicted to travel at the speed of light and to have zero rest mass and charge. A graviton is the gravitational equivalent of a photon. It is this anti-photon-being-a-gravity - motion by just merely swapping direction and all is proved that I find not very indigestible in modern science. One of the main issues that I wish to protest by my writing of this is my argument that if the Universe can be compressed back to the size it had at the point of 10^{-38} seconds after the Big Bang the daily outdoor temperatures of 10^{27} K will also come about once more. The expansion was the result of compressed space, which then formed into heat and in turn resulted in finding a Universe with all the insufficiency of space less ness prevailing throughout and wherever space was needed. By that it forced space-time to come into being. Space-time came about at the time of endless time duration without space availability, which brought about the period of the Big Bang wherein space growth was the converting of such heat to space. If the Universe was in a vacuum as big as being available now then what was the temperature of the vacuum while it was empty before material filled it later. Then I presume the vacuum was there present as it is now in this present day. If the Universe then employed the space of say one atom, the impression comes through that from edge to edge and from

Universal border to border the space occupied was the same as one atom will claim in our present day and age. Normal gravity started at 10^{-43} seconds. The Universe was the size of a neutron or somewhere in that vicinity. The Big Bang began and GUT, or the grand unified theory, produced the attempt to describe the strong and weak nuclear forces and electromagnetism in one single mathematical theory. Somewhere before 10^{-12} seconds of counting the Universe cooled to about 10^{15} K the electromagnetic and the weak interactions acted as one single physical force. Science reckons that unification may come about at temperatures of 10^{27} K, which was the temperature of the day at 10^{-38} seconds after the Big Bang. This statement echoes my viewpoint but one has to look carefully for that to surface.

In the suggestion the presumption claims that all the space that the Universe made available at that time was the total space one atom might take up today. If that might be the case then where was the rest of the space that now fills the Universe? Or was the rest of the space we now find in the Universe and what is now explained away as the vacuum, also available back then. Did the Universe only have that one tiny hot spot it filled with huge volumes of heat? Was the rest of the space vacant being out there all along during all the time running to the present date but filled with emptiness standing around as a big vacuum with no better to do than sucking on the Universe while the Universe was exploding at the speed of light. Then that statement suggests that in this hot Universe there were light-years upon light-years of vacuum waiting to be filled by the intense heat soaring in the smallest spot. If that is the case then why did the vacuum not fill in the blink of an eye by all the exploding expanding material growing at the speed of light? Was the Universe overall bitterly cold where the vacant space was locked in with one spot of the vacuum filled with temperatures so hot we can only produce it in numbers suggesting a value but never claim to be able to digest the reality thereof in the human mind? If so what happened to the natural consequence that heat flows in the direction of cold and equalise between hot and cold. Was the space being available at present available then or was the hot space the only space available at the time. If so what prevented the heat from instantaneously filling the eternally cold vacuum because with the rules controlling vacuum in affect, it should have filled in such a manner in less than a heartbeat? I believe that singularity formed space-time and space-time developed from the overflowing thereof at the time it was extending. With time marching onwards and outwards to this day space-time developed. Space-time developed another product that everything in the cosmos has to have. It must be in such large quantities everything imaginable in the Universe has to have it and that is space using time to move about. I suggest that it is space that is holding heat in a quantity providing density and ratio to space available and in relevance to the space being available to quantify the presence of the heat and which then proves to form the time factor. The container and contained all together mixed by motion. From that very first separating of heat and space, which is what formed from singularity to produce space-time. The Universe was full... It was overflowing by the speed of light in the beginning...so where and when did vacuum or nothing enter the Universe as a factor if and when the Universe was so full.

The answer to that is absolutely crucial because how did the Universe decide to fill some parts with a variety of something and decide to fill some parts within the in-between with nothing? If that is true why did gravity not prevent the vacuum filling because no gravity that came about since can beat the force that gravity had back then? This leads to another question following the previous one in asking why did gravity at the time when it was so strong with r^2 so much compromised not fill the nothing immediately as it entered with something that could absorb the nothing. At the very beginning the mass that was pulling on the mass by force was immeasurable and none quantifiable. Even more to the point is the question to be asked in how big was the radius between the materials with the immeasurable mass placed in such a little space. This is all the more important in the light that the smaller the radius is the bigger the force will become from the immeasurable mass pulling.

With the immeasurable mass that was producing the first gravity between the particles divided by an almost non-existing radius the gravity produced had to be in gigantic proportional quantities and with the separation of the radii being in the infinite measure that it was at that point then how did the Universe establish the chance to expand. It did expand, as we all are witnesses to in spite of this contraction of gravity that had to have been compromising the expanding factors. Still the expanding filled the unknown part of the unoccupied Universe, which at the time was there or was not there and where it was not there it was then filled with nothing. If the nothing was not "nothing" then the nothing that was not being nothing was also filling the rest of the vacant Universe that was or that was not because if it was it was filled with nothing and if it was not then it was nothing.

This is then taking into account that then all the reducing that is resulting from Newtonian contraction and that was going about in the space available at that time was something filled with nothing and surrounded by more nothing. With everything in the Universe being that much crowded and crammed where and how did nothing enter the Universe and fill the rest that was unfilled? What factors introduced nothing into the picture since the entire Newtonian concept finds its base on the principle that matter reduces using gravity by force which then brings about reducing or the removing of the many nothing between particles, which will then lead to nothing that has to vanish even before nothing can enter the space. This question may seem small-minded belonging to the mentality of a child or to that of the mentally impaired with not much factual appreciation developed yet. Please do not see it that way. If you think in those lines it will be because you do not have an answer to challenge these silly questions. Beware, silly as they are they represent official backing by the Wise-and-Informed. If the space is nothing and if the space was as large as it is at present then there was no need for such a small area to fill with something leaving only the rest filled with nothing at first since all the space we know about was there present and by being present it was there then for the taking. What ever filled the Universe had to start at the centre of the Universe and fill the entire Universe all over from a centre as it moved outwards filling from the inside outwards. This is a natural human instinctive realisation but is beyond proving by using Accepted Scientific policy. But that leaves Newtonian science with a massive unsolved problem: where is such a centre at the present time and where does the centre produce the limits or border it apparently has to form as it expands.

By expanding there is an additional contribution too that which was when that was, it was receiving more than there was before the addition increased that which was and by then becoming more than there previously was it had to be improving the border from where it must have been before the adding took place to where it was after what that was added was added. When that was less than it became when it was added, it was at the limit that was there before it was added too and that limit there was, was a limit that is the limit that I am referring to as a border being there. The cosmos is filled with unrecognised borders. The expanding has to be an ongoing filling that is at the same time expanding from the inside towards the outer limits of the Universe. Since nothing can enter from the outside where nothing is, the filling of nothing as a substance that would take up vast quantities of room had to fill from the very centre spot where all other filling came from. This filling of nothing with material has to be well mixed. The truth about cosmology is that space forms no borders but by using any Newtonian centre from where mass is attracting we must find a point where there has to be the ultimate Universal centre which is the cardinal point in the entire Universe and it is the first, the prime position to locate coming before any other concept one wish to put forward because all concepts has to start with locating that cardinal centre. There has to be the ultimate r^2 radii located precisely between the ultimate mass drawing the other ultimate mass closer. If there was a Big Bang then there has to be the spot where from the Big Bang developed therefore there has to be such a centre connecting the past to that ultimate centre with the line of development flowing onwards to this day. The fact that science is Newtonian proves that in the meantime Mainstream science is still of the opinion that there was the specific centre in the Universe that is nowhere to be found as it was filling the unknown with nothing coming from nowhere, but which somehow is still somewhere in the centre of all of that which is something. On the opposite side of nowhere there is an outer border in space producing a limit to nothing and serves nothing with a specific point to stop being nothing because that point is precisely where nothing ends and forms a beginning of a Universal border or a Universal end. How one will stop vacuum being no longer nothing was a question everyone comfortably missed to ask therefore no one ever seemed to deliver any form of answer.

One night some years ago very close friend of mine had a meal at his restaurant and as the conversation progresses he asked me about space and where it must end. I tried to explain to him what I believed in comparing to what Mainstream physics believed, but soon saw I was not gaining in his understanding. Then I decided to jot it down on paper and he could read it at his leisure as he saw fit. That led to the first book written by me (in Afrikaans my native language). What I tried to explain to Johan Boonzaier that night, is that if the Universe was the size of say even a tennis ball with only the size of a tennis ball being the very all of space there is available, then yes, it must take time to expand from that having the excessive heat there was back then in all the space we have at present. It then is converting heat into space bringing about the expansion. But one will find most expanding within the atoms, as the atom must grow since the Universe in all was the size of what one atom is today. The space in the atom pushed the space outside the atom but there must be plenty more to the growth. Something outside the atom contributed in it own right because there is more expanding than there can be blamed on coming from the atom. But the space then also developed as the Universe developed and if space developed then it cannot be total vacuum filled with nothing because "nothing" cannot develop. You the reader must judge whom is correct between my view that space developed with the Universe as part of the Universe and reject the official view about space being nothing or otherwise you the reader must then decide that I am wrong, but should you do that, then find a reason why the Big Bang started out small and filled all the available vacuum or what is contemplated as vacuum that we have with the motion of time. When Mainstream science accepted the Big Bang as the principle that will take science into the future the view about such a Big Bang concept unlocks a different door to another view on the cosmos from birth to end. It calls for the revision of all aspects of the entire history on cosmology and change that what is dead wood and that which needs to be chucked out. Most of all it was my following the lead I got from Kepler that unlocked the doors I now present to you. I claim there is no gravity - motion as there is no gravity forming weight or forming mass. I hope the sketch contributes to my explaining effort:

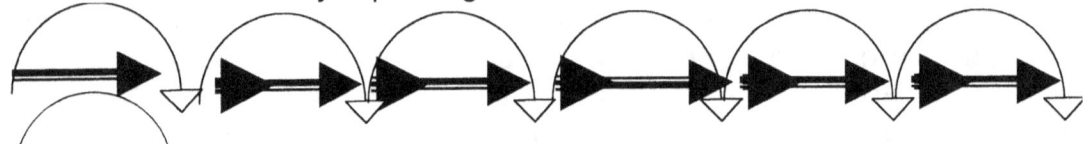

∇ The duplicating frequency the Earth shows as k_1

$\blacktriangleright$ The frequency of motion duplicating my body maintains as k_2.

k_1 minus k_2 ➡ The frequency of motion difference my body has minus to what the Earth has where that difference in motion becomes my mass. It is the sum total of the reducing of the motion that my body has ir comparison with the motion capability of the Earth that is the mass value. In k_1 as well as k_1 the symbol represents motion, however in the case where k_1 minus k_2 that shows an incapability of motion, which is motion frustrated or a more commonly used name would be <u>mass</u> is created

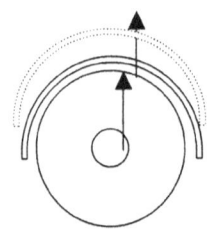

The new k that is applying the relevance, must link the space a^3 being equal to the motion T^2 to singularity k^0 in order satisfy k^0. The flying of the aircraft is then unequal to the motion in the previous relation that was in place where it was part of the sun and the Earth motion relation and the new motion will bring a correcting in the relevant distance k to put the motion in balance with space. Space will always demand a correct establishing of the miss-interpretation of the equilibrium that is needed to sustain the effected singularity because of the space-time factor.

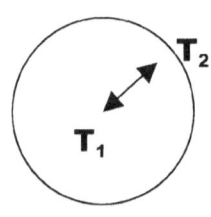

There is a point where the two points forming the relevancy unite in shared singularity. It comes because of shared motion

In all space-time, one finds at least two relevancies where one is at the centre.

That part Newton saw and formulised

He missed another part. Crossing a limit of inclusion is the limit of division and such limits are in distinction by motion producing the gravity, which is parting the two objects. Motion brings about a relevancy where two positions no longer share a common point in singularity. That is what Newton missed.

That is the gravity aspect Newton and all other Newtonians miss.

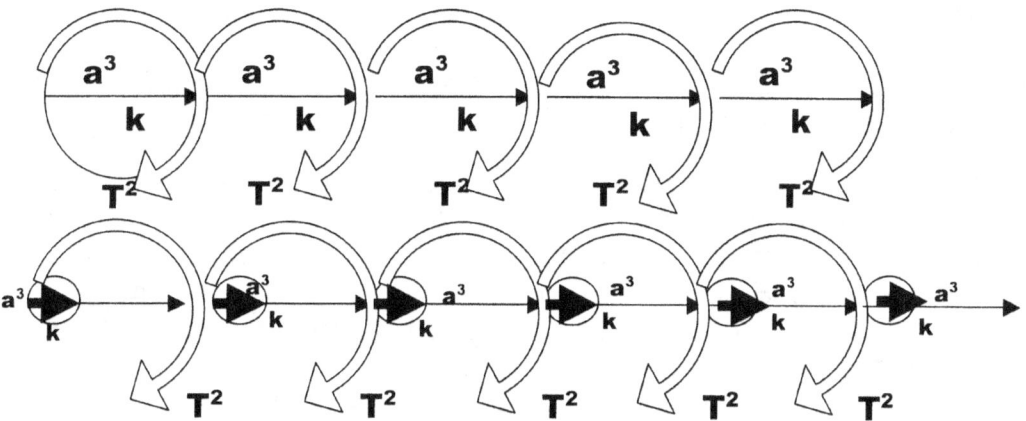

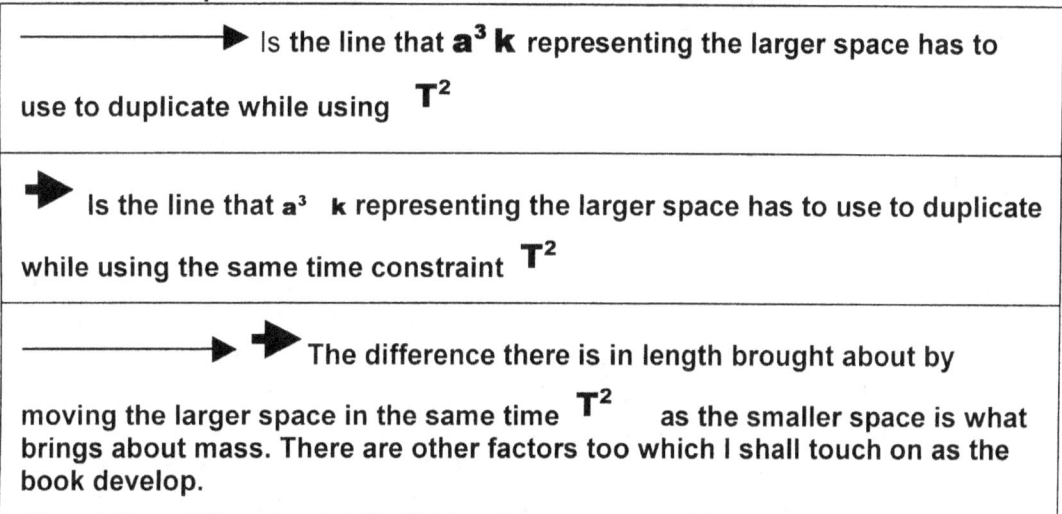

Two objects of substantial size differences are travelling at the same time but one has a space, which it has to move when it travels that is considerably different from the larger space. The larger space will produce an extending line equal to the space it moves while the smaller space will also produce a line in ratio to fit the space holds relevant and that it has to move.

⟶▶ Is the line that $a^3 k$ representing the larger space has to use to duplicate while using T^2	
▶ Is the line that $a^3\ k$ representing the larger space has to use to duplicate while using the same time constraint T^2	
⟶▶ The difference there is in length brought about by moving the larger space in the same time T^2 as the smaller space is what brings about mass. There are other factors too which I shall touch on as the book develop.	

Two objects of substantial size differences are travelling at the same time but one has a space, which it has to move when it travels that is considerably different from the larger space. The larger space will produce an extending line equal to the space it moves while the smaller space will also produce a line in ratio to fit the space it holds relevant and that it has to move.

Mass has precious little to do with the whole affair except to be an obstacle intended to restrain the motion of the hosting space. The difference in size between the one in circular motion and the space in contracting motion must bring about that the smaller object has to move about a circle much closer to the centre because the larger space form the centre hosts it. However there is no large or small in the cosmos but only those better developed or those poorer developed. By duplicating there is more to duplicate in the better developed than in the lesser developed. When the lesser-developed space is duplicating the lesser developed space would hold have a lesser extending relevancy from point to point forming a shortfall by distance in comparison. The motion being extended needs less extending and should therefore be closer to the centre in relation to what the better developed space would need in extending by a duplicating effort. This is the principle we find that is behind the sound barrier. The motion the aircraft produce forms an increase in the duplication of the aircraft which extends the duplication splitting the Earth producing an extension and the aircraft producing an extending that goes beyond the attempt of the Earth by duplicating material.

I know this may sound barely believable but please hear me out. While we use gravity, the use of gravity as such makes us part of the Earth. We see gravity as some influence or force producing mass and that mass is forcing us down on the solid ness and onto the Earth. By having the mass we become a semi unit with the Earth. That is how we on Earth see gravity but when investigating gravity in outer space we must come to a basic question: Is that what we experience as gravity on Earth truly gravity? Much of the proof about gravity is part of our perception about gravity because we experience certain conditions with gravity while we find ourselves bogged down on Mother Earth. But are our perceptions about gravity truly correct? We experience mass but are the mass the result of gravity or are the mass the product of gravity. We only experience gravity, as a factor from the position we have on Earth and the conclusions we form is a product of a perception we formed while we are being forced to be part of Earth. It's as if we are upside down and have to decide on which route we should follow. I want to make a suggestion, which I aim to prove in the following pages. My personal being on the ground and having mass that is keeping me on the ground comes about because of the speed that I travel through space being the very same as that which the Earth has.

By me not applying a speed difference I then inherit the speed the Earth places on me. But my space which I use $a^3 = T^2k$ to travel and the space which I use tot travel through is much smaller than that which the Earth burdened with to move and to move through. By me having a smaller space to move $a^3 = T^2k$ the space a^3 being moved k in the time it would take to move T^2 will produce less space a^3 to shift k and therefore a smaller distance k to replace all the space a^3 that is moved in the time T^2 the space a^3 needs to enable it to move k. To duplicate by motion the smaller space requires a smaller distance to shift the space but the motion will take up, as much time to complete than would the larger space take to complete though the space the larger space has to duplicate will require a longer distance to complete the total duplication of the larger space. A large space a^3 will produce a large extending k when using a^3 the same time duration T^2 when using the same time factor as that which the smaller space is required to use when under obligation to use same time constrain. Behind this is the most basic principle hiding which allow us the fortune to be able to fly using a flying machine. It is all about motion supplying relevance and forcing on time constraints.

Because my body that I have is travelling so much slower than the Earth is travelling due to my size in relation to the size the Earth has and although I am using the same time as the Earth does to move, such a speed difference is not in the time differences it takes to complete but in the space differences that has to be completed in the same time but is unable to fill and the space is trying to crush me into the Earth where I am forced toward the centre. If I were able to penetrate the soil solidness I would reach a point where my speed as zero would equal my space I occupy.

The space I duplicate by moving from one position and placing the space I hold in the next position while keeping my space I move as it is identical in the next spot but located in the next position. Such moving by duplicating takes a certain time to move from one spot to the following spot and it will use a certain frequency that will have the same ratio in bridging the gap from one point to the next point as that which the Earth has. My speed of duplicating by motion has to be even in frequency because I am within the duplicating space, which the Earth is duplicating and as part of the space that the Earth is duplicating but the duplicating of my space I do myself. But in size there is a massive difference between the space I hold and the space the Earth holds but to duplicate will take me as long as it takes the Earth. Notwithstanding this common factor the Earth has to use equal time in duplicating its massive space, as I have to duplicate my small space when we both have to share a frequency that will keep us duplicating evenly. Therefore the frequency of duplicating using the same time period will be a lot different to my much shorter frequency of duplicating space.

The difference is between me being in mass and me being in the correct position in the space-line the Earth has will place me in the correct position but the heat that then will surround me will fry me into non-existing. Fortunately for life the soil forms a barrier through which I cannot fall any further as to correct my location. But being where my position would have no mass would allow me to float there in that location in the same manner as I would float in water. I would be buoyant. It is because I do not harmonise the displacing frequency as I should that I have mass.

$$\frac{a^3}{k} \rightarrow T^2$$

My having weight is what Mainstream Physics use to give me my gravity. Science purposely switch my having mass and confusing my mass with my having weight to explain what is beyond explaining. It is said that while I float in outer space in state of suspending hanging above the Earth in the weightlessness I still have all the mass that I had on Earth. But in order to prove that those in science will give me a mass even in outer space whether I deserve it or not. By that token science first has to cheat all logic by reasoning in some bizarre way that I take my mass up there to where there is only micro gravity. They firstly claim that all of a sudden I take my mass to outer space and in their next argument they say I have micro gravity in outer space since my body is floating as if it is in the sea. But if I stop floating and start falling to the Earth my body and I did not gain any mass. My falling then comes as the result of my motion being much smaller in relation to the space I claim and my motion then is being less than what is required to keep me in the position I have which in I maintain my orbit up there. By moving to slow I fall. I do not fall because my mass grew. But science has been proven wrong by their work without any of them aver admitting to such a defeat. All the satellites fall if the satellite motions are not reset. The satellites do not gain or lose mass. They gain or lose motion. By amplifying either my space (using a Hot Air balloon) or by accelerating my motion that I have in relation to that which the Earth forces me to have, I will break free from my weight or mass. I shall become airborne and float as if I am in outer space. By pretending my mass can be multiplied many times over in using a process, which then is called not gravity but momentum. But motion and gravity is all the same because motion is gravity that is redirected, which then forms another part of gravity where gravity again is also only motion applying. Science maintain the argument that when I am in outer space and am no longer part of the Earth I then will only have mass. But since there is only micro gravity I will be in a state of weightlessness. My mass is what gives me gravity and while being up there I take my mass long with me. But with my mass up there I will only have micro gravity. I am floating with my mass and it is my mass that is responsible for my gravity and I am floating above the mass of the Earth, which is right down below me, but still I have micro gravity. That is true if I wish to incorporate the dubious use of double standards by separating mass from weight. The mass my body will have in a Black hole will be a billion times (at least) more than what it is on Earth. With that the Black hole destroys the fact propagated in science that my mass will be the same everywhere. That is more than permitting double standards. Because our motion is much slower than the Earth is spinning, we place a breaking effort on the velocity the Earth has and that breaking effort we accept as the mass we have. The truth is that my mass comes about from the lack of motion I have in relation to the space I occupy and has nothing to do with any gravity - motions pulling me down. If I increase the motion I have there shall come a point where my motion will be sufficient to pull me into the air, as I then will have the required velocity to lift from the ground. That motion being in excess of what I have and is complimenting the motion that I receive from the Earth counteracts the motion of gravity that is containing me. The motion I adopt then release me from the motion containing me and if motion can release me by only becoming more, then gravity is my motion not being enough in the first place to keep me onto the Earth. Nowhere and at no time does my mass ever gain by having more protons that will get me back to the ground as if I am bigger or carrying more material or does my mass reduce to get me into the air as if I am smaller or carrying less material. Please note that this is my way of explaining to you about the fact of bodies having weight or mass. It is not mass or the lack thereof or any means to measure occupied space within the atmosphere of a larger body that pins me onto the ground. My body is claiming space by motion in space. Gravity is the result of motion because it is in the motion that bodies have that gravity affects them. This is proved because by adding motion the mass does get more but the body never gets bigger or hold more material, and in defiance of that statement by increasing the motion my body lifts and flies. The reality is that my body in motion has more mass being momentum but still my body lifts when motion allows my body to lift. This statement confirms Kepler that a^3 becomes more (massive) when motion T^2k becomes more (moving).

Mainstream physics admits all along that nobody, human or otherwise knows what gravity is. While investigating Kepler's work with employing much motivation and detail in order to give his work the much duly credit it deserves it will also serve a valiant purpose when by the same token we try to establish what gravity is, because I believe Kepler possibly answered that mystery. We have to start with the person that introduced gravity or so does everybody acknowledge. Newton saw an apple fall from a tree and he subsequently realised there is some force pulling the apple to the Earth.

Allthough he still was a student he announced his findings and became a genius on the spot. The concept he introduced as gravity gave him instant admiration from which he became the legend he is today and that reputation he gained there at that moment would last him from that day he instantaneously unveiled his mastermind, and that same genius still serves him in his honour to this day long after his death. He found that this force has to have some thing to do with the weight and the mass of that particular object and the mass of the Earth. There is some force pulling that apple as much as the force is pushing the apple and the same goes for the Earth because the mass the Earth has is doing the same to the apple. Between the two objects facing gravity there is a force that develops where such a force is pulling the apple on a constant basis towards the Earth even after the apple is already in a steady state on the Earth. That forms the mass and the mass forms gravity. He concluded that the mass is responsible for the pulling. Remember this observation came three point five centuries ago when knowledge and brilliance carried a much different defining than what such defining of brilliance is worth today. He realised the pulling on that apple brings about weight that brings about mass because the apple departs from its location and arrives at its end location when the falling is completed. Then he went out convinced all that was in line of finding the needed convincing because no body before Newton thought of what Newton thought quite in the way that Newton thought about gravity.

Newton succeeded because he found a way in presenting science with the fact that objects move closer because of some force. He went one step further and named the force he fathered as gravity. But there it stopped! Any and all other further defining the matter or going into any possible observations of whatever magnitude concerning the topic never realised any motive to go further. Inspiration to further commitment just flew out the window as the essence to do so immediately expired as far as the rest of science is concerned. What might he have missed if he missed anything? We all fall down when we are unstable and out of balance. He never realised that balance is more crucial than brutal gravity because that part is the defining part about gravity. No one ever gave a thought about the balance part even centuries later even as we grew into all the sophistication we now enjoy. What brought about the balance that secured objects in an upright stance and supplied some form of control over the managing of a position? Any other position than being flat on the floor would have a better defining than being just at the mercy of the force gravity. Standing tall is a stance that defies gravity so there is another force other than the pulling of gravity. Admit tingly the force would first and foremost have to aspire to the rules of gravity and then comply with other demands. True enough is the fact that that position would ultimately and firstly by all accounts have to satisfy gravity before any further motion could commence. Yes but then by balance motion defies gravity by changing gravity's force of pulling everything straight down towards a visionary centre between the objects. In affect this means somehow there is control over gravity and gravity does not leave objects beyond outside control. Gravity is manageable and can be controlled; we just have to find a way…

Years later some one came up with the novelty of hot air ballooning. Ballooning proved that there is antigravity but that part was missed by all even to this day. Some people speak of antigravity as if that is some mystifying mysterious concept that is so well hidden in the secret annals of the hidden Universe that only Ali Baba and his magic words can reach it. Please consider the following statement. If gravity was bringing the object down, because of the affect of gravity which is that what we experience as the gravitational sensation and that is what we interpreted as gravity by our sensation and observation, then that is only coming about by our bodies that is in a state of being dragged down. The dragging down of the body is in the direction of the Earth centre. That sensation of being firmly locked onto the ground constitutes to what we believe we experience as gravity.

When some influence brings about the very opposite effect, which then results in establishing the opposite result, it deserves to be anti. In example we feel dragged down but anti will be the lifting of the body into the air. Anti will be going in an opposing direction of the motion that gravity inflicts. It will counter the influence that gravity applies. Such motion has to indicate antigravity. The counter acting of the mass dragging us down must be anti gravity pulling us up into the air above the ground. Antigravity must come from such an opposing influence that will bring about the lifting of my body. If hot air ballooning gave the object an opportunity lift, then ballooning must be antigravity. The balloonist and the entire balloon found a manner to counteract the pulling of gravity enforcing weight. The balloon can lift what gravity depresses and if Newton said gravity is the falling then later Newtonians must agree that the opposite of falling is flying or lifting. A balloon is lifting-and- flying. If gravity is pulling down objects in the direction of the centre of the Earth then flying is antigravity. Moving away from the Earth by means of motion and in particular flying is using whatever means to defy gravity where the lifting can also be the hoisting of a body by a crane. Lifting by ballooning in a hot air such balloons escape from gravity where the balloon constitutes to bring about the effect of establishing antigravity. Climbing up mountains must fall into the antigravity department because parachuting down the mountain definitely falls in the gravity department. Nevertheless it still does not answer the question of what gravity is.

Let us look at antigravity because the antigravity is releasing the object from the gravity that controls the object by an Earth fed force. The balloon starts flying when the confined space of the balloon is veraciously and violently heated in access. The balloonist shows us that in order to overcome gravity we have to introduce heat. That is the only manner in which we can defeat gravity. Even by an engine driving an aeroplane such flying can only result if an engine combust solid fuel by creating motion as the fuel mixture is turned into heat. It is heat that makes the difference. That is the very thing that Kepler said. Expand the space a^3 and the motion T^2 will move further increasing k. Blowing hot air into the balloon is increasing space within the balloon a^3 which then results in providing the balloon with a larger distance k from the Earth centre k^0 that still holds time with in the Earth atmosphere with the Earth T^2 within the space of the Earth k. Using Kepler provides us with insight and the ability to see what gravity is by showing us what antigravity is (a^3 gets bigger and that will bring in a larger k). But moreover the larger space in enough compensation to bring about extra motion that will defeat gravity by the extending of k. If that is not antigravity then we can forget about Ali Baba and his magic rhymes too.

The balloon assists us to escape the Earth's hold on our body, because there has to be the force producing motion countering the motion of the Earth gravity. The balloon shows that releasing enormous quantities of heat into an inclusive area excluding space such as that which the balloon canvas provides, which is establishing the release from the gravitated containing force on the body giving the body a means to escape by floating about above the ground. The motion is at that point breaking free from the containing gravity by moving in a specific direction, other than the direction the Earth gravity inclines the body to travel. By concentrating the releasing of heat into the balloon, the direction of motion starts to contradict the enlisting of the Earth gravity and the heat breaks the balloons confining properties while the balloon is released from the Earth as the balloon and us lift up into the air and away from our confining to the Earth.

At the point of explaining we arrive at the point where we can say what we think the difference is between the balloon floating in the air above the Earth and a body suspended in outer space floating above the Earth's atmosphere. The difference is the heat that is in the confined air per volumetric ratio favouring the heat being more in the space than what the heat is outside the confined space. If we had any method to put the required heat we need to escape from the limits of the Earth to outer space into the canvass of the balloon there was no canvass left to contain the heat. The heat is available to do the job but the means to do the job with the tools in hand is unavailable as far awe can use the balloon. By having more heat in the one area than there is in the other area beats of the pulling of gravity. Obviously it is antigravity that keeps the balloon in the air and what keeps the balloon in the air is having a larger volume of heat per space unit than what is in the atmosphere. The balloonist shows us that by applying more heat we can defeat gravity more. Someone took the advice, because the next minute the Germans had rockets. The launching of rockets brought about the ultimate defeat of gravity but it involves almost the ultimate releasing of heat.

In antigravity we find heat more concentrated in one definitive area than the heat concentration is elsewhere. The more the heat is that we release into space the more the antigravity is that we achieve and the more release such antigravity can produce. But what connection can gravity have with heat and if there were any connection between heat concentrated and gravity, what would such connection be? The history behind Carl Benz should bring the answer but more so would be the story behind James Watt and steam although the James Watt story may not be that thought provoking because it is much less filled with the ever popular cheap thrill only sensational gossip can provide…Still both stories cover the same principles. In the Carl Benz story a housewife leaves a pot of benzene fuel on a coal stove. The pot with benzene heats up where the pot with benzene becomes hot and under pressure. This performing of heat increase, such increasing expands as space and releases the heat as newly creates space, which then removes the housewife with her house from the neighbourhood she used to regularly frequent as her residential address. Afterwards almost the entire neighbourhood is not there to tell the tale or ask why...
It was a stupid tragedy that brought about the end of steam and the rise of the internal combustion engine and on Earth billions on billions of human souls are in torment not to please or suffer for the advantage of coal Barons any longer but now they are dying and suffering in agony to please the wishes and desires of oil Barons. How much did the world not change…While it is no longer the coal Barons shackling us in chains and telling us democracy broke our burden of slavery, we have now the pleasure of the oil Barons enslaving us with democracy and telling to be happy because we are the fortunate slaves, there are others circumstances in which they can enslave us that will leave us worst off. All this came just because the pot of fuel created a houseful of space that was enough to remove the house from the address the house previously enjoyed. But

Mainstream science neglects to appreciate this. They see the heat, they see the antigravity but they fail to add the heat, the anti gravity and the space that no longer housed the house of the naive and rather impractical thoughtless housewife. They call the tragedy an explosion but then again everything that expands while using a noise during the expanding is an explosion. Adding of new space to the space holding the house at first altered everything that was previously proportional positioned in the space where the house was. Such exchanging of heat to accumulate and introduce more space in the process referred to as an explosion was bringing in more space that came directly as a consequence from the explosion which was producing more space where the increase in space brought disorder because the well organised material distribution and placing was before the event filling just enough of the required space arrangement that was holding every object in a prearranged order of tidiness.

Then suddenly out of the blue the space which held the house in a tidy arrangement had to accommodate more space therefore the ratio of material per space volume increased dramatically many times over in the favour of the space in the balance. That part no one ever acknowledges. However the losing of the house was not much surprising to Mainstream science back then and even today because who cares about old news. All of Mainstream science was at the time, as they are today, very familiar with all explosions because of wars and bombing that leads to maiming and killing and all the unspeakable monstrosities we associate with war so that the dirt poor can suffer and die to leave the disgustingly rich even richer. The poor has not the means to pay science to be clever and devise methods to save their lives, so the rich does the poor the favour of paying science to find methods whereby more poor could be killed as long as the rich saw it as a good investment with great capital gain on the part of the rich. Therefore science is well established in the method of creating more elaborate and destructive explosions that the rich pay them to invent. In the explosion caused by our housewife no one put up money to investigate what happed during the explosion but money went to why the explosion happened.

That inspired an investigation in connection with the fact of the finding more about what takes place during the carnage as more money goes to finding means to create more carnage per money unit spent. At least that is why the poor were invented and that is why wars are invented. It is invented so that no money goes wasted on saving the poor people except if the poor has the money ready and available to pay the rich for medicine to enable the poor to stay alive. So science goes out and develops more fuel for carnage but fails to find out why the housewife and her house are no longer part of the neighbourhood she used to frequent. With the loss of the presence of the ignorant housewife with her house her neighbourhood and all were a normal way of leaving us with a new way of tapping and harvesting energy and untold riches which was born with the death of the absent minded housewife. But according to the mindset of science they saw not what the incident presented in space producing for to their view nothing new came about since it was just another exploding of fuel...so no body bothered as to finding out how. What they missed was the part the coal stove played in the whole tragedy. Without the intervention of the coal stove producing the heat that turned the liquid fuel to liquid heat liquefying the space that turned the liquid space into a gaseous space where the liquid space revealed its true incentive in nature by turning out as space and the newly created space that was in fact liquid space that went on to become more space, well that space was providing the one main factor in space-time relevancy. The stove's heat was producing space by transferring the heat from the stove to the pot filled with volatile fuel and the transfer of the heat to the fuel brought about the expanding of the space that the fuel claim to need in the pot filled with the heating and volatile fuel. The fuel space requirements became more as the heat filled the space that was filled with fuel and that took up more space, which the pot could not cope with since the pot had no room to allow such an increase on the demand for more space. The fuel expanding as such was claiming new space to sustain and accommodate the growing requirement for more space to be created in order so that the volumetric increase of heat added to the fuel could be accommodated. At one point the asking for space became a claim on new space, which we see as an explosion. The heat transformed to new space by an exhilaration of breathtaking increase in heat forming space and this increase of newly formed space was transforming all other surrounding space within the room, the house and the neighbourhood in general. By increasing the volumetric quantity of the space it rearranged all space, which included some of the space held by material that was in a solid form and scattered the rearrangements as fragments in all other designated places far away from each other. This meant there was excessive and all around rearranging and relocating as well as re-allocating of space in general. This to science sees as shock waves resulting from an explosion but is merely heat expanding space to set new required standards. It is rearranging every aspect that contains space or that space contains. It will bring a much different looking end. Everything about this concept is missing from Newtonian science because Newtonian science failed to investigate Kepler. Kepler said space a^3 is equal to the motion T^2k thereof and then that says without Kepler directly saying it, it says that if space a^3 goes bigger as a result of the explosion then such increasing in space will constitute to more space a^3 which has to produce an increase in motion T^2k where more motion T^2k will bring about faster displacing space. This is one small fact that Newton robbed the world of realising with his ignoring of Kepler's work.

We are now serving time in the twenty first century. One Professor once told me I must realise that Newtonian science took man to the moon and back several times and in such a view I am rather annoying presumptuous to criticize Newton. The Professor missed the point. I criticize Newton on what he did not give us, which he gave us as incorrect by his own admitting that it is mostly guesswork on his (Newton's) part and his guessing about the facts where later that guesswork became institutionalised facts believed by all concerned to be correct and to be proven to a degree of correctness that is far beyond doubt. Newton gave us gravity but Newton never gave us the explanation about gravity. At the time Newton met strict opposition from his colleges and piers because others felt his introducing of an unexplained force was taking Science back in time, which of course it did. Many scientists at the time accused Newton by name of dragging science back in the wrong direction of progress by introducing unexplained forces acting in a superstitious and mediaeval manner. I went one step further by asking myself the question: If space becomes more when heat becomes uncontrolled why can space not become heat when space is under control? If space becomes more as we see with every explosion of every kind and such heat forming space releases energy, then why would space being managed not form heat being under control and produce energy. We only have to see what Kepler said gravity is. Motion gives us energy.

Where space is the least, which is in the centre of the circle, gravity is the strongest. The gravity located in the circle's space less centre holds not only the sphere together but all that is in the surrounding of the sphere outside the sphere as well. It is from there in a giro action that gravity bonds all atoms forming the structure of the sphere as one unit together in a unit as well as distributing a specific alliance in shape and form. How the atoms manage that we will get to in a while, but there is a law allowing for that to take place.

Gravity is the strongest in all cosmic structures holding the form of the sphere and gravity controls all around from that very centre where space is the least, therefore the more material there is to generate motion within a star the more secure would the generated centre be in any star where such centre produces gravity. It is not the material but the motion the material accomplishes that becomes the factor of gravity. The smaller the star is as far as volumetric occupation goes, the stronger the gravity is that is coming from such a centre. The less the space there is the less the motion is and therefore the stronger and more deliberate the motion is evoking gravity. From the centre in the middle where space is absolutely at a premium the gravity grows stronger as it draws all material.

Nothing in the Universe is without motion because "nothing" as a factor is "not any" factor in the Universe. Other than nothing everything else is in motion because motion brings separation and points boundaries that establish individual space. Therefore what is in space is in motion. However, as Einstein pointed out motion is a relevancy between factors of different motion in space through space. When looking at the moon on a semi cloudy night we can place the relevancy on the clouds being without motion and the moon travelling at a dazzling speed or we can put the motionless relevancy on the moon and find the clouds travelling at a large speed.

Space consolidating by claiming independence by as space 1

Space contracting to centre operating as a holding space

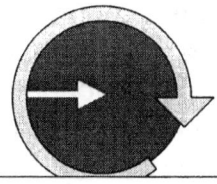

In the normal flow of events there is a certain ratio of liquid time that stands related to a specific gravity that is generated by the motion of the liquid in regard to the space that binds the solid.

$$k = a^3 / T^2$$

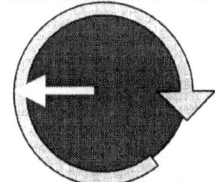

The solid on the other hand confirms the motion whereby it acknowledges the boundaries where the motion increases the limit on the space ending. It is a reflection on the motion providing the space with gravity that steak off boundaries.

$$k = a^3 / T^2$$

When an explosion occur the flow of liquids amplify by many fold. There is then a much bigger concentration of liquid heat forming time that increases the gravity to which the space has to respond.

$$10k = 10 \, a^3 / 10T^2$$ The singularity reacts within by counter measuring the excessive gravity established by the massive increase in liquid flow that establish much more space than with which the governing singularity can cope with.

Since there is an excessive demand on more space because of the liquid becoming more and thus demanding more space the singularity that is generated responds to the increase in gravity. The liquid demands enlist more space within the boundaries of a^3 with the increase of T^2 that brings on a larger k in both directions.

$$10k^{-1} = 10T^2 / 10a^3$$

However the gravity bonding the material are unable to compromise for the increase in space that the increase in liquid demand by providing more motion.

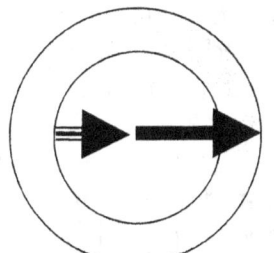

The sudden increase in space that the motion demand by providing a much accelerated gravity relevance k can not be accommodated by the sudden expanding that the material has to provide since the space the material holds is insufficient to solidify the space the motion requests.

The expanding calls for a stronger gravity in more space, which the material cannot provide, and the material relinquish the space it holds because it cannot withstand the call on a bigger demand of space by more liquid motion.

The motion is one of confining the space to a centre by the moving or trying to move the flow of space and whatever is in the space into the centre where the space is least. Take the Neutron star and the Black Hole as an example and compare that with the Sun and the answer is simple. I claim that gravity is all about reducing space and not attracting matter but that I explain a little later on. Therefore the matrix of gravity must be permanently located in the location where space is the least. Looking at a sphere we find that what holds the sphere true to form is placed in the centre of the sphere, which then has to be the most intense point of gravity. Gravity is confirming the round shape without favouring any specific point. Such evenness of gravity comes from what is applying at such a centre and is in control of the surroundings. The centre that secures all of the space and material in the space holding the specific form has to be round if it is anything. That shows that in the sphere one can see that the sphere as a form is dominated or controlled from one specific location in the centre. The explanation about the reason there is control coming from the centre, has a very childlike simple answer.

The Big Bang was where gravity held the Universe in the least space there ever was. To find the original gravity we therefore have to reduce the sphere to the circle and reduce the circle from there narrowing the circle down to as far as one can go. The Universe is a magnitude of spheres constructed by a complexity of circles. This is because everything sprouted from singularity. To narrow any circle down will be the same as narrowing down the Universe. In our reducing of the Universe we must first acknowledge that the Universe constitutes many spheres, which is giving the Universe gravity as a combining unifying part which is the part of the sphere giving the sphere form (or gravity) and that confirms that the sphere is a circle in many times over multiplying the positions from where gravity secures form. If we wish to go back in time by taking the Universe back down the same route and at the same time maintain some coherency we must concentrate on a single circle because a sphere is a circle by millions of possibilities linked together by just a name that changes the concept. When one takes this accepted route in thinking that by reducing the connecting line to the connecting circle point in the centre of the lot, it must take us back in time at the same time as the circle reduces to the time during the Big Bang.

During the Big Bang where all circles were as small as they can get, we run into an unknown substance we came to know as antimatter. This theory is propagated according to Mainstream science but what is most surprisingly is that I do agree with this part of the statement. All material produces gravity. I go one step further and say all material applies motion where some motion may be to contain by using gravity attributing to the contracting that leads to the reducing of their space. Then as everything in Creation has an opposing to restore and maintain balance, there had to form another or other material that did not by our lamentable standards produce gravity because those material produce antigravity, a concept beyond human discernment. Antigravity must be the expanding in counteracting contracting. A counter action to contracting is where expanding provides growth of space to that which has then reduces the gravity effects. Forming pappy provides more space by losing density to the advancing of their space. Material either have gravity by solidifying or concentrating the space they hold in ratio to the material within the space they hold whereas others lose their solidness by entertaining more space within the ratio of material to space where such material becomes liquid and in more extreme cases they become gas. Being a gas they float which gives that material a high degree of antigravity being airborne. It is however not clear if antimatter produced gravity as it did when it went to lunch on and ate up all material in the immediate surrounding. It was cannibalistic but the unanswered question is this: was it a gravity producing predator or a non gravity-producing carnivore. Did material find a comrade in their gravity forming of form or did the gravity it produced bring on the demise that subsequently followed the event as is reported by the highly informed.

The Accepted statement on antimatter reads that matter composed of anti particles where such subatomic particles that have identical rest mass to corresponding particles of ordinary matter but opposing charge and are opposing in other fundamental properties. One example given is that an electron would have a positron, which then functions as the anti particle and has a positive charge compared to the electron's negative charge. That is put bluntly in its utmost simplistic form. Unanswered and tough questions arise from such a statement. What kept the electron bonded to the atom since the protons must by implication produce expanding or by definition be repelling the atom and surroundings instead of the normal contracting or confirming of form. What is a positive compared to a negative charge, because it is human concepts that put the directional qualities of material into a positive or a negative contexts as we did with hot and cold. It is human standards that humans brought about to make all human inadequacy by lamented human understanding better but it is not applied cosmos principle. If there is extracting electrons performing in the capacity as antimatter, then there better be protons by other name in service to the anti electrons, which then of course serves the anti electron in the capacity of an anti proton with an equal but negative charge to that of the proton. When matter and anti matter meet, the two opposing particles annihilate each other until one vanishes from the Universe. I have to add that at the time this theory was devised the first computer games became a crazy fashion played by young and old, those wise and those foolish all alike. This game was called the packman and the packman ate up all the skulls and after eating left nothing as evidence. The theory about antimatter has some very striking similarities to that packman game. It still does not answer the most ardent questions: What makes a positive electron different from a electron in the working place each has and can any person show such an object found in nature.

Can people take a positive electron to an investigative bureau and be acclaimed for bringing about such evidence? It is unwise to substitute nature with human concepts just to further mathematical equations. This was apparently presented as normal as nature was when nature developed with the Big Bang and nature then did behave this oddly just after the Big Bang came about. But one huge misgiving in this argument is declaring that everything the antimatter had as a meal vanished and even moreover then antimatter went and vanished too. Where could the combination that was produced when the matter and antimatter collided go after it disappeared and did it form the by-product of antimatter science is talking about, which since then apparently vanished too. What a bloody none-intellectual fairytale that is on the in addition as well as one of those made–up-as-they-go-along stories, which is told by persons that supposedly should know of better. Since there is no place other to find a location to be within than being in a place inside the Universe it is hardly possible to vanish from the Universe except in fairy tales because for one simple fact: there is no other home to have but the home we have which we call by the name the Universe and we have no where to escape to but within the walls that the Universe provide for such a purpose.

There is one Universe containing all and preserving the lot. Mainstream physics is accepting this fact. But then by the same margin they accept a principle that allows property that once was part of the Universe to leave the Universe and go somewhere outside the only Universe. They create a loophole whenever it suits them to misplace what they cannot explain readily and logically. In Creation to their and my thinking there can be no hiding of anything but in the Created Universe. This they admit and confirm although with the same breath those very same intellectuals also admit that there is another place outside of what we are able to find in the Universe. When someone comes up with the marvel where such a person can declare in all honesty that the product of antimatter or singularity escaped from the Universe to God knows where that person should leave the field of science and go for fantasy writing such as fairy tales or reporting about politicians inner deepest chastity and integrity. That is what we can find outside the spectrum of what the Universe can deliver. With such a statement on anti matter or the loss of any Universal product disappearing from the Universe then alarm bells should go off in the mind of the trained and professional Scientist working with such matters.

Yet those in charge do not once belabour the question on the validity of a statement that involves stating losses occurring of substance being in the Universe going lost or being removes from the Universe. Their surprise of pressruns stating the **loss** of factors and declaring the possibility that there was a possibility of such losses where those factors now are outside of what once was part of the only place there ever can be. They can read mathematical calculations and agree on an outside the Universe without stating it in an explanation what happened to the lost and found or they're ability of introducing the concept as a reality, which they claim it is. That such factor can go outside the Universe and leave the Universe by causing a Houdini vanishing act of never –to-be-repeated-again status. Science would have us to believe this antimatter went into hiding in a manner that is out of the Universe.

They applaud this thoughtless presumption while fully knowing that at the time they do this acknowledging that there is no other place for anything wishing for a place to be within then having to be in another place other than inside the Universe. If it was ever anywhere it still is within the Universe merely because there is no other place to go than to be inside and part of the known Universe! There cannot be some factor and then misplace it as if a valid factor calculate the value can prove the disappearance and by disappearing it no longer is. If it was in the Universe it must still be in the Universe somewhere. Then we better start looking for it.

Another big issue is that what ever the Big Bang produced must be in equal terms everywhere. The Big Bang was a process that had the Universe act as a high-speed cocktail mixer of no repeating ever again. Whatever the Big Bang was of all that it was, the most it was in the beginning was that it was one massive mixer mixing everything in it at the speed of light. The relevancies might change slightly and balances may change favouring opposing ends...yes and known appearances did change...yes. But in the end all the factors must always be present everywhere through out the Universe. By this lacking of a fundamental explanation about what antimatter will look like when found Mainstream is incredibly poorly judged by scientific standards. Those mathematicians calculating physics suggest that science should take antimatter as a cosmic fact and then in disregard of other realities they dispose the truth by discarding its properties onto the unknown.

That hardly suggests plausible science by any one's admitting. In the cosmos is, was and will be all the material there can ever possibly be. Our concepts we put forwards can be faulty but nature cannot ever be at fault. Our arrangement of our ideas can be at fault, but we cannot pull a vanishing act on certain cosmic products and in doing that then dismiss the existing of such a factor or factors, which we then claim, have vanished in the further developed Universe. Our concepts of what they became may be at fault and by changing some basic principals such changing may produce a better understanding about what we think we read into mathematics. Mathematics is purely a language and mathematicians are purely translators. Mathematicians translate from the language they read to the verbal equivalent they speak. As in all translations made, certain concepts may become misinterpreted.

The terminology used to explain this is "lost in translation" Mathematicians must see what there is in the translation and try to incorporate what there is available in the cosmos to what the Mathematician sees in his mathematical calculations. The Universe was full of heat and it was full of material but it was not full of free space. If that is the case then where did the heat come from and where did the heat go? Hiroshima and Nagasaki taught us many things about the horror of human nature but most of all it taught us that material is heat secured in atoms and atoms are heat tightly wrapped in a cocoon, which we named the atom. Heat in any form cannot have anti in another form. The package holding heat wrapped can unwrap as it does with nuclear atomic demise. But the anti to heat is cold and cold is space.

The undeniable fact about the Big Bang theory is the accepting of a growing state in which the entire cosmos seems to be in. With all the expansion that went on we came to the point where we now are at and in such growth all aspects in the Universe must grow in relation to quantifiable progress in all different aspects, which takes us to that which is seen and that is unseen and which came along as products in the Universe where everything took everything on a growing spruce by unveiling space. That is where we now are. Such expansion include all there is including everything and not just with outer space growing. The dynamics of outer space alone cannot grow by leaving the growth of material behind. Should we wish to see where we came from we have to reduce that which we now see in our surroundings to apply to the measures that once applied in all aspects of the cosmos. Mainstream physics is over pronouncing the growth of space and with that suppresses the part matter must play in such growth by simply ignoring the issue. That is the reason why they prefer to ignore the evidence that material is growing notwithstanding that material is growing or that their disbelief about the matter of material growing do not change that material is growing in any case. Because they cannot find any reason why material should grow they refuse to admit that material does grow.

This is hiding from the truth by hiding the truth. If space grows and the Universe is getting bigger then all space grows to allow the Universe to get bigger. That includes matter and space not in matter. Space can only grow if materials that also hold space also grow within the space that is growing with the growing space. It means that stars get bigger by the cosmos growing from the Big Bang onwards and outwards to the moment in which we are at the present. But if stars grow then the atoms forming the stars are doing all the growing as they secure more space within the space they claim.

If Hubble saw space grow, the growth of space must include the growth of space holding material as well. In studying the Hubble's expanding theory we come across evidence that makes it clear that all material expand in a manner as if the expanding comes from the centre of each and all particles within the expanding space and the expanding grows outwards from every particle centre. It is using every star centre to grow from in all directions proportionally in all directions evenly. This leads one to believe that gravity is this securing of space in the material just as Kepler showed it to the world. It proves a connection with deliberate implications coming from every as well as in every specific centre. It proves that the centre $k^0 = a^3 / T^2 k$. It becomes apparent if and when separating Kepler from what Newton thought about the work of Kepler which Newton accepted as being inferior and all incorrect.

That is making the Universe small and as man grows man allows the Universe also to grow in relation and corresponding to man's ability to comprehend. We see the cosmos as a circle and we accept the circle because the circle is what gravity implement when the choice of form is coming from material that has all options to freely choose from. By taking the circle back one will follow will trace the rout of the cosmos to where it then started.

All stars are many circles in many dimensions, which form when all circles join into what we call a sphere, but that leaves us only with the circles in the plural. Taking the cosmos back can only lead to one point and that Kepler told us we will find singularity $a^3 = T^2 k$ which is $k^0 = a^3/T^2\,k$. We can only reach $k^0 = a^3/T^2 k$ if we repeat $1/k = T^2/a^3$ in a continuing manner indefinitely. When one does the effort of reading this correctly, it says that when distance k brakes from singularity $1 = k^0$ that is then $(k^0 = 1)\,/k = T^2/a^3$ where the space a^3 produced a time T^2 equal to singularity k^0 and singularity k^0 is equal to eternity which was where all was equal to a never changing cosmos that was holding the single form into one dimensional space that included all the filled and vacant material filling in from all sides.

This is one way of looking at the issue and by doing that I am about to prove that singularity is Π. I am about to prove that not only is the planets adhering to the Titius Bode rule of seven over ten and ten over seven in relation to the Roche limit but that the Roche limit explains the very, very first instant the Universe experienced outside eternity. The atoms relates to space in the very same manner of seven singularity positions to ten points and from this motion of material interacting with space is securing material on the inside as well as on the outside. By that motion gravity comes about finding the value of Π^2. Gravity uses the relation of the Titius Bode seven on ten and ten on seven as well as the Roche factor to form gravity and gravity is always Π^2.

Once again one must look at the process from that which we find the cosmos uses in the atom. There is a solid centre, which the body of the Earth takes on which is covered by the atmosphere. The atmosphere is the liquid or neutron and the moment that the independent object moves that object then becomes the neutron. The role of the neutron is to establish the fringes of the electron. That will then be the exact role the moving object takes on. However when the moving object start moving faster that what the atmosphere being the neutron moves, there is a huge discrepancy in time relation between the neutron and the moving electron this discrepancy we call the sound barrier.

However when breaking the barrier and the satellite enters the atmosphere there are laws applying that changes

k_1
k_2
k_3
k_4
k_5

The Coanda effect starts

Any object entering the Earth atmosphere is starting an Earth wide Coanda effect by establishing a space link through motion.

From the centre running through occupied space singularity positions space-time at a point that connects by the seventh dot and in this, a straight line forms from the centre out to the edge of space. This is then coupling the object with singularity controlling everything within the realms of the Earth by using the line running straight but abides the 7^0 points. When the object accepts the duplicating and the dismissing, motion of the earth such an object maintains a perfect relevancy with the singularity in the centre of the earth.

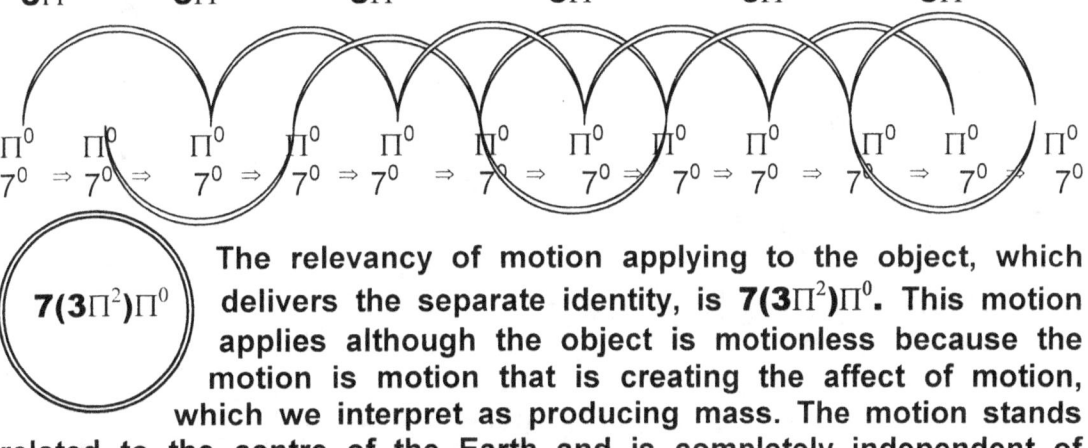

$3\Pi^2$ $3\Pi^2$ $3\Pi^2$ $3\Pi^2$ $3\Pi^2$ $3\Pi^2$

Π^0 Π^0 Π^0 Π^0 Π^0 Π^0 Π^0 Π^0 Π^0 Π^0 Π^0 Π^0
$7^0 \Rightarrow 7^0 \Rightarrow 7^0 \Rightarrow 7^0 \Rightarrow 7^0 \Rightarrow 7^0 \Rightarrow 7^0 \Rightarrow 7^0 \Rightarrow 7^0 \Rightarrow 7^0 \Rightarrow 7^0 \Rightarrow 7^0$

$7(3\Pi^2)\Pi^0$

The relevancy of motion applying to the object, which delivers the separate identity, is $7(3\Pi^2)\Pi^0$. This motion applies although the object is motionless because the motion is motion that is creating the affect of motion, which we interpret as producing mass. The motion stands related to the centre of the Earth and is completely independent of whatever may present the mass factor.

This I see by reading Kepler's work as Kepler produced the work and introduced the work as $a^3 = T^2 k$. With this formula $k^0 = a^3/(T^2 k)$ must also be true because $a^3 = T^2 k$ is a relevancy that has to be in relation to singularity and therefore singularity must be $k^0 = 1$. Where will we find $k^0 = 1$? When an object is within the singularity alignment of the Earth, meaning it is either stationary, or free falling on pure "gravitational" momentum it holds the $4 \times \Pi^2$ in relevancy to Π. What I try to say by that is that the Earth holds all objects within the atmosphere to a displacement value of $3\Pi^2$ within the seven are where it is using 10 as the liquid in space. Because the aircraft does not leave the Earth's atmosphere the 6^2 and the 10^0 does not come into effect.

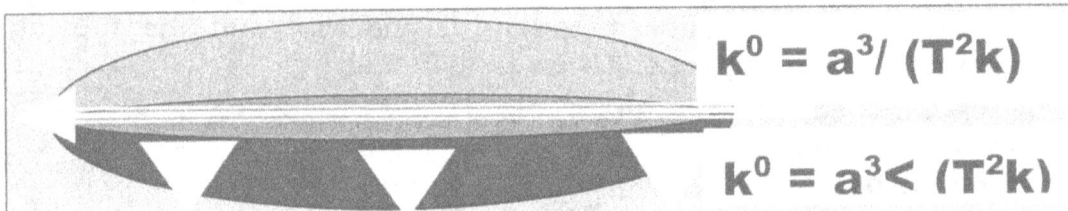

$$k^0 = a^3 / (T^2k)$$

$$k^0 = a^3 < (T^2k)$$

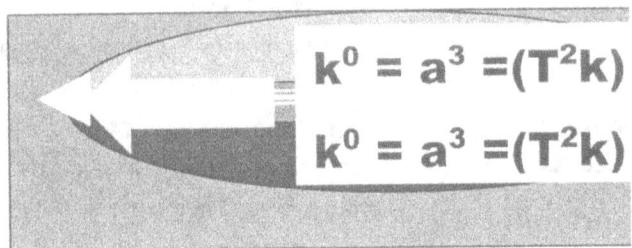

$$k^0 = a^3 = (T^2k)$$

$$k^0 = a^3 = (T^2k)$$

It is when the motion exceeds the mass the aircraft has the ability to break the sound barrier. Galileo proved that no mass is present in falling, which is also matter in the process of flight and because of that can the sound barrier become some form of constant.

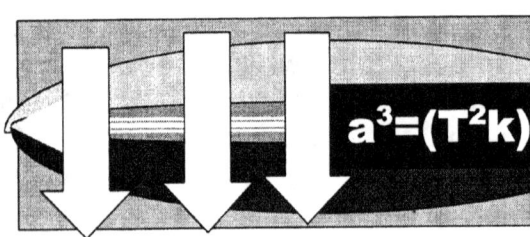

$$a^3 = (T^2k)$$

Mass is the refusing of any object to dismiss the form it has and to join the Earth solid structure. Mass cannot and does not contribute to the establishing of gravity except by depleting space through motion and such numbers of the protons in a space forming an exclusive unit.

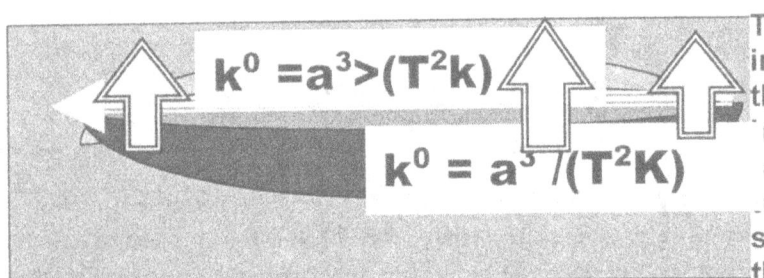

$$k^0 = a^3 > (T^2k)$$

$$k^0 = a^3 / (T^2K)$$

The establishing of independent motion of the craft secures an individual gravity and such individuality leads to the breaking of the sound barrier because the one gravity can no longer subdue the smaller motion, which is producing gravity.

When the aircraft stands, still the sun provides such a pivoting centre to the Earth and that also include the aircraft. When independent motion of the aircraft comes about the point of relevance then shifts from the sun to the Earth centre where there is a line contact between points in the singularity that the Earth holds.

$7\ (3\Pi^2)\Pi^0 = 207.2$ km/h. $\mathbf{R^2/T} = 7\ (3\Pi^2)\Pi$ and $\mathbf{R/T} = \Pi^0$

The Roche limit also changes to accommodate this change and becomes either Π^0, Π, or $(\Pi^2/2)$. Falling "free" will then mean that the object holds a position of $7\ (3\Pi^2)\Pi^0$ to singularity. In the half circle applying two of the quarters in time, the position is in the triangle of singularity placing the half circle in the first quarter. Any further linear movement will follow the triangle second line by multiplying the line by the value translated as a number of Π^0.

$7\ (3\Pi^2)\Pi^0 = 207.2$ km/h or $(3\Pi^2)2\Pi^0 = 414.5$ km/h. $\mathbf{R^2/T} = 7\ (3\Pi^2)$ and $\mathbf{R/T} = 2\Pi^0$

This is the first barrier but we only see the linear part as a value and peaks at $5\Pi^0$

$7\ (3\Pi^2)(\Pi) = 651.13$. In this $\mathbf{R^2/T} = 7\ (3\Pi^2)$ and $\mathbf{R/T} = \Pi$ where then the linear component Π becomes the second line in the triangle. triangle of singularity. The linear component or negative displacement can go as high as 2Π but then another barrier would come about and in this we star to locate the principle behind the breaking of the sound barrier. There are definitely TWO barriers to comply with when breaking the sound barrier

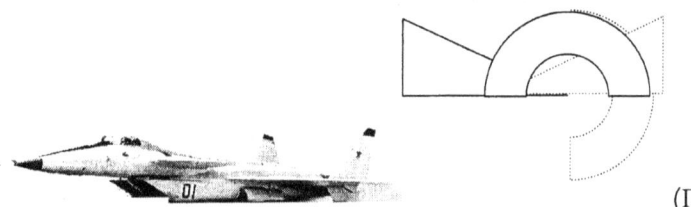

$(\Pi^2/2)$

$7\ (3\Pi^2)(\Pi^2/2) = 1022.79$. In this $\mathbf{R^2/T} = 7\ (3\Pi^2)$ and $\mathbf{R/T} = 2\Pi$ where then the linear component Π becomes the second line in the triangle. triangle of singularity.

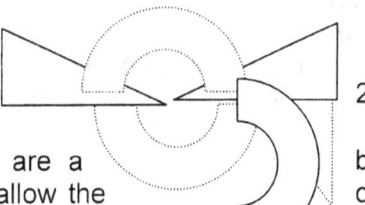

$2(\Pi^2/2)$

$7\ (3\Pi^2)2(\Pi^2/2) = 2045.59$.
From this point on and above there are a
can cross because singularity will not allow the
quarter by any object.
boundary no one
crossing of the third

Stars can and stars do **overheat**, sometimes and the **Polar Regions** where **the Titius Bode matter-to-matter applies** holding the square matter (7+7) in relation to the square of space (10) and **other times** in a double relation to the **square of space** 10 to that of matter in a half square (7 /10 or 7/ 10). Saying that one has to differentiate between heat and overheating because a star represents the coldest space in the universe and not the hottest space.

Everything that is in the Universe is heat. There is liquid heat that can flow and there is solid heat that cannot flow. The solid state or liquid state has no bearing on human perception but is in motion. When the body moved in relation to the body that does not move the body that moves are liquid and the body that is relevantly stationary is a solid. Movement is liquid and immobility is solid.

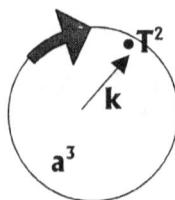

$a^3 = [T^2 = \mathbf{7(3\Pi^2)}]\,[k = \Pi^0]$.

The Earth holds a specific size in relation to the motion of the individual object being the Aircraft $T^2 = = \mathbf{7(3\Pi^2)}$. Since the craft is stationary the distance between the craft and the object is $k = \Pi^0$

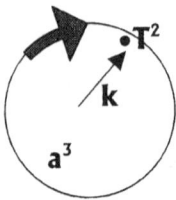

$a^3 = [T^2 = \mathbf{7(3\Pi^2)}]\,[k = \mathbf{2\Pi^0}]$.

As the motion of the aircraft accelerate the Earth still holds a specific size in relation to the motion of the individual object being the Aircraft $T^2 = \mathbf{7(3\Pi^2)}$. However the connecting flexible link being the atmosphere which plays the role of the neutron has to extend in order to compromise to being $k = \mathbf{2\Pi^0}$

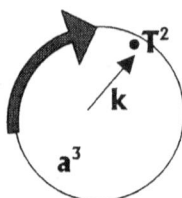

$a^3 = [T^2 = \mathbf{7(3\Pi^2)}]\,[k = \mathbf{2\Pi^0}]$.

The Earth holds a specific size in relation to the motion of the individual object being the Aircraft $T^2 = \mathbf{7(3\Pi^2)}$. Since the craft is now in motion the first beacon to arrive at in relation to singularity Π^2 extends the distance between the craft and the object to $k = \mathbf{2\Pi^0}$ then $k = \mathbf{3\Pi^0}$ and so on.

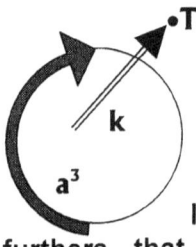

$a^3 = [T^2 = \mathbf{7(3\Pi^2)}]\,[k = \mathbf{2\Pi^0}]$.

As the motion of the aircraft further increases the Earth still holds a specific size in relation to the motion of the individual object being the aircraft $T^2 = \mathbf{7(3\Pi^2)}$. However the connecting flexible link being the atmosphere now has to extend to being the most furthers that the neutron can possibly extend and such extending can compromise up to being $k = \mathbf{5\Pi^0}$

$a^3 = [T^2 = \mathbf{7(3\Pi^2)}]\,[k = \mathbf{2\Pi^0}]$.

When the motion of the aircraft further increases the Earth expands beyond what the limits will allow being the Roche factor of $\Pi^2 / \mathbf{2}$. This puts a cap on the specific size in relation to the motion of the individual object being the aircraft $T^2 = \mathbf{7(3\Pi^2)}$. In this the neutron of the aircraft parts in a dimensional time value from the neutron the earth holds and the connecting flexible link being the atmosphere then cannot extend beyond what the neutron can possibly extend and such extending breaks down at $k = \Pi^2 / \mathbf{2}$

Stars can and stars do **overheat**, sometimes and the **Polar Regions** where **the Titius Bode matter-to-matter applies** holding the square matter (7+7) in relation to the square 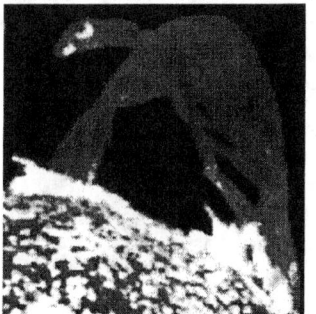 of space (10) and **other times** in a double relation to the **square of space** 10 to that of matter in a half square (7 /10 or 7/ 10). Saying that one has to differentiate between heat and overheating because a star represents the coldest space in the Universe and not the hottest space. **Heat and cold are relevant dynamics** forming **in appreciation of singularity. The sun is the coldest place in the solar system** and that is fact. Looking at evidence the sun provides it contradicts everything science wishes to believe about cold and hot. Science wishes to see the cosmos through the eyes of what fits the needs sustaining life on Earth and what benefits maintaining surroundings in support of life as one finds on Earth whereas life has no part in the cosmos except for the speck of dust we call earth. Looking at the cosmos impartially to life the evidence supports another view. Every aspect in **the cosmos is the very opposite of what science believes** it is. The sun is **not a ball of gas but** a **giant sea of liquid**, frozen **without any** form of **gas or air** in the interior. Having a liquid interior **the sun** has **no pressure** but has the **very opposite of pressure** to which there is yet no name given. **The liquid comes from singularity freezing** space-time within the atmosphere of **the sun**, and such is the case with all stars still in the shining phase. **Stars more developed than the sun is frozen solid causing fusion.**

In **the picture to the left** we find not withstanding whatever name we attach to the **red liquid substance flowing from the sun** into space and back to the sun, **that liquid is heat** in a very direct form. **If outer space was the coldest place in the solar system** the heat **should** immediately **escape to outer space** and **not return to the sun** as it clearly does. If **outer space were colder the heat would not return to the sun**.

All elements forming matter in as much as the heat forming **an atom is** as much a liquid as it is a gas and a solid. **There is no hot as there is no cold. It's about storing energy in space or in heat, which is another Cosmic equal being opposing similarities.** Hot and cold are  relevancies **brought about by singularity valuating space-time** and during **the Big Bang** the Universe was **freezing cold** at **three billion degrees C**. It is the relation matter has with heat that provides the form the particle has at that moment. Increasing or decreasing the heat will alter the form of the element. Therefore all elements forming **matter is as much a liquid or not than it is a solid or a** 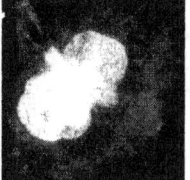 **gas.** It is the space surrounding the atom that provides the state of relevancy in forming a liquid, a gas or a solid which the atom find its relativity in relation the rest of the atoms it share space with. Hydrogen is as much a solid as tungsten is a gas depending on the heat in relation to the space matter is within. Should **you reply** that it is **the gravity pulling the heat back to the sun**, then that **confirms** my theory that **gravity is all about collecting heat onto matter** with outer space being the hottest place. **It is the concentration of heat in space being relevant to form. When overheating a star turns its liquid to gas whereby it merely transforms it's interior to a relevancy it has from pre- to post- Big Bang.**

One thing we humans have to realise and that is that heat expands and expanding is producing while claiming more space. It is having what there is just more of it proportionally than the rest has or more than what was. If it contracts it is colder notwithstanding our human mentality and when it expands it is hotter notwithstanding our human perceptions. When it moves it is hot and when it does not move it is cold.

We humans on Earth think that hydrogen is a liquid at -259^0 C but that only applies on the earth. The picture clearly shows the **heat in a liquid** flowing **from the sun** and **back to the sun**. In the **sun the hydrogen holds enormous quantities of heat in a liquid at a temperature of 6500^0 C.** When a star has its singularity secured the star is bitterly cold because it has heat in a liquid form flowing back to the point of singularity although

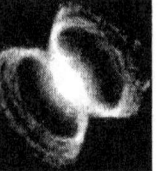

we may regard the star to be rather on the hot side. The sun (fore instance) freezes hydrogen in a liquid form at 6500 ^{0} C. If hydrogen remains a liquid at 6500 ^{0} C, just think how cold it must be as the star's interior approaches the point of singularity. Therefore fusing protons comes from cold and not from heat or pressure.   By allowing the singularity to overheat the star overheats and heat within the star flows

from singularity to outer space freely. In such an event outer space is then colder than the star because the heat is released into outer space with no intention of returning whereas in the sun it returns as soon as it leaves.

There are two ways to reduce heat; one is to bring about expanding space, as the photographs clearly show. The second one is where heat will reduce when in motion by spin. When withholding or retarding the motion in which the matter is moving, matter will overheat. Gravity is the motion of unoccupied space through the dimensional transformation to occupied space. Motion and space therefore is the anti-, the opposite the negative to heat being the positive. With singularity overheating the expansion of the singularity drives heat into space, creating space to compensate overheating **That is a natural phenomenon**. The only reason why **heat will** rather **flow back** to the star than **escape to outer** space once the star released it into outer space is **if outer space present more heat than does the star,** because **heat always flows from hot to cold** no matter what influences may arise. **Outer space must hold more heat than does the star but the accumulation of space in relation to heat makes it seem colder bringing expanding of heat to become space. Space and heat directly relates being in one or other form**.

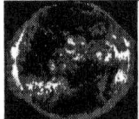

The cosmos is all about **converting space to heat** which we see **as gravity** and **returning heat to space** as a **control mechanism** always **keeping** a very delicate **balance** which we see as **a star shining or being normal.**

The purpose of the converting of space to heat is to supply the core where singularity is with heat. **It turns space to heat** sustaining matter but sometimes singularity overheats and then matter converts to heat allowing heat to convert to space. That we call many names amongst others exploding into super

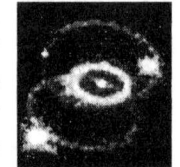

nova. Whatever names are used, it is not important because the **process rests on space and heat interacting to form energy**. That

was what **the Big Bang** was and **the Hubble Constant** is all about where **matter converts heat to space.** I show that **space and heat is the very same thing** and there **is no such a thing as pressure** but releasing **heat produces space** and **concentrating heat reduces space** with the two interacting on singularity demand setting time to space with time being the spin or motion of heat in space. **Heat and space form the second singularity** caused by the **fragmenting of singularity to compensate overheating during the pre-** Big Bang matter forming era. That is what we see as **light and space,** which again is the **same thing and is fragmented singularity forming radiation and heat, where the star re-transfers heat back to space due to an overload.**

When I make controversial comments nobody finds reason to listen to me. Everyone finds an incoherent novice trying to make sense of some incompetent view that strays from the accepted. Nobody takes the accepted and compare that with the n views I have.

Every position in the universe either holds singularity in a form, or relates to singularity. There can be no position unrelated to singularity therefore every aspect of the cosmos is space-time in various forms under the provision of singularity connecting. Matter cannot be if not surrounding singularity

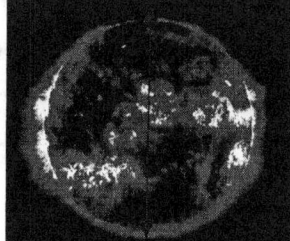

Singularity is as close as any spot can ever come to zero **BUT IT CANNOT EVER BE ZERO.** From singularity diverts space-time and there cannot be space without time as much as there cannot be time without space, not withstanding the size of space or duration of time.

Through space-time singularity connects as much as relates linking the universe into a network of influences beyond what ever we can ever conduct. There can be no spot that does not participate in the curvature of space-time. From the point of singularity runs space holding time to the prescription singularity dictates.

The spinning top is all the evidence any one needs to come to such a conclusion. In the past when I have acknowledged the fact that I have no academic background in the field of cosmology, this admitting brought about discontent, or rather dismay and I might even add some scorn from academics. In my opinion this trend of behaviour is uncalled for as every person is entitled to be opinionated. Every person has an opinion and it is the opinion that has validity, not the persons social standing. It is the way one absorb and reflect a personal view about matters that are important and not the manner in which the person arrived at such conclusions but it is the validity of the conclusions that should carry importance. The cosmos contains matter, space and time in heavens or dimensions standing in relevancy to one another. Whether you refer to dimensions as dimensions or heavens it is no different because it is the same thing. Singularity brings about heavens as it brings about dimensions and singularity is a mathematical fact. A straight line cannot start at zero and still be a straight line because zero extending to wherever brings about a full zero. A straight line starts at the point where the pen point meets paper. That point may be any distance from infinity to a measurable dot, but it cannot be zero.

Any straight line is also half a square because the line forming the square cannot start at zero for the reasons I just mentioned. That is singularity pointing an eternal direction from a point of infinity and that is the basis of the cosmos as much as that is the basis of mathematics. To escape from nothing one has to become something and by doing that one could not have been in nothing in the first place. If one holds a point in nothing one cannot become something because of the nothing value.

—————————— 180^0 X 2 = 360^0 ⊏————————⊐

To back this argument that no line can ever start at zero is to ask the simple question: what will the length of the shortest possible line be. It must be a line where the starting point is so close to the ending point the distance parting the two is incalculable yet there is the line therefore the end and the start is apart still sharing the same spot.

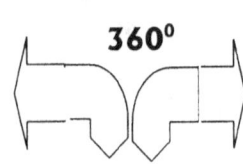

360⁰

SINGULARITY

To this end any shortest line will start run and end in infinity Should a straight line start from infinity and never be able to reach a point of zero (because that will bring in the factor of no line (0 X 1 = 0)) that means the line dips into infinity and it has to come out on the other side leading in the opposite direction of the first line in an attempt of avoiding zero.

The length cannot be zero because zero means no line. The starting point and the ending point may be inseparably the same point with virtually no space between the two points but neither of the three points can be zero simply because there is a line (be it infinitely small it is there). If the point is zero, then the line will be shorter than the shortest possible line. If there is a line and the two points starting the shortest possible line and being the continuing of the line it may still be the same point and even by sharing the spot as the point ending the shortest line possible the line must be there and the line holding the start and finish is next to zero but can never be zero otherwise there is no line.

If that is the case then the one side of the square also presents the point of singularity to the other line holding the other dimension. That then makes the straight line not 180^0 because it is half that of the complete straight line and indicating the point of singularity of the other line and that results in any form holding only one aspect of the full form. But because the two lines are in a relevancy to a common starting point, both will enjoy the remaining distance of flow still holding the relevancy in comparison to each other and that brings about a triangle that also has 180^0 in number. That means any straight line is also a half square which is a triangle bringing about 180^0.

The co-ordinates of referral will only hold three references as surface contact areas pointing in one direction of the possible six surface areas available. That is mathematics and mathematics cannot be bias or lie. Every object in the Universe holds three points of face value to any direction possible. You can only see three sides of any six-sided object. Because the lines has a starting point in infinity that starting point has to represent singularity outside the sphere because at some point the lines cross the border of being the shortest lines possible to being normal lines. That means every starting point represents singularity by measure of r instead of Π. Every point is also the point indicating singularity at another point pointing in the direction of a point holding singularity as the pointing line as well as the supporting line.

I explain r later on but r does not in any way refer to radius, as does the r that I use for that purpose. By the same measurement, I also provide the third line with a singularity as reference and a singularity of direction. The prominence of this will become most apparent when explaining the Titius Bode law in relation to matter relating to space and matter relating to matter claiming space, but as usual, I am getting ahead of myself. What I have just indicated is pure mathematics moreover the most basic mathematics there is. There are always two lines running in supporting but opposing directions claiming space from that one point of singularity and a third line confirming (controlling) space in a third direction.

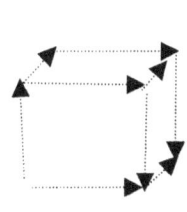 **Every line is as much part of a square than it is pointing in another direction relating to a position holding singularity. That brings about my conclusion in changing Kepler's formula of $a^3 = T^2 k$ to my own which is $R^2 / T \times R / T = 1$ where any of the three components can form a square in relation to a mutual point of singularity where each one is the starting point on that lines individual singularity**

That is space-time holding one square of space in time and directing time in space in a specific direction of flow. You may not agree with me on this issue at this point but my reasons for such a claim comes in other parts of the book and there I substantiate the claim with much more explicate mathematical detail. Space-time is matter-claiming space in time while directing space in that same duration of time from any point forming singularity. Now one arrives at the question of positioning singularity. Singularity cannot be in space because space is claimed from the point of singularity and space is confirmed from that point of singularity. If space is claimed and controlled Einstein must be at fault looking for singularity in outer space.

To find such a point one should once again look at the possible dimensions available. A square is six sides holding three to any direction. A circle is a square without edges. That is what we are looking for. A square without edges can only be a circle with a definite un-varying point of singularity indicating a specific location. A square can place singularity at any position because space holds no specifics where as a circle holds specifics at a defiant point. Space provides no specifics because an astronaut can and does drift in any direction when not being attached to a sizable object. That throws outer space out as a possibility.

Take the top with some astronaut to outer space and ask him to spin the top in outer space. We do not even have to bother such a busy person because we know the answer. The spinning top will not be able to turn because there is no stability factor in outer space. The top needs support from the needle running upwards, therefore the top will then have thrust running downwards, and the accumulative effort will bring about a point where the opposing points will allow a spin to occur. In outer space there is no such a point therefore the point we seek is within matter. A top cannot spin in outer space. We already established the fact about singularity being the position where two points originate in a square and the third will show direction. The top is a circle.

The difference between the circle and the square is in the direction the indicator follows and a square cannot spin, but through contact with a sphere as a circle cannot be motionless The factor of Π indicates eternal motion and NOT zero motion. There is a massive difference in that concept. If no line can have a zero point to start with where will the circle get the zero to indicate motion! This principle is the most basic mathematic rule and even I the ILL EDUCATED can see this. When the end of the rotation arrives the end rotation also announces the beginning of another rotation and not nullifying of the previous rotation because the rotation will have a line showing the effort it made and as it forms a wave, the wave will be there forever. The pitch may decline to a straight line, but the line remains. When calculating the motion a triangle does the honours.

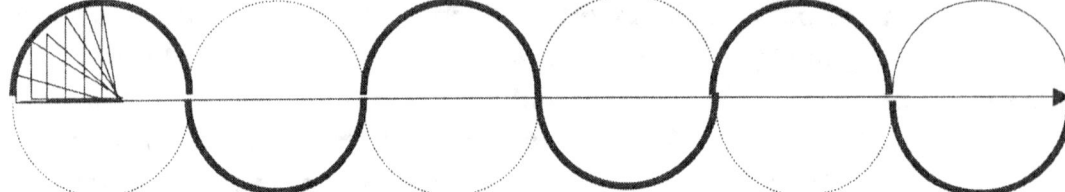

The wave confirms rotating directions followed by the circle as it spins. By stating that a wheel has a relevancy of zero by completion of a rotation such a claim denies the wave its entitlement of existing. The wave going flat, as it becomes a straight line also has an indication to singularity. All spinning matter has the point where the spin is still there but the radius is to small to measure by any means. That point is standing still in relation to the rest of the spin. In relation to that logic I do not accept that science holding the radius of a spinning object unaccountable in the spin, whether the spin is applying or not. It is in the very fact that the spin places the reference factor in the graph that makes the graph that astonishing accurate. It is mainly mathematics through the graph, which affect day and night and not the Earths standing in relation to the Sun that brings about climatically and weather changes and Earth development conditions.

Nothing said so far is high tech or mind bending complicated. All the above arguments from the first page to this point reached are simple and there's only ordinary primary school mathematics involved that every scholar should know. One does not need a brain fitting Einstein to come to these conclusions but just thinking about everyday issues.

All circles have something in common with squares. It has a surface but where the square holds a surface pointy the circle holds no points. To calculate a square there has to be a point where the line starts and that line will hold a value of at least infinity running from that point in two opposing directions. Arithmetic presents the possibility of zero and mathematics excludes such possibility. That is the difference there is between arithmetic and mathematical science. It is where mathematics departs from arithmetic and is as basic as counting is in arithmetic. There is nothing outrages about that which I have mentioned. Neither does one need the brainpower of a person like Einstein to come to such basic conclusions nor yet it completely destroys the claims.

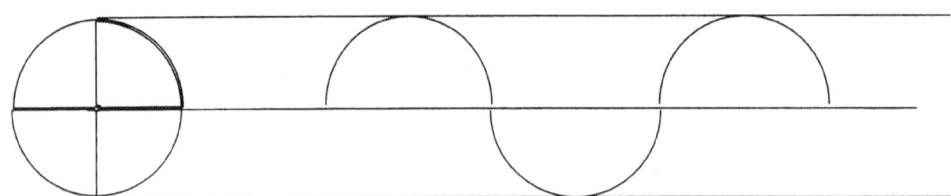

From the graph one can establish the link in the circle's rotation around a conforming unit being singularity.

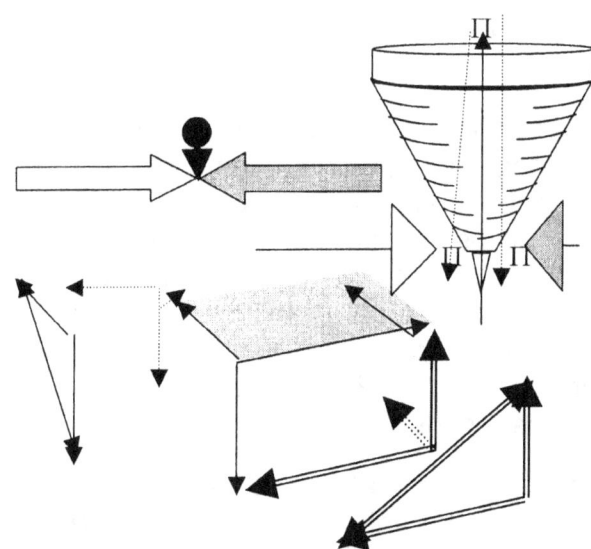

Being a circle means the thing, must be round and spinning. In that case, let us take an example well known to all, the spinning top. The top spins on the thinnest of points, and still maintains a balance.

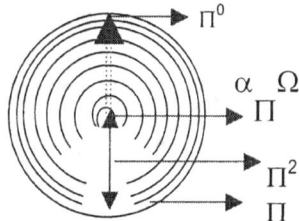

In the scenario depicting the square there is always some three pointing triangle everywhere and the common denominating figure is therefore three.

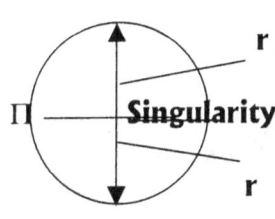

In the circle, there are also two lines where each line holds one point to singularity. Splitting the line in the two opposing directions, so in that way it is the same as a square, but the third line indicating direction brings about a difference that distinguish the circle from the square. The circle direction indicator is always Π placing the pointers at $r \times r = r^2$

Squares rotating will find a point where it goes into self-destruction through faster motion. This action is imbalance but in principle it has a point where Π forms a meeting with r and Π at that point forms r.

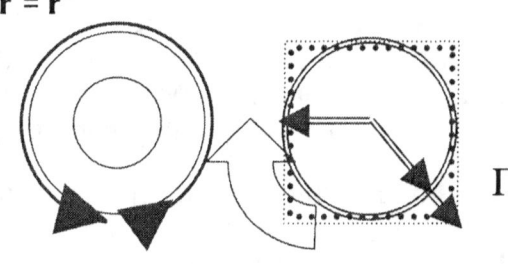

Applying the second law $F = m\,a$ one arrive at the formula $G(M \times m) / r^2 = m(\omega^2 r)$ claiming a zero influence between the radius and the orbit of planets. I shall later again return to this issue but firstly I would like to take your mind to one other thought that seems to have escaped every body. It concerns the medium science rely most on for the gain of facts and information about the Universe. It is the influence of light.

When realising the error of science in accepting a value as zero to be legitimate in mathematics, one can establish from that that the circle does not employ zero as a value after the completion of one rotation therefore $F = G(M_1 \times M_2)/r^2$ is invalid, one has to return to Kepler's $a^3 = T^2 k$ and establish a value from that.

From the graph one can establish the link in the circle's rotation around a conforming unit being singularity.

Saying that one therefore has to admit that the smallest spot has to hold space because the most insignificant dot can transmit light and being able to accomplish that, one must accept it to carry a value of something. If that spot had the value of nothing, it means that spot was not there to begin with. Holding space-time one should return to the original formula indicating space-time in as much as $a^3 = T^2 k$ where a = R and T = T. Being time it has to alternate positions and that can therefore only apply to **k** where **k** will indicate a relation to the space-time in question or the relevancy to singularity being $k^0 = 1$. This reality has serious implications on the speed of light we take as a constant.

Have you as you sit reading this part at this minute sat back and gave a thought about the light enabling you to read? Such a thought brings to mind the most simplistic answer one can imagine. The light hits the page bounces from the page and contact the lens of my eye where the lens conveys the photons becoming electricity to a part of the brain that translate the electricity to an understandable message and that makes one read. It is as simple as that! Ever gave a broader thought about light streaming across the night sky, coming from ends of the Universe we do not even realise is there? How does the photons manage to convey one complete picture coming from as far apart and as wide an area as it does? With a few photons connecting the eye or lens no one ever noticed the wonder of light. The photons reflect a view that seems as if coming from all the billions upon billions of stars. But most is coming from darkness covering an area no man can measure. Yet how many photons can actually connect to the lens of the camera or to the eye? Still a few photons coming from a single direction directly ahead eventually tell the entire storey. It is very simple to take the process of seeing by means of photon conducting very lightly and I have never heard one of the Brainy Bunch really in sincerity dissect the process to its potential. It is impossible that light from such an array of assorted sources can simply come together at the eye lens and show a picture of objects spanning across a Universe as wide as our mind can receive where the objects they reflect is beyond human measurement and the quantity is inconceivable many.

Light is much more than the medium science takes it to be. Light connects the Universe in a way we cannot contemplate. Light being far apart originating from regions not in the same time or universal space connects in a way that present us with a picture holding the Universe in an understandable content. From the point we stand and we watch the Universe the significance of what we see surpasses the sense of understanding of what we are experiencing. How can the few photons that our lenses catch coming from such a vast area covered by the night sky cover transmit the complete picture of what we see. Take a few seconds and inspect the picture of the night sky then rethink the picture applying the full content in the picture to what the size of your eyes. Think how big the picture is that your eyes take in and translate that area to the size of your eyeball in an effort to determine a ratio. One will be forgiven if one thinks of the ratio as eternal to nothing. Yet a few pages back I showed that according to mathematics there couldn't be anything as nothing. Consider the path the light followed from the source connecting to light from all other sources where all particles of the other light may come from and bringing a full picture to the lens one use to look through. In your mind connect a line from every atom producing light and connect the lines to your eyeball and see how you can manage to fit all the lines, as small as the lines may be.

Scientists think of outer space as geodesic zero, with nothing in outer space but space. Geodesic zero means the light travels in a straight line from where it originates unhindered all across space to where the light connects the eye. Such an idea by itself is outrageous because the stream of photons reduce in space to such a minute quantity, that taken the area the photons travel and the space in vastness it covers, the chances of one photon coming across many hundreds of light years through billions upon trillions of cubic kilometres of space and selecting my eye to convey the electricity is less than infinite. Yet such conveying takes place every second of every minute. The position of the location of the second singularity, which is the precise duplication of the first singularity but in a diminished capacity, is obvious to miss when one is not applying a detective mentality, as one should in scrutinizing the cosmos. Culture will have us believe that when one sees a colour shining from an object the colour is associated with the object. Logic tells a different tale. A yellow dot is all the colours in the spectrum but yellow because it is disassociating with the yellow. That goes for red blue and all other colours we may visualise. I think the norm accepts this as scientific fact with very little argument or substantiating proof.

If light came as individual streams of photon flurries our visage would translate that as such shown in the fragmented picture above. It would be a picture unconnected bringing across some photons in the manner where every object stands apart not being related in any way and that will be what we see, if it is anything that we see. That we know is not the case but that means geodesic zero is as much rubbish as anything Scientists regard with simplicity and with careless thought. Geodesic zero means nothing and how can I see nothing as darkness because "nothing" is not darkness, nothing is "nothing" and the darkness I see is darkness showing the darkness as something.

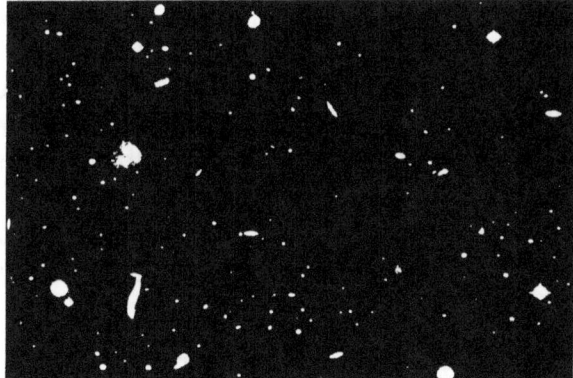

What then about colours that are technically not colours as is the case with black and white? White is simple. By spinning all the colours in the spectrum the colour white shines through. Black is quite another matter. A friend of mine whom is one of the best painters I have ever come across told me that one couldn't paint black but have to make black a dark blue to show shade on the canvass. That apparently is his success in achieving the realism.

He also went on to explain how many variations of dark blue form the shadows in one simple tree. This remark set my mind in motion. One cannot see black because black has no colour to show, but black is the colour most prevalent in the

Universe. One can see only by colour and since black is not a colour we should not see black, but we do.

If the darkness was the representation of "nothing", then that should be exactly what we must see, nothing but the stars. Some stars taken from the top picture and leaving the rest to nothing is what we see in the picture below. A blind person sees nothing but when we look at space, we see something that we think nothing of as we see as space. One cannot have the ability of sight and see nothing except by closing your eyelids and then you see nothing. But in that case you do not see "nothing" in contrast of "something" you see "nothing" without it contrasting to "something".

 **Nothing is all about not being and not "not seeing".**

By the ability to see the darkness renders the darkness something other than nothing and that changes the acquired value of the darkness from nothing to something. There is an eternal difference between something in infinity and nothing.

The arguments introduced up to this part of the introduction prologue only touches the most basic aspects of my work and by no means can such an introduction secure an opinion. Yet, not once through all my long investigation in the past thirty or more years have I found any other person claiming such views that I have brought about even in this skimpy way as I do in the prologue.

The arguments introduced up to this point of the introduction prologue only touches the most basic aspects of my work and by no means can such an introduction secure an opinion. Yet, not once through all my long investigations in the past thirty or more years have I found any other person claiming such views that I have brought about even in this skimpy way as I do in the prologue. As it applies with all things, so it does in this case as well that when delving deeper into any issue. The complexity of the issues truly comes to the fore ground when analysed in more detail. I wish to advise the reader to treat the seven books as seven different works and in that light I have separated each work in volumes of seven separate books with individual I.S.B.N. numbers with adding one part, the one you are reading, with one sole purpose and that is to bring about an academic introduction to clarify a quick perspective. Then the next three parts being of a general introductory nature there are overlapping in some sense but each highlighting issues in a different manner as to clarify facts used in the last three parts bringing conclusion to different cosmic perspectives. Yet the work is seven parts of one thesis and as such it serves.

I WISH TO DEFINE THE CATAGORISING I USE AS PART OF THE BOOK.

I have the utmost admiration for Scientists and I shall never dream of placing me in the same category as academics mainly because of their intellect and achievements. To substantiate this segregation I use some referring to place distinction between the highly schooled super trained academics that spent most if not all of their lives in preparing to further their minds. Because I tip the opposite of the scale and spent as little time in an official capacity on learning and education I have to be on the "other end". From where I stand and admire you, I can only see intellect: being the academic's common denominator. If that is the common denominator on the one side, the joining factor on the other side *"my side"* must then be the class of stupidity. To you and your class such a remark would be an insult but to me and therefore my class it rings truth and that makes it not an insult but a norm we should accept and learn to live with. It is rather a pity that while the SUPER CLASS will never say it to our faces; the SUPER CLASS is strongly of such opinion that we on the one side of the Universe have no minds to think in any way, and it is our duty as much as privilege to accept what you the ones occupying the other side of the Universe inform us to accept and you live by that idea. As I said I have to live with it too and if I am the illiterate, then the SUPER CLASS must be the SUPER–EDUCATED; where I am the class amounting to stupidity the SUPER CLASS must be the Brainy Bunch. It all comes from the fact that there is such a huge differentiation between us. You consider one with merely one Baccalaureate degree a stable boy; think how low your opinion must then be of my type including me that never even came that far. To distinctly point to grouping or class or whatever you wish to consider the division there is between you and me I refer to your side of the Universe by the names I use above. Further more when I refer to mistakes that I do prove to be mistakes in the book as we go along I refer to it as Xepted mistakes to clear another distinction of necessity.

Introducing the book **Matter's Time In Space** written with facts about the creation in mind produces a problem because the complete picture that I introduce has nothing in common with current accepted science.   The issue remains comprehensive even by using it in a very simple form. In spite of this, I shall explain three of the four unrecognised phenomena I use in proving my statements in this very book aiming at a theme of simplistic introduction being "an Academic Letter". Then in the following books I go into extensive detail proving all of the Cosmic Pillars. With my introduction of the phenomena, which I named the four cosmic pillars you will find it obvious why science do not accept them even if it is documented throughout the Universe and is quite commonly found. Applying them totally annihilates Academic's formula of the basis on which science rests in the formula being $F = G (M_1 M_2)/r^2$. The four cosmic pillars are the following:
1.   Roche-Lobe
2.   Titius Bode principal
3.   The Lagrangian principle
4.   The Academics Gravitational Concept forming the atomic relevancy.

The Universe is a combination of many material formations holding positions in space. Some of such material was covered in the blanket of heat, distributing into more spacious surroundings as the material expanded from the centre flowing outwards. Hubble's constant is proof that the space between cosmic structures are departing from many centred positions between such objects and this is a trend being located between all the objects throughout space but also indicating a definite growth in the radius and such radius growth follows a pattern where the growth seems to flow from any such a centre point away from the centre. With out the absolute and undeniable proof coming from the Hubble constant bringing proof beyond any possible doubt in any one's mind that expanding is very much and a very big part of all Cosmic activity. The accepting of the Big Bang would not be in place. $F = G (M_1 \times m_2) /r^2$ is in essence a big issue about contraction while

Hubble showed the space was not dividing. The space was multiplying. The stars are growing apart and so is the galactica. This then brings in the question of space available.

With me not whishing to go into the formation of structures at this point in the book I would like to point to the fact that my following referring to the solar system is actually referring to a similar solar system that is somewhere and is now a part of a galactica we do not know about. I bring this in to disqualify any academic loophole that may come about from an argument about the solar system coming into place at a later stage of the cosmic development and therefore the argument I am about to present that such an argument does not apply. To avoid such a loophole we now use a hypothetical but real solar system in space, which formed as the Big Bang took place. There are those who avoid admitting to inconsistencies by arguing that my argument about growth is invalid because the solar system was not in place at the Big Bang. To them we now present a solar system that is identical to the one we know in precise duplication thereof. But it represents a precise duplication of our solar system and was in place ever since the Big Bang. That means with the solar systems being apart in millions of kilometres there was a time the planets and the Sun were apart by the measure of kilometres. The Big Bang shows a growth in space. Then there must have been a time when the planets were between fifty-nine and fife hundred and ninety kilometres away from the sun. How did the planets being the size they are at present fit into such a space and still be apart? Material too must be part of the growing. This line of arguing I suppose is much below the Academics pursuit of matters but since I am much lesser in mental standards of development than they are, such reasoning prompted me to go on some investigative journey. Light journeys through out the cosmos and it will be sensible to follow the travelling of lights.

With objects being apart at some distances and light flowing in straight lines between them it must take light a straight line to travel between cosmic objects. The distance the light has to cover depends on the radius there is between the objects and as such the Universe is then about structures claiming space and space setting objects apart being the radius standing between those objects. The objects are circles by dimensions and the space is also dimensions that are crossed by lines travelling through the dimension. With light being a line and the Big Bang coming from a situation that was a lot more cramped for space than at present, the correct path to follow if I wish to trace the steps of the Cosmos back to the Big Bang is to reduce the straight line between the structures and find where such a line will no longer be a line. The same procedure will apply to the material structures all being in a sphere form. A sphere is a lot of circles forming a unit but not repeating any occupation of the space, which any one particular one claimed. Such a circle also applies a straight line only known by another name but still serves the same purpose. Reducing the line will lead us to the beginning of time.

 By dividing the radius r by the half of the value that then reduces r to a point where the left edge of the line reducing will be at the very same place the right hand edge of the line that is reducing will be. At one point the spots that formed the two ends of the line will be at the same spot. Any further dividing will land the left hand spot past the right hand spot

in the opposing half where it then will grow once again but in the opposing direction. All possible dividing then ends on one spot where such a one spot shares a location with all other possible sides. The centre then physically is in the single dimension applying as one spot to share a location for all sides. At such a point there is no further dividing possible. On several occasions in the past I have been accused of manipulating the argument to produce none-existing or overrate facts. That is not the case. I am not manipulating facts to create an argument as so many accuse me of. What I am talking about is a mathematical fact that any one can prove by calculating following a very simple procedure. A child is capable of using the two times table and dividing by two every time is the most simple form that mathematics may be used. It is a mathematical fact that a line will reach a point where all sides are at one spot and as such cannot divide any more. At that point all sides share but all sides prevent zero becoming as factor since the sides share on spot. While the different sides are in one place the factor and value is one to all.

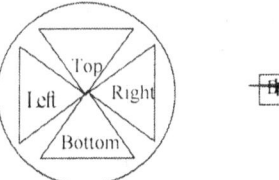
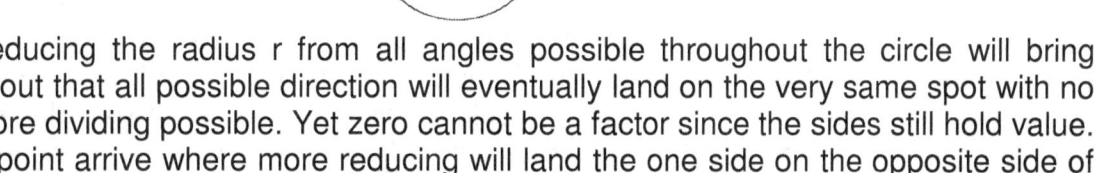

Reducing the radius r from all angles possible throughout the circle will bring about that all possible direction will eventually land on the very same spot with no more dividing possible. Yet zero cannot be a factor since the sides still hold value. A point arrive where more reducing will land the one side on the opposite side of the line but it will not bring about zero in the equation

What this argument further proves is that the circle reducing must then come from all points because the radius might be a line but that line represents a circle through 360^0. Taking that into account it is important to recognise that notwithstanding the size of a line, which any radius of any size is there is another line (or dot) eternally bigger as well as eternally smaller than the line in question. While we are in the third dimension being part of the third dimension then allows that all parts of the third dimension forever can be divided once more until the line in the third dimension is no longer part of the third dimension. When such a line leaves the third dimension it is still dividable because it might not be part of our dimension any more but it can still reduce further as part of the second dimension. By that time it has left our scope by miles. It does not mean that it end there because from our perspective that is where it ends. Yet it can still reduce infinitely more until it has left the second dimension and then at last forms part of the first dimension. Only then when the line reaches the first dimension, no further dividing of that line is longer possible. We can never grasp the size of a line that forms the utmost or the least of possibilities and therefore size belongs to the human mind forming conceptions of big and small, but it has no place in the cosmos at large. This concept not only applies to size, but to all limits and divides we wish to create forming borders that we can appreciate. When looking at the circle in the conventional manner, we persist with errors brought about in culture and not by applying some significant modern logic. Take a circle and reduce such a circle constantly to where it no longer can reduce. Reduce it to a point where only form remains part of the circle because the radius has gone beyond human measure and becomes so small it is not noticeable with what ever tools man may use, then what remains is pi since pi does not indicate size but indicates form, and form is all that then will remain. In any circle or sphere the size only depends on the fluctuation of r, as a component to the circle or sphere but that does not affect the form by indication of Π in any way there may be. The conclusion I drew from following this process is that from this line no start can be at zero because that will be a mathematical impossibility since no line can ever reduce to zero. A line will forever be able to reduce further becoming smaller but it can never reach zero because zero is not on the scale of lines. If a line cannot reduce to zero it then cannot start at zero. A line or spot starting at zero would therefore be shorter than the shortest line possible. For obvious reasons can no line, or any line grow or extend from zero because such a line must then quit zero and become something, thus abandon its original value. That would mean the start of the line has a different value to the end and a line holds conformity through out. When any line is starting from point zero it can never leave zero because of the influence of being zero disqualifies any possibility of growth. If the line then had to grow in all directions at the same pace the line must then become a circle or being three-dimensional, then forms a multi circle we name a sphere. Since the Universe is about circles and lines connecting circles, I came to conclude that flowing from this fact is that in the Universe there can be no zero improvising as a filling ingredient for the space of a point or be unfilled space. In the case of the growing sphere the value of the circle is Π, and that is where creation must have started. That gave me the clue where to start looking for singularity. One would find singularity in the value Π and the value Π will be in all things rotating in a circle. As usual I am again shooting the gun before the hunt started. Lines in mathematics do not start from zero and that is no discovery on my part that was a realisation I came to.

UNIVERSE
Everything that exists, including space, time, and matter. The study of the Universe is known as cosmology. Cosmologists distinguish between the Universe with a capital 'U', meaning the cosmos and all its contents, and Universe with a small 'u' which is usually a mathematical model derived from some physical theory. The real Universe consists mostly of apparently empty space, with matter concentrated into galaxies consisting of stars and gas. The Universe is expanding, so the space between galaxies is gradually stretching, causing a cosmological redshift in the light from distant objects. There is growing evidence that space may be filled with unseen dark matter that may have many times the total mass of the visible galaxies. The most favoured concept of the origin of the Universe is the Big Bang theory, according to which the Universe came into being in a hot, dense fireball about 10-20 billion years ago.

UNIVERSAL TIME (UT)
A worldwide standard time-scale, the same as Greenwich Mean Time. Universal Time is the mean solar time on the meridian of Greenwich. It is defined as the Greenwich hour angle of the mean Sun plus 12 hours, so that the day begins at midnight rather than noon. It is closely linked to Greenwich Mean Sidereal Time (GMST), since the mean sidereal day is a precisely known fraction of the mean solar day. In practice, UT is determined by a formula from GMST, which in turn is derived directly from such observations of the meridian transits of stars. The

version of UT derived directly form such observations is designated UTO, which is slightly dependent on the observing site. When UTO is corrected for the variation in longitude due to the Chandler wobble, a version of Universal Time, UT1, is derived which has genuine worldwide application. When UT1 is compared with International Atomic Time (TAI), it is found to be losing approximately a second a year against TAI.   Broadcast time signals use the time-scale known as Coordinated Universal time (UTC). This is TAI with an offset of a whole number of seconds. The offset is adjusted when necessary by the introduction of a leap second, and UTC is always kept within 0.9 s of UT1. On this issue there is much more to explore than the meagrely mentioned. Time stands related to the position an object holds to a centre such an object refers too while in rotation. Kepler found for instance that T^2, which holds the orbit to a rotation specific, is directly dependent on k to value the space a^3.

Einstein proved that in the presence of a strong gravity time slows down. Surprisingly, with that evidence being around this long, nobody since then in science took those statements and made any further progress from there. It was left in some drawer to dry. Science still sticks to its change that time did not change slightly since the beginning of the time and holds the same pace ever since. With the entire Universe including all the gravity now present and not excluding one Black hole or dust speck pressed in an area possibly the size of a lepton the gravity extending from that must have been beyond what words can ever describe. If the gravity was that high and Einstein already proved gravity slows time down, then there is one logical conclusion and that is that time was n fact standing still. Mathematically it is incorrect to allow gravity to compress the Universe into a spot smaller that an atom and exclude any other factors and relevancies to change. But before coming to the mathematics I would first like to bring your attention to the practical side. I am promoting a theory in which I am able to prove there is as much contraction (moving in the direction of the Big Crunch) taking the cosmic Universe back to the size it had during the Big Bang as there is expansion (moving apart by Hubble's Constant) and the contraction is as much part of the expansion. By contracting the Universe is expanding and everything is based on gravity providing both actions. The Universe rides on a balance and we have to locate such a balance. To prove my theory I firstly had to locate the centre of the Universe. Even admitting to such a notion sounds like madness or in the least a tasteless joke, but please give me a chance to explain in more detail. I realised that my effort to locate the point holding singularity only stood any chance of success if the reducing of the line enabled me to backtrack the exploding Universe to its origins. By applying some basic effort I have located the position from where all movement came and the direction it took moving forward in time…and yes, during my search on locating the centre of the Universe I also stumbled on time as such.

Let us find the smallest possible line first. Reducing the line will eventually leave all sides on the same spot. Such a spot must be round in form. The line being the smallest line will start off as a dot. A line so small it has reached a point not dividable any more will have all sides literally on the preside same spot, and I have located singularity in just such a spot. I came to the conclusion that the spot I found had to be singularity purely on the grounds that that spot holds only one side to serve as a start to the starting point of all directions possible. There in that side is only one spot is only one side applicable and one dimension present. With all the factors given one can only come to one conclusion and that is that there can be only singularity. In such a case more dividing by two will land further positions on the other side of the divide. That point serving as a position for all point and cannot allow further dividing is the smallest line or spot there may ever be. This spot is the result of a most basic process of reduction as the Hubble constant is a most basic process of doubling up during a matter of time. By reducing the line constantly the only value that will eventually remain without dispute from any party arguing about the facts is Π. By only having Π and a radius as one square (the radius effectively becomes one holding any and all sides on one point) of any significant measure as the radius it will be an evenly spaced dot. From the smallest ever possible dot will grow a line in every imaginable direction relating to a prospect of Π not favouring one direction that puts all directions at equilibrium meaning that any form of what ever might develop from such a spot will have the end and the start being in the same position, which will also have to be a sphere as the flow outward will be equal in all directions. Please think clearly, is that not precisely the commitment we find in gravity, where gravity is flowing from singularity outwards but never favouring any side? This reasoning prompted me to look for singularity in such a spot because if the prime spot from which all came was a spot holding all, then the spot must hold the shortest line but more prominent it will hold the smallest form including the smallest circle or for that matter the smallest sphere. With gravity always being in the centre of a sphere where the space is least available in the entire structure (there is not even space left to fill) one finds a flow of gravity from that centre spot outwards in all possible direction even-handedly. The fact that the original gravity will begin as a circle or will be a circle is the direction it will take when being the first spot created. All progress will be evenly in all direction because no direction will stand out or be in favour above any other direction at first.

The spot forms a full circle, but the line running through the circle is forever present because that is the future radius of the circle that will one day develop the circle, which is equal to the present diameter. The fact of the presence of such a possible line in such a possible circle dividing the possible circle into two parts makes the centre line equal to the half circle. The line forms the half circle but not only that the line presents the half circle as much as the line is the half circle. The line then is 180^0 and the half circle is 180^0 because in singularity the two factors are the same. The same value is of course $\Pi^0 = 1$.

In this half circle of the future, which is no half circle as yet because of a lack of space there are three future points indicating the space less ness that will go on to become space filled with something. On top of such a circle to form must be a marker indicating an awaiting boundary or future border and at the bottom of the future circle there also must be a similar marker that is no marker as yet. Between the two possible points that are not there yet is a future line running that is not there yet. Then indicating the possibility of a position to come that will bring about the half circle being a future distance apart from the future line indicating a diameter that will one day be there a third such a marker must be established for the future. That forms a triangle with two more sides being connected by either a line being one or half pi being one. From singularity comes about that the line is the same as the half circle is the same as the triangle and all has one value being

180^0. From this come the most basic principles in as much as forming the ground rules of the law of Pythagoras.

When drawing a line such a line then starts off with a dot serving the spot that holds all sides equal. That means the line serving as the future radius will be equal to the half circle which is then Π. The only aspect of the point that stands in for the end of the single line forming the radius of the circle is that we then mathematically reach the single dimension. We decreased the line to where a circle being Π formed on the single dimension. This dimension also hold the circle dividing line because from there the radius must once again generate a value and by such a gesture that the extending would form the circle that forms the sphere that eventually leads to the formation of particles. This leaves a problem to investigate.

With no line possible there had to be another dot that formed since the Universe has many dots that formed lines. But let us not to get confused and lost in the range of possible diversions but let us stick to two dots. One dot was next to the dot next to the dot, but as I said we stick to one dot next to the second dot. M X M $/ r^2$ is the first step gravity began with. That leaves us with a huge problem in as much as when $r = 0$ then $r^0 = 0$ and 0 dividing any value will leave 0 as the answer. If the particles were inseparable at the start it must bring about that gravity would not be forming since the distance will not permit any dividing. By allowing the distance separating the particles to be zero, the particles melt into a unit.

Again this is Mathematics and not my incoherency as some Academics dismissed my work. Let me run through the argument one more time because I have been insulted by Academics in the past telling me I am bending mathematic rules with my applying double values to try and produce some argument. The two particles formed by an inseparable unit separated by a sharing of a spot. We know that at least two spots formed because there are many more than just two that remained to become part of the visual Universe. Let us name the spots because that is what humans do best if they do not know what to do with what they have to do. Let us call the one em and the other one spot next to em we then call emtoo. Between em and emtoo there were nothing because em and emtoo were inseparable. By they're being inseparable we would naturally be inclined to think that the separation value should be nothing or at least zero. But putting zero in that place is a mathematical excluding procedure leaving future mathematics excluded. With m multiplying m_2 and then dividing $\div$ r with zero (r=0) such a procedure will leave the lot at zero and with that nothing is going nowhere. That means although we think the space between the two parts are nothing the non-existing space has to be at least one to be a future factor.

Every part of the argument is sound but was never yet used. I repeat once more if my argument reflects on inconsistencies those inconsistencies are not about my work. In order to disprove my argument replace Mass one and Mass two with any number possible, then divide such a number with the square being zero. If there was no space then the value of the particles had to be one. If there was no space between the particles the particles then had to form a unit. But if there is a mathematical possibility of reducing a line to the single dimension then there had to be a factor representing r as a factor of one. Take $(M_1 \times m_2) / r^2$ and substitute any of the factors with zero and the result coming about has to be zero. The factors in the equation have to have any and all the elements at a value of at least one. Only if r was a factor of one can gravity bring about any mathematical equation developing from this argument.

That means the mass on both sides must have a factor of one being a limit, which does not allow such further reduction of r and any further reducing of r beyond the limit will not be tolerated. Only if $r = 1$ then r^2 can be 1 and mass can be apart. Like it or not but believing in the Big Bang must also bring about the accepting that the cosmos moved apart somewhat. The fact that r brought increase in the space separating the mass produces a problem that was solved already. About a century and a half ago Roche found just such a limit. Once again I were confronted by zero becoming growth. There is a huge hole that needs filling when bringing into a relation any forming of an alliance between a cosmos coming from nothing and filling with nothing and a cosmos growing spontaneously through balance shifting prominence. Mathematically the fact of applying nothing serving as a factor applying in the cosmos is not a strong and convincing argument. The minute one brings in zero as a multiplying factor forming a definite value working into the calculations of the cosmos, growth disappears. If growth was not a factor, the zero factors could be involved with some form of maintaining stability and where then further growth will accept the responsibility of zero

The Roche limit is:

The region surrounding each star in a binary system, within which any material is gravitationally bound to that particular star. The boundary of the Roche lobes is an equipotential surface, and the lobes touch at the inner Lagrangian point, L_1, through which mass transfer may occur if one of the components expands to fill its lobe. It names after the French mathematician Edouard Albert Roche (1820-83).

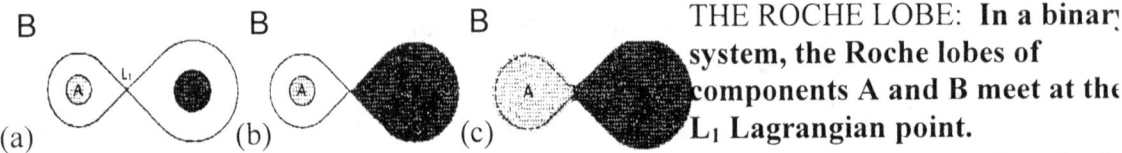

THE ROCHE LOBE: **In a binary system, the Roche lobes of components A and B meet at the L_1 Lagrangian point.**

(a) **In a detached system, neither star fills its Roche lobe. (b) In a semidetached system, one massive component, B, fills its Roche lobe. (c) In a contact binary, both components overfill their Roche lobes and share a common envelope. Lets explain the importance of this Roche limit and how the Universe used the Roche factor to produce the Big Bang. That is where it all started...**

The closest encounter worth noting we ever had with this law in the modern age of news and Television was the Shoemaker-Levy 9 incident during the previous decade. At the time and even in the present no one drew any similarities but after completing this book the reader should find why I could draw such a similarity, which there is between this incident and the Roche limit. Even the phenomenon called the Sound Barrier became clear when applying the Roche factor with the laws governing the influence of singularity.

singularity k^0 by creating motion in spin and duplicating space by reducing space.

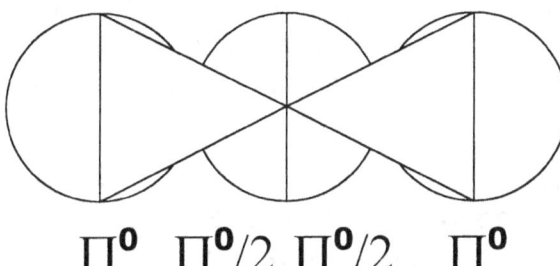

$$\Pi^0 \quad \Pi^0/2 \quad \Pi^0/2 \quad \Pi^0$$

Let us start telling the story as it was.

At the very first sign of any of the sides departing from the centre shared by all, all other points must also show signs of a willingness to depart. There will be one point where r still is one coming in as a factor but pi moves out from only being a factor of Π^0 = **1** and at that point, pi will become a full factor of Π.

But keeping Π as one (Π^0 = 1) we keep the Universe in the first dimension.

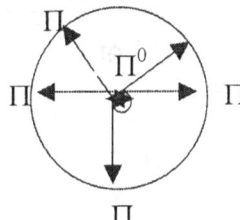

This point, which I now am referring to, is the point where Π a fully appreciated value while the diameter D still remains a dimensional factor of one. His is the dawn of the second dimension where space was there but space was sparsely shared in some cases. It is when Π^0 shifted to become Π for the very fist time.

The point without movement, the point holding singularity must have a value of Π being the eternal dot but since the dot has no dimension in having form the Π that indicates the dot must be Π^0. From such a point there has to be to the side of the centre point be a point where space do start. That point will then receive a diameter but that point will have form only in being a circle. In that point there is a shift from in relevance from Π to the centre Π^0 and for the first time it brought about two separate values for Π.

Em   B₁ B₂ Emtoo

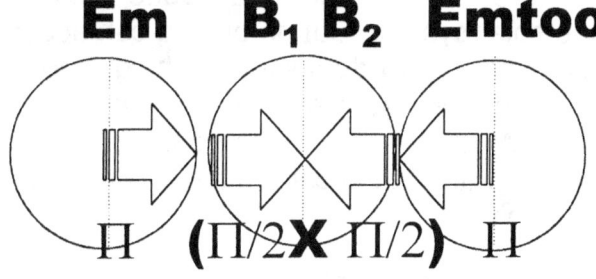

$$\Pi \quad (\Pi/2 \, \mathbf{X} \, \Pi/2) \quad \Pi$$

By allowing the distance separating the particles to be zero, the particles melt into a unit. Again this is Mathematics and not my incoherency as some

We return to the fact we established before that Em and Emtoo was divided by r and then r had to be one since r could not be zero. Such a centre would then carry the same value as Em and Emtoo. That means whatever value Em and Emtoo receive has to go in equal measure to r with Em sharing half of the divide and Emtoo sharing the other half of the divide.

Academics chose to interpret my work or rather then to find grounds on which to dismiss my work without bothering them with the effort of reading my work carefully.

Because the three points existed on equal terms in singularity sharing a same spot the coming out of singularity will enforce that equal value comes to all. That means the circle gets to become Π, the diameter becomes Π and the distance setting the structures apart will also become Π. This is what the coming from one point brings along. Only when being part of the second dimension can there start being separate values.

While the form was still being in the single dimension from the one side of the form the dots had to establish identities apart but not separated yet. The one circle had a factor of $\Pi^0 = 1$ and the centre had to have a value of $(\Pi^0 / 2)$ extending past the very next object but also cutting such an object into a square double half value that was going to come about as soon as the other dimensions came into form.

The only definite place one will locate zero is in between the starting point of the lines going in opposing directions in the position the lines hold before there was the least of directions applied, but that is only because there is no such a position, not because any line is coming from there. The two lines are still one holding the opportunity of parting as an option but have not yet parted and therefore are on the very precise same spot. The line coming from there is already there because it already has the choice of going in any and all opposing directions and when it starts running it will place filled space in that location because the space was already filled with a line starting and not with a line not there at all. When reversing a line we might find a better idea of what is in place and where it is in place. Gravity is officially a force without limits going past and through borders and has an unlimited reach. It seems to remain even and this is conflicting with the flow of perceptions about mathematics.

The formula $\mathbf{F = G\ (M_1.m_2)/\ r^2}$ is unable to explain the principle discovered by Titius and later by Bode and in contrary to all statements to that effect made by Accepted Science policy makers the Titius Bode principle is not coincidental. In fact it is one of the four most adhered and important cosmic pillars holding the cosmos structural in place. From the two above comes gravity. In a few pages I prove how one can arrive at the facts that prove how the Titius Bode Principle leads us in the direction the origins of the solar system. But before we can accept the influence of the Titius Bode Principle we have to deal with "Nothing" and as such dismiss nothing from science. "Nothing" in the Universe is coincidental; "nothing" in the Universe does not apply. Where mathematics connect to lines nothing disappears. Should any principle not match an accepted theory or change the accepted theory, the theory does not apply.

The content of my work holds a new view about Cosmology, which I have been working on for the past twenty-seven years and exclusively for the past six years. I always had a problem with the idea that space constituted of nothing, while I came to realise that lines mathematically couldn't start at zero because there is no evidence of zero as a factor in mathematics. Should you disagree with my statement the question in need of answering is this: What will the length of the shortest hypothetical line imaginable be and moreover, what would the total overall length be in that case? The shortest possible line (hypothetically) must be so short it must have an initial and ultimate point sharing the same spot. The two points must be one and only then can further reducing of any line not occur. If it used zero as a start, the zero part would not count, because the line will only start at a point past zero where the line then will start forming an infinitely small dot. I press this point in urging the understanding because there is such a point, but in an attempt to recognise the point, I have to convince the reader to abolish four or five thousand years of accepted and practised mathematical culture and that is no easy feat. Taking the line back as far as possible brings a dot because of the equilibrium that will stem from such a position. The dot is in infinity, however small, it is not zero. Zero ultimately means not existing and then that point, as a start does not exist. The smallest line has a beginning and an end at the very same spot located in infinity, and infinity may be beyond human scope, though infinity is still not zero. Infinity may constitute of something we do not yet understand, but we may not define our human misunderstanding not present in our minds and therefore as nothing. In this aspect lies the difference there is between arithmetic and mathematical science where arithmetic can have position such as zero since arithmetic excludes the cosmos calculating numbers only. Cosmology is not about numbers because no one can calculate the number of stars. Cosmology is all about lines and angles positioning objects, and in those there features no zero. No line can be zero long and forming a position of zero degrees in relation to another object.

A man may have that many oxen or so many sheep and even this amount of wives, (in Africa) or not have any therefore having then a total of nothing, but there cannot be nothing between the Sun and its orbiting structures. The having and have-nots are part of arithmetic. Light will indicate a line flowing between the Sun and whatever planet, following dot after dot thereby proving the existing of the possibility of something going about by a straight line, and any straight line in relation to other straight lines will be under the law of Pythagoras in as much as obeying the rules of trigonometry. There is no possibility of a straight line not forming in space. If there is space, there can be a straight line. The mere fact of two spots having different positions in space gives the two dots different values. If the line has the length of zero it is not present. If the triangle has one angle at a value of zero it is no triangle because the zero would dismiss all the other angles. Mathematics converts the values of integrating lines according to Pythagoras and arithmetic is about numbers to be added or subtracted. By mathematically excluding zero from cosmology a new Universe opens to the human mind. With the distance between the Sun and Pluto being roughly one hundred times more than the distance between Mercury and the sun, the distance must hold something more than pure vacuum filled with nothing except one atom here and there occupying the vacuum between them and the sun. If space supposedly comprises of nothing how can nothing then become plural forming more or be multiplied by a number as to indicate a growth in something not even existing. As the one becomes one hundred the one cannot substitute a value of nothing but then must be part of something. If the one substituted the nothing, all laws of mathematics will go in disarray because when one multiplies any number by zero it becomes zero placing both planets in the sun. If Pluto was one hundred times closer than it is at present was it one hundred times nothing closer? $100 \times 0 + 0 = 0$. That is mathematics!

By allowing the three hundred a value the nothing must form one making that which is between Pluto and the Sun not to be nothing but there has to be something. This argument follows mathematics to the letter and in precise detail. With Pluto and the Sun being apart that being apart has to have one of something in place forming the being apart from each other's cosmic position one time multiplied by the many ones we find in that space standing relative to other space regarding whatever the space becomes what is between the Sun and Pluto. That factor cannot stand in for not one, which is the same as nothing as that is because one cannot take the place in the position that zero secures. By excluding nothing from the equation space becomes something bringing in a value lying inside the realms of the infinite that must form singularity. As the zero becomes a dot, something else becomes clear about the dot. Looking at the night sky we find darkness overwhelming the space in relation to the stars bringing across light.

My approach to cosmology shall prove to be somewhat unconventional but through the abandoning of the accepted, it enabled me in locating the precise location of singularity that forms the connecting basis of the Universe (and this I say with some degree of confidence). There **are two locations** but I shall **first concentrate** my explaining effort on **the prime singularity**. Singularity did not vanish into the unknown after the completion of the Big Bang development but is in a place science incorrectly valued and classified incorrectly and in that, there is something hiding the truth. If singularity was or is where the beginning is we have to go back and see just where such a beginning was. I cannot accept that the Universe started at zero and neither does anything else in the Universe start at zero. My excluding the possibility of zero includes that the Universe is not filled to the top with nothing and neither is nothing part of outer space. The Universe is about lines allowing light to flow from one point to another point and in following that line it has to continue in the line as the line has to represent something. The Universe is all in relation about lines indicating distances between cosmic structures. The cosmos is in short about lines connecting points in space being apart. It is about a line starting and continuing from such a start. But science advocates their opinion that such a start of a line flowing between any and all objects can hold zero because according to them the Universe are full of nothing. If the Universe in as much as outer space is a container filled with nothing at the present moment, and there is no place anything that was part of outer space previously could release to and there was no emptying of what ever filled it before, then it could not get rid of what was in the outer space when it started with what it started off with. We must then accept from what is not in the Universe was not in the Universe at the time during the start that at is present at present according to science because it then still must contain the same nothing and must have that same filling from the start to the present. If it was nothing it still must be nothing and that same substance being nothing is what it also used to grow using it as it grew because it filled outer space with nothing growing from and growing to nothing. Is that true? The filling of the Universe could not go anywhere so one has to presume it started off from nothing and from there it kept filling with nothing since what ever was in the Universe at the start had no place to escape to or no place through which to escape. That is only applying if it is nothing filling the Universe at large. Can nothing grow as much as a line is growing from a start of nothing? The answer is that such lines not only indicate a distance but since the Universe came from such a small space as science propagates with the theory of the Big Bang then all particles in the Big Bang Universe were rather cramped for space when the Universe started from that small line between particles and is now the same line but is now so big. In the past it seemed being so small and showing the space between particles to be awfully short at one time. It was short but how short was it? Did it start off as nothing? Is the line starting at nothing as science wishes us to believe? If it does then all lines must start from nothing so we better investigate this trend with the start of a line. In this following I show my argument with which I hope to prove the counter part of what science believes. I am about to prove that which science sees as nothing in space and in material is the very location of singularity.

Lines mathematically cannot start at zero because there is no evidence of zero as a factor in mathematics. Should you disagree with my statement the question in need of answering is this: **What will the length of the shortest hypothetical line imaginable be and moreover, what would the total overall length be in that case?**

Locating zero

Zero in place

Let us duly test my statement by taking the line back as far as possible. The shortest possible line (hypothetically) must be so short it must have **an initial and ultimate point sharing the same spot.** The line that **cannot reduce** any **further** must be **so short** that **directions flowing away** from each other **are located** in the **same position.** Any theoretical line being the shortest possible line cannot have the line holding the initial starting point at point zero and advance from there. Mathematics simply will not allow it. If the point had zero as all it had to offer, such a point is not present. The zero means there is no such a position. If it used zero as a start, the zero part would not count, because the line will only start at a point past zero where the line then will start. Zero ultimately means not existing and then that point, as a start does not exist and where the line then stars is a point in existing. When the line **has a beginning and an end at the very same spot** and it wishes to extend the position as to further the possibility it has, which direction should it favour. Extending the line in any one direction will favour one direction without any clear reason not extending in other directions. The fact of direction being present only proves and is proved by another point established, which is placed in relevance to such a second pointing a position already established by the relevancy of two point located in a direction to one another. But if one point starts one line there is no favour of direction since there is no established direction yet. The only mathematically sensible option about extending any line starting at a pre-designated point without any other point to establish a pre determined direction will be non-bias progress in all directions equally in order to give a meaningful flow of mathematical equilibrium. Not one direction stands superior to other directions and all directions are equal with no bias anywhere. Of this statement the Pythagoras mathematical principle is proof of and that I explain later.

Let us dissect nothing, as we find nothing in the presence of the cosmos. The distance between the Sun and Pluto is roughly one hundred times more and if the distance between Mercury and the sun, but both has nothing between them and the sun. The space filling the distance from the Sun to Mercury has nothing more than the space between Pluto and the sun. That means the distance between the Sun and Pluto is as equal in relevancy than the distance from the Sun and Pluto since both is the measure of nothing. If the one substituted the nothing, all laws of mathematics will go in disarray because when one multiplies any number by zero it becomes zero placing both planets in the sun. The distance between the Sun and Pluto **is Pluto is 5900 X 10^6** kilometres of space, but in that statement we take it that the one of a kilometre is present in such a multiplication. The one constitutes the presence of fact being a statement of a value. By saying the distance constitutes of nothing we have to substitute the one factor with a factor of zero. Then the calculation must read **Pluto is 5900 X 10^6 X 0 = 0.** Including nothing as to state the presence of that part contained by the calculation delivers the total of zero. By excluding nothing from the equation space becomes something bringing in a value lying inside the realms of the infinite that must form singularity. Applying this logic to the Lagrangian system and interpreting that information to the law of Pythagoras a clear pattern comes about.

The reaction responding from my argument is that it is silly, but should that be your personal opinion too then test where the silly part applies. Bring the zero into the calculation, the zero that science so eagerly places in outer space and see the mathematical result. By applying the distance one accepts automatically that the figure become calculated with one as it represents one in being a calculating part of the cosmos. The calculation as all calculations normally are is in order to calculate something and the something will at least stand in as one in relation to the rest being part of the calculation. But saying that the factor of one in fact represents nothing since nothing is so much the part in the calculation being calculated, then the zero has to replace the one as the fact of being calculated.

The claim becomes obvious when observing the connection between the half circle, the straight line and the triangle, which could also promote all the qualities lurking behind the pyramid. Consider the connection between 180^0 sharing three different forms all part of mathematics where each is different in form, but equal in value and then one may realise in considering the very basic in mathematics being the Law of Pythagoras on which all mathematics are focused. The triangle stands in for one factor represented by one at a value of 180^0. So does the straight line become a factor of one and the half circle also becomes one where the factor of one equals all 180^0. All three are most seriously part of shapes in the cosmos. Revalue any one form to zero and the rest too must follow and share the same value. The Law of Pythagoras is about angles in relation to lines and not one angle can represent zero because that will reduce all the lines also to zero. The measure of angles between stars at a distance uses parsec as the indicator, but the parsec between the stars indicating an angle has to represent an angle whereby one may measure distance and such a distance cannot be zero because then the parsec will be equal to zero. Again it is multiplying the factor with the measure but if the measure is about a factor of zero, then the factor too becomes zero. That is as basic mathematics as I can present.

If the argument seems ridiculous it is not my mentioning such a fact that is ridiculous but the mere fact of the reasoning also becoming a recognising of an argument accepted by science making it as such ridiculous. If space is nothing then it has a number to use indicating just that value being zero or the capitol O indicating zero. Try and indicate what is measured and calculated in space, but not by simply not thinking about the fact and therefore simply ignoring that what is measured forming the sole value of space, but put the value of nothing as part of the distance in calculation because that is what is measured. When stating the distance between the Earth and the Sun place on paper what will allow the kilometres measured to represent the factor that is being measured. If represented by one being the total of one by hundred and forty nine million kilometres of nothing put that language in the International language of mathematics that spans all dialects spoken on Earth. Put it in mathematical terminology by saying there are 149 000 000 X 1 (multiplied by the kilometres) multiplied by what it is being measured which is 0 and what will the total come too... a full zero.

149 000 000 X 1 (km) x 0 (indicating what the km are made of) = 0

 Mathematics says it. If there is something to be measured then the least value the measurement can have in relation to what is used in the measuring has to be one. It cannot be zero and be measured...and we do measure outer space! It sounds as if something here is at fault. It is not with my mentioning the inconsistency one should find fault but the fault is with the fact that it is there and no one noticed! I am not to blame just because I am mentioning it, but the blame must go where it belongs.

I think it is by now little understood although I imagine not nearly accepted that by adding a million of nothing to one nothing there will remain one nothing and that is still nothing. Nothing cannot accumulate therefore I cannot accept anything holding the vastness of space being able to constitute nothing as the major component.

When reducing the circle in size one have to reduce the radius or the diameter because the pi is the indicator of the form as being a circle. Divide the r until there can be no dividing any further and that cannot in the end indicate zero because no matter how small, in that will forever be a value in place.

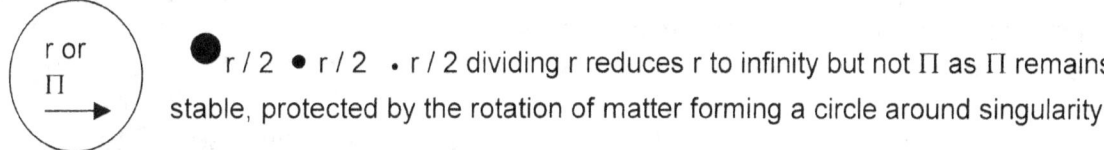

r or Π →   ● r / 2 ● r / 2 ● r / 2 dividing r reduces r to infinity but not Π as Π remains stable, protected by the rotation of matter forming a circle around singularity

Taking that into account it is important to recognise that notwithstanding the size of a line, there is another line (or dot) eternally bigger as well as eternally smaller than the line in question. We can never grasp the size of a line that forms the utmost or the least of possibilities and therefore size belongs to the human mind forming conceptions of big and small, but it has no place in the cosmos at large. This concept not only applies to size, but to all limits and divides we wish to create forming borders we can appreciate. When looking at the circle in the conventional manner, we persist with errors brought about in culture and not by applying some significant modern logic.

By reducing r indefinitely to the tune of half each time, r would become infinitely small, beyond human calculating means, however as mentioned in the case of the smallest dot holding one spot, r would become insignificant beyond human comprehension even, but never reaching zero and still Π would remain intact and dictating form. I believe one can begin to see where my suspicions are heading because the flaw comes about in the manner mathematics are practised for thousands of years. But before coming to the mathematics I would first like to bring your attention to the practical side. I am promoting a theory in which I am able to prove there is as much contraction going on in the cosmic Universe as there is expansion and the contraction is as much part of the expansion. The Universe rides on a balance and we have to locate such a balance. To prove my theory I firstly had to locate the centre of the Universe. Even admitting to such a notion sounds like madness, but please give me a chance to explain in more detail. If I wish to achieve success that would depend on my ability to convince all that outer space comprises of material and as such we can locate such material even if we are unable to see such material.

To find the invisible I had to locate singularity. I realised that my effort to locate the point holding singularity enabled me to backtrack the exploding Universe to its origins.

By applying some basic effort I have located the position from where all movement came and the direction it took moving forward in time

The reversing of the circle radius is not alien to nature at all. An observation coming instinctively to mind one may recognise is that the form reminds rather explicitly of natural phenomena such as hurricanes, water whirls and even the shape most commonly favoured to express the cosmic object referred too as a Black Hole. The similarity may be more than coincidental. Let us consider the statement in the reverse. In our calculating of a circle we apply two formula methods. The one use an r to indicate the radius and the other use a D to indicate the diameter, which is double the radius and therefore needs to be divided by a four to eliminate the Newtonian inverse square law amounting to the difference there will be between the two. The one using the radius is Πr^2 and the other formula is using the diameter is $\Pi D^2 / 4$.

In any circle or sphere the size only depends on the fluctuation of r in the square as a component to the circle or sphere but that does not affect the form by indication of Π in any way there may be. The conclusion from this is that no line can start at zero because that will be a mathematical impossility. A line or spot starting at zero would therefore be shorter than the shortest line possible. For obvious reasons can no line, or any line grow or extend from zero because such a line must then quit zero and become something, thus abandon its original value. That would mean the start of the line has a different value to the end and a line holds conformity through out. When any line is starting from point zero it can never leave zero because of the influence of being zero disqualifies any possibility of growth. If the line then had to grow in all directions at the same pace the line must therefore be a circle or being three-dimensional, a sphere. Flowing from this fact is that in the Universe there can be no zero point or unfilled space. In the case of the growing sphere the value of the circle is Π, and that is where creation started. That gave me the clue where to start looking for singularity. One would find singularity in the value Π and the value Π will be in all things rotating in a circle. You might wonder how does that apply to the cosmos and moreover to gravity? In my search I stumbled on two accepted but not intergraded laws and when I found and located singularity the two laws became very much plausible and factual. Take a circle and reduce such a circle constantly to where it no longer can reduce. Reduce it to a point where only form remains part of the circle because the radius has gone beyond human measure and becomes so small it is not noticeable with what ever tools man may use, then what remains is pi since pi does not indicate size but indicates form, and form is all that then will remain.

I believe one can begin to see where my suspicions are heading because the flaw comes about in the manner mathematics are practised for thousands of years. But before coming to the mathematics I would first like to bring your attention to the practical side. I am promoting a theory in which I am able to prove there is as much contraction going on in the cosmic Universe as there is expansion and the contraction is as much part of the expansion. The Universe rides on a balance and we have to locate such a balance. To prove my theory I firstly had to locate the centre of the Universe. Even admitting to such a notion sounds like madness, but please give me a chance to explain in more detail. I realised that my effort to locate the point holding singularity enabled me to backtrack the exploding Universe to its origins. By applying some basic effort I have located the position from where all movement came and the direction it took moving forward in time...and yes, even time as such.

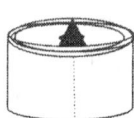 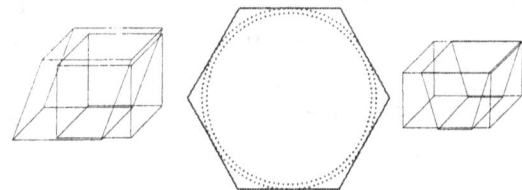

Anything occupying space in the cube will apply r and by r I mean just a distance not using Π because Π serves as a form indication while the collective product of r will determine form as well as accumulative dimension total. Notwithstanding the name used confirming the shape or r named as length width or height, it is all just a straight line bringing about the cube with all its other names that may find attachment to specific form but nevertheless still remains only a six-sided cube with connecting lines applying different angles changing in some cases. The normal perception is that any circle growing spontaneous would grow by the radius, which is r. In mathematics that may be true but it is not true in nature. In nature that cannot be the case because r is an indication of a straight line. By growing with the aid of a straight line from the centre to circle the influence that that would have on the circle would result in many circles following one another and not a continuous growth.

Gravity is the dimensional changing of space holding r as reference in the cube as to the sphere holding Π as the reference. In order to generate spin producing time in matter occupying space, therefore creating dimensional change, Π has to be a factor indicating the possibility of spin because implementing Π the circle sides will follow one another without establishing separation. The answer must be in finding Π, and thereby locating singularity. If singularity is in affect the original point of the cosmos birth, the reducing path we should follow will indicate the whereabouts such a point must be.

There are two standard formulas used to calculate a circle. The one uses an r to indicate the radius and the other uses a D to indicate the diameter, which is double the radius and therefore needs to be divided by a four to eliminate the Newtonian inverse square law amounting to the difference there will be between the two. The one using the radius is Πr^2 and the other formula using the diameter is $\Pi D^2 / 4$. However one looks at the mathematical expressions and Kepler's formulating of space-time, there is an exceptional difference between the two scientific uses. When investigating Kepler's formula one do find it appreciably differs from the normal Mathematical equation like $a^2 = r^2\Pi$ and $a^3 = 4/3\ \Pi r^3$. In the normally used mathematical expressions such equations tend to concentrate on the volumetric aspect. In the case of Kepler's expression it is something else that wants to surface. It is another idea that is coming to mind. In Kepler's formula a^3 stands to symbolise the third dimension and such a third dimension becomes equal to two other dimensions grouping and sharing value to equal a^3 efforts. It is not the circle of the rotation because with such a normal circle the radius is in the square and Π evaluates form. Here there is no mention of a factor Π, which one would suspect to be somewhere applying since the circle is Π and Π is the circle and the two are inseparable, but not in Kepler's a^3, where there is no mention of Π at all. The fact that there is a radius of some sorts used to indicate a position cannot hold the square as it normally does in the case of the normal equations. In the mathematical equation the factor indicating the position of the circle edge has the square value being called the radius or in some cases the radius doubles and which then is the diameter, and the circle indicator is Π. But in this event the formula value will bring about a square value to the answer one receives. It will bring a value to the surface of the circle. In Kepler's formula it specifically does not.

I realised before starting my quest that one possibility that the shortest line or smallest spot can never have is having a starting point on the zero mark. If the mark of zero holds the start it must also hold the end because the end and the beginning has the same position. If the position of zero then is the beginning, the end will also be zero leaving the line or spot without an end as well as without a beginning. Such a spot will constitute all of nothing. Any line starting from zero would inevitably start from a point where it ignores the zero mark because the fact of zero does not implicate a start or a size of value, but only the not being there of that position. All lines would form a duplication of another line sharing value since there will always be a possibility of yet another line in the realms of singularity lying between the two lines in question reducing the size infinitely to either side of the divide we humans create. Boundaries therefore are human and as man made substances it does not belong to the cosmos outside the influence of man and must be discarded. No mathematics will ever measure the thickness, because as the line that is standing still it cannot have a width at all. The moment a width appears which one can measure or calculate, the line will become part of the factor forming the divided and not the divide. The instant when space connects, the spin direction will produce the partisanship of space and spin. Any form of space (even in the most minute) will expand as it favours a direction but changing the direction is by rotary motion. The moment there is an area there is a measurable rotating brought about and no longer a non-interfering divide. Such a line holds space in a position that runs far beyond the boundaries and limits of the three-dimensional. Another factor of such a line would be that the radius (let us substitute the radius r with the using of Kepler's k), k would be immeasurably small. The factor k cannot be zero because infinitely close to that first k is the start of the third dimension where time plays the part as the fourth quarter. The presence of k is undeniable and recognisable yet it is not visible. The fact that k is there albeit stripped of any influence, disqualifies it from being zero and therefore not being there. With k already beyond any measurable space, leaving a^3 as a factor of one and not being able to pin any volume measure to that one k will have to be to the power of 0 being k^0. In Kepler's formula $a^3 = T^2 k$ the area a^3 would be one because of the dimensional non-existing of measured sides in any direction. If $k^0 = 1$ and $a^3 = 1$ the only alternative T^2 could possibly have is also one. The factor of T^2 identifies the time in the formula and when the formula indicates time as one, the time component must therefore be eternal. Only time in eternity does not change

The real formula applying when the calculation of the sphere volume is calculated is $a^3 = 4/3 \, \Pi r^3$ where it places one third of the dimensional (but lesser) factor in direct relation to another third dimensional relation held by the radius and all aspects about the factors being in relation is to acknowledge the form that is applying and serving as a sphere. However there is no criss-cross matching of dimensional accumulating. It places time in the square directly in relation to space in the cube in association where time shows two distinct qualities. The one factor is time in the circle rotating while the other is in the linear or the straight line implicating the position that the other would have. In all instances of measuring the distance the orbit travels around the Sun as the space displaces or space covered by travelling in the time it is covered and dividing such a ratio one find the distance of the orbiting object from the Sun the in relation to the other factors form one or very close to one. It is relevancies carried from the Sun and the Sun is the governing singularity representative for the entire solar system. This is about relevancies applying throughout the Universe. This balance is much, much more than what the figures say. It underlines and it explains gravity as a life form in the cosmos other than what we consider our life to be.

The German mathematician and astronomer KEPLER, JOHANNES (1571-1630) German mathematical and astronomer became Tycho Brahe's assistant in Prague in 1600 A. D. where he undertook to complete the tables of planetary motion Tycho had begun. Kepler first calculated the orbit of Mars. He spent much time trying to reconcile Tycho' s accurate observations of the planet with a circular orbit, but concluded (in Astronomia nova, published in 1609) that Mars moved instead in an elliptical orbit. Thus, he established the first of his laws of planetary motion. A theory that the Sun controlled the planets by a magnetic force led him to the second and third of his laws, which were published as part of his treatise on theoretical astronomy, Epitome astronomiae Coernicanae (1618-21).   The Rudolphine Tables (named after Tycho's patron, the Holy Roman Emperor Rudolph II) of planetary motion appeared in 1627 and were still in use in the 18th century. Kepler also wrote De Stella nova, on the supernova of 1604 and Diptirce on optics and the theory of the telescope. The overall view followed in this book **Matter's Time in Space** places the true significance of his work in true contents. In KEPLER'S EQUATION is the equation that relates the eccentric anomaly of a body in an elliptical orbit to its mean anomaly. The equation is $E - e \sin E = M.$, where E is the eccentric anomaly, M the mean anomaly, and e the eccentricity of the orbit.   It is important as one of the mathematical relations enabling the position of a planet about the Sun, or a satellite about is planet, to be calculated from the orbital elements for any time. However this only relates to the solar system, and KEPLER'S LAWS only apply in the contents of the solar system. The three laws governing the orbital motions of the planets, discovered by J. Kepler is as follows: The first law states that the orbit of a planet is an ellipse with the Sun at one focus of the ellipse. The second law states that the radius vector joining planet to Sun sweeps out equal areas in equal times which as it says refers to time and not the circle. The third law states that the square of the orbital period of each planet in years is proportional to the cube of the semi major axis of the planet's orbit. The first law gives the shape of the planet's orbit; the second describes how the planet must continuously vary its speed as it follows its orbit, moving fastest at perihelion and slowest at aphelion. The third law gives the relationship between the planets' average distances from the Sun and their periods of revolution. Instead of placing the true value to Kepler's laws, I. Newton placed his own interpretation to Kepler's laws, and in doing this he wilfully destroyed the principle working of the Creation. Through Newton's tunnel vision, he applied his own miss interpretations to the correct presumptions of Kepler. Newton reduced the implication that Kepler's findings hold, by using Newton's variation of what Newton wished Kepler's work would provide Newton when Newton was introducing his interpretation the law of gravitation. He then went about and changed it to three laws of motion. I. Newton generalized Kepler's first law, verified the second law, and showed that the third law should be amended to the form; $4 \pi^2 a^3 / T^2 = G (m + m_p)$. In this, the value of T and a are the period of revolution and semi major axis of the orbit of a planet of mass m_p about the Sun of mass m, and G is the gravitational constant. The major aim of this book is to correct these misgivings of Newton. I shall return to the statement about $4 \pi^2 a^3 / T^2 = G (m + m_p)$

What Kepler saw was more of a dimensional nature than the practical mathematic symbols and values. On the one hand was a value to the third dimension, which equalled two-dimensional values one the second dimension, and one to the first dimension.

In the argument Kepler made he had hiddee so much more facts into one formula than what I think even he realised. Well, it is much more than what the Accepted Policy Protectors Of Science ever came to realise. He officially formulated space-time, he officially coined not the name but the origins of the Universe being the Big Bang and he was the first to put the speed of light in relation to cosmic development…and all of that with his rather simple formula. He said the space a^3 not the circle (a) or the circumference a^2 but in the circle a^3… where such a circle represents a factor in the third dimension.

The formula he compiled was not rather but very specific about the area being a third dimension area and to prove it beyond doubt he placed it in the relevancy of the formula in a ratio of presenting the third dimension in space. He said a^3 is equal to $T^2 k$. Newton and Newtonians came afterwards and played with mathematical toys as to challenge their mental capabilities. Newton introduced a $4\Pi^2$ to indicate the presumed circle on the one hand and on the other hand he brought this lot equal to $\{G (m + m_p)\}$ which he then presumed to be the general Universal gravity constant (G) and the sum total of the two structure mass. Newton saw a ring circling around a centre having $4\Pi^2$ to indicate such a ring outside a centre and he positioned $\{G (m + m_p)\}$ where the two mass factors combine the gravity effort in the general grand gravity constant in space. I have had so much resistance in the past from all Academics but that is not what I see what Kepler saw. I shall trace this back to the centre of creation.

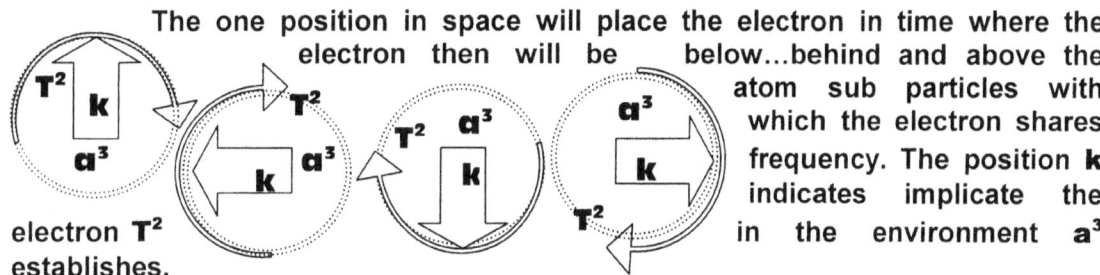

The one position in space will place the electron in time where the electron then will be below…behind and above the atom sub particles with which the electron shares frequency. The position **k** indicates implicate the in the environment a^3 electron T^2 establishes.

In their eagerness to calculate they calculated a formula to measure the circumference a^2 of a circle being Πr^2. I have seen an Astro physics examination where they use $4\Pi r^3 / 3$ as the formula to calculate the Sun and other stars volumetric space! They formulated the measuring procedure of the circle being in the third dimension that will show how big the volumetric space is of a sphere at a^3 being measured with the procedure being $4\Pi r^3 / 3$. This too was a fanciful devise allowing mathematicians to be much superior to the rest of the commoners and to dictate to the lowlife how and what they should think when they think and if they indeed can think of anything to think of. Then some Mathematician and an Englishman of Substance came onto the idea of gravity. Being a mathematician the Englishman placed the Universe at the feet of mathematicians. He saw circles where Kepler saw three dimensions. He saw three dimensions where Kepler saw nothing. He knew time had to be somewhere as something and then covered it by denouncing the circle as nothing.

What then is it that Kepler saw as he formulated $a^3 = T^2 k$. At the normal flow of time it takes the electron a certain time to spin around the atom. The atom uses space a^3 and the atom is a certain length **k** that forces the distance the electron has to travel in one cycle period T^2. The atom a^3 connects the electrons travel **k** to gravity T^2. The relevance **k** produces to support a^3 is to point T^2 to two positions the electron will be in the duration of one specific time. The electron travel will be cyclic and periodic in relation to the space the atom holds. The space stands related to the gravity with which the Earth confines the space of the atom to the space and speed with which the liquid heat confines the atom space.

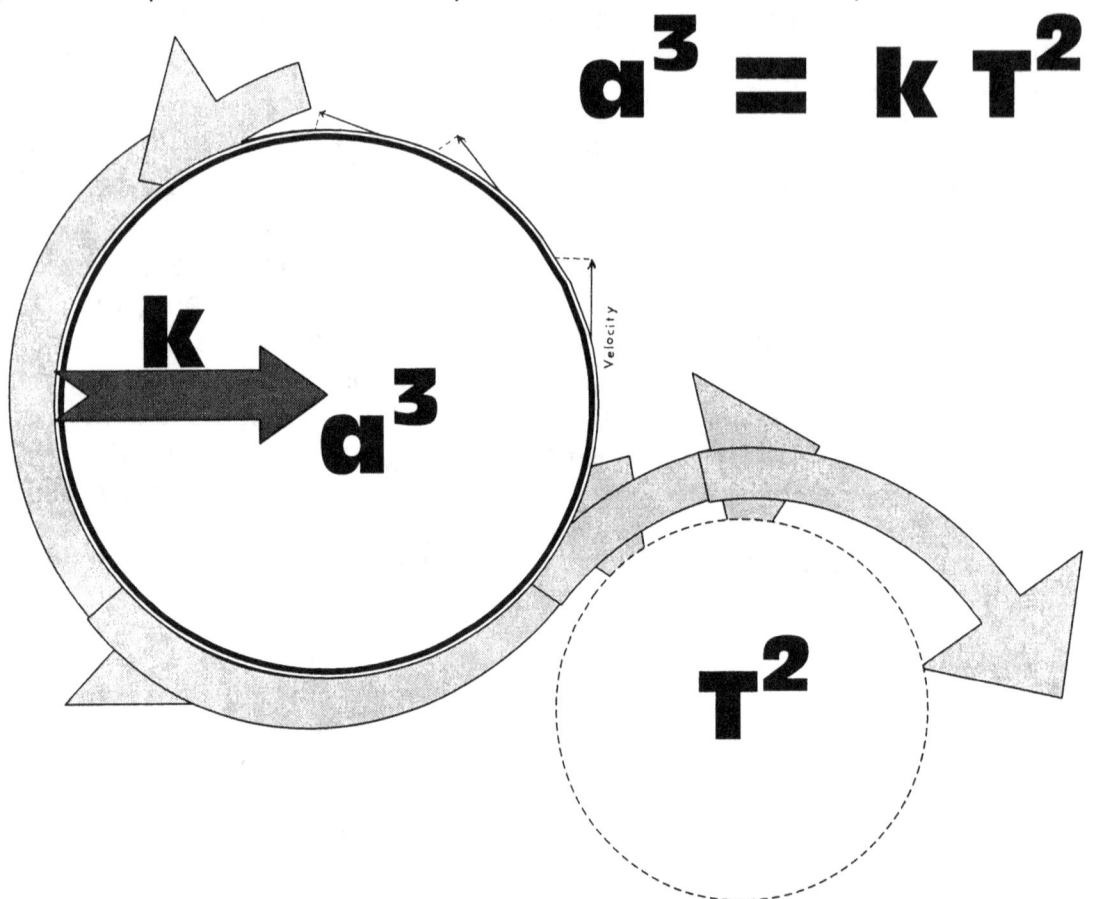

$$a^3 = k T^2$$

Planet	Period T years	T^2	Distance k	Space a^3	Ratio
Mercury	0.241	0.058	0.39	0.059	0.983
Venus	0.615	0.378	0.728	0.381	0.992
Earth	1.000	1.000	1.000	1.000	1.000
Mars	1.881	3.54	1.524	3.54	1.000
Jupiter	11.86	140.66	5.20	140.6	1.000
Saturn	29.46	867.9	9.54	868.25	0.999
Uranus	84.008	7069	19.19	7067	1.000
Neptune	164.8	27159	30.07	27189	0.999
Pluto	248.4	61703	39.46	61443	1.004

At the first glance Kepler's formula seems to be numbers and positions applying between the sun and specific but different planets in the solar system.

The time frozen on paper in a single t is effective in remembering the viewer of an event but that is not the event in the present any longer. That was how the event occurred during the time from where the camera shutter opened T_1 to where the camera shutter closed T_2 and the time frame T^2 was then during the open period of the camera shutter. But afterward it represented **t** when looking at the picture and the looking of the picture became an event during a specific T^2 that went from where one is taking the first look to where one is looking away from the paper carrying the first dimensional image of an event gone by and that is at that stage a representation of **t** in another milieu of $a^3 = T^2 k$. The **t** in the single is when mathematically presented as only **t** indicating a mathematical single flat dimensional view of time and is then correctly applied because it represents a reminder of a four dimensional event $a^3 = T^2 k$ that went single dimensional because the moment in the fourth dimension was then frozen in a single dimension on paper while the fourth dimension $a^3 = T^2 k$ soldiered on and time will always be representing T^2 as Kepler stated in the square allocated to space having a cube $a^3 = T^2 k$ at a time even before gravity got a name.

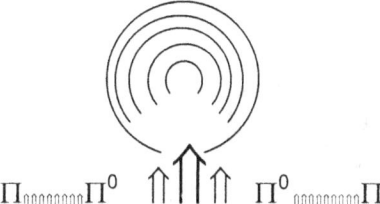

With singularity placed in infinity within the centre of every rotating object every atom and its relation to its surroundings including other atoms form space-time diverting from the point holding singularity as far as rotation goes because every object holds three relative positions in as far as where it was, where it is and where it will be in relation to singularity providing time. I elaborate on this else where.

Newton said a sphere is $a^3 = 4/3\ \Pi\ r^3$

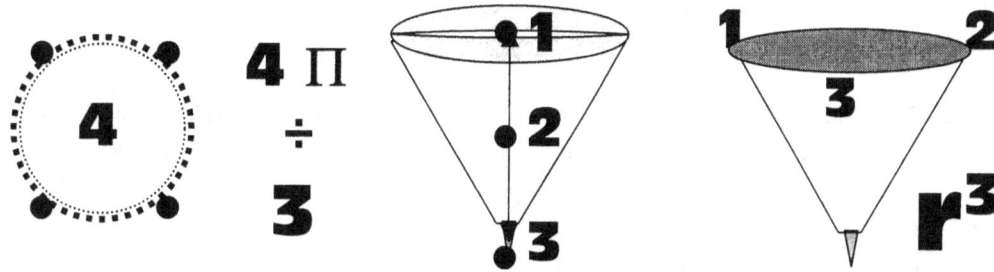

In $4 \pi^2 a^3 / T^2 = G (m + m_p)$

$a^3 = T^2 k$

$a^3 / k = T^2$ but at the same margin is

$k / a^3 = 1 / T^2$

$k = a^3 / T^2$ singularity

$a^3 / T^2 = G (m + m_p) / 4 \pi^2$

and $a^3 / T^2 = k$

then $k = G (m + m_p) / 4 \pi^2$

But I showed that $k = a^3 / T^2$ and Newton's claim is that $a^3 / T^2 = G (m + m_p) / 4 \pi^2$

The only definite place one will locate zero is in between the starting point of the lines going in opposing directions in the position the lines hold before there was the least of directions applied, but that is only because there is no such a position, not because any line is coming from there. The two lines are still one holding the opportunity of parting as an option but have not yet parted and therefore are on the very precise same spot. The line coming from there is already there because it already has the choice of going in any and all opposing directions and when it starts running it will place filled space in that location because the space was already filled with a line starting and not with a line not there at all. When reversing a line we might find a better idea of what is in place and where it is in place.

In the action of the inseparable drawing closer and moving closer gravity finds the dual value of linear and circular gravity. There is no separation of the two factors acting as one but both have different applications and values in the unit. This is the result of singularity having three parts acting as one but giving three distinctions in application. Gravity is as much part of dismissing space as it is about making contact with space in time.

But since the connection comes about as a circle, the connecting points will relate to Π as the value.

Due to the spinning nature of such a point with all surrounding the point will be alternating direction favouring change every second and in that the value of such a point can only be Π because of its constant changing. Using r would specifically oppose another r from every angle because the use of r will bring about a static relation to the previous and following instant and therefore it will cancel the constant spin flow.

The new direction pointing to a new location in relation to the previous point will oppose the previous point it had in relation to direction considering the centre point.

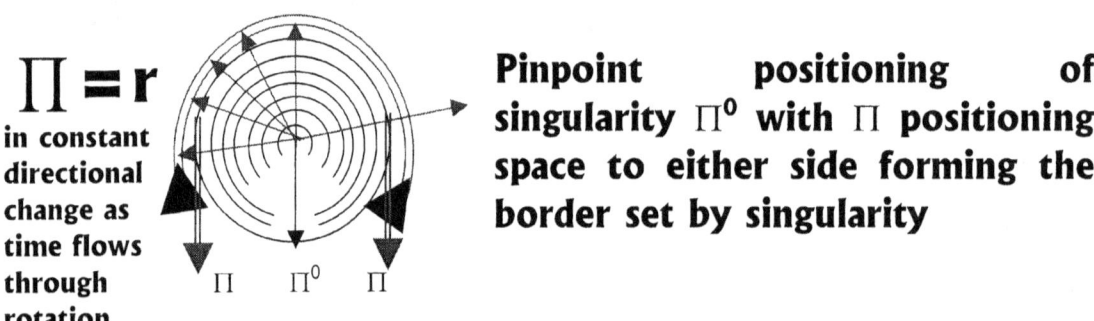

$\Pi = r$

in constant directional change as time flows through rotation

Π Π^0 Π

Pinpoint positioning of singularity Π^0 with Π positioning space to either side forming the border set by singularity

The motion of a liquid confirms a centre and the confirming of the centre provides a flow of space-time in either direction, which produces the gravity. Without establishing and activating such a centre the centre is not active. It is there but it is inactive in being present.

By taking the line to k back to where the line or k cannot reduce further $k^0 = 1$ establishes such a value where k then finds a position in the single dimension. But in that case a^3 also is equal to one and so is T^2. In fact k still has to produce a line and we find that k represents a^3 to the full as well as T^2. The point where k forms the most slightly distance the area a^3 establishes a value outside the single dimension because T^2 adds a value. The fact that T^2 comes in as a factor in the presence of the first sign of k appearing indicates the start of motion taking a^3 from one location to another specific location. It indicates the travel of the planet during a month or a day or an hour. It does not indicate a circle except at the end when completing one cycle. T^2 is the distance in time a^3 will take k from indicating one point to indicating another point. The formula points to a referring of the very time space was indicated by position location and time. The astonishing part is not as much the way Kepler formulated his formula to cover the movement and the position of the electron in relation with the rest of the atom, but the brilliant way the mathematicians neglected to see the fact. Kepler saw a three dimensional a^3 something in a specific position in time T^2 relating to a specific density k of the atom. With space in a cube as it cannot ever be otherwise the time too has to be in a square because placing time in the single dimension of t the time then becomes part of a single dimension such as one may find in a photograph picture. One can justly use the same formula to implicate the electron taking time to complete the distance between two points indicating the area from the centre of the atom.

In the sketch below the circle to the right would come about from a straight line r growing influencing the appreciation of Π, but to influence Π would lead to a breakdown in r as Π and r are different entities. The circles to the left shows a continuous growth by extending Π every time and since Π is the same part as the previous Π, only extending that billionth of a millimetre or many times smaller each time, the circle will be truly continuous without any signs of a break. In the context of dimensions one finds coming from the centre $Π^0$ an established eternal flanking of Π to six positions since $Π^0$ forms the centre to the six sides and all six sides not having a diameter yet must apply Π to indicate specific value. In the very centre, which I am referring to, rotation must end or start depending from what vantage point the relevance is placed.

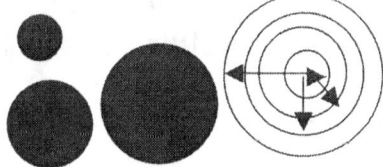

One should not try to focus on an image of such a spot or dot because there is no image. The line dividing the cosmos and that run through every particle, no matter how large or small is beyond our vision. Such a small line, so small it is not even noticeable is large enough to part the cosmos into sectors. It splits the biggest there is into particles and we are not even able to notice the precise location of such a split. In truth there is no top or bottom that we living in 3D can see. We shall have to use a general conception brought about by intelligence. Your intellect tells you about such a spot, but that is all because that spot is on the other side of the Universe (quite literally). From the centre of the dot there is a top and a bottom spot. From those points there is connection with four quarters. That produces six connecting points that are all aligning to the centre.

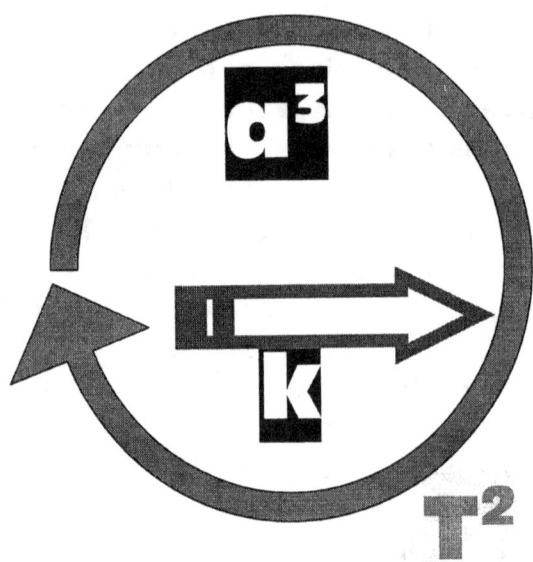

a^3 forms the space the atom claims while travelling in the Earth spinning all the way and travelling with the Earth around the sun

k positions the electrons travel in the space relating to the space the atom holds while travelling with the Earth in the Earth around the sun in relation with a specific position k will indicate in relation to the sun.

T^2 is the time it takes the electron to be relevant to the position the Earth places on T^2 while the Earth captures the space of the atom by providing the space for the atom to be within while the Earth travels from one point T_1 to another point holding T_2 in frequency to the atoms T^2 relating to the Earths T^2 in perfect harmony with the sun having another T^2 relevant to all the other factors we call cosmic particles. Big or small it is only about cosmic particles holding space in time in relevancy.

Looking at the effect of gravity it shows the precise quality of no distinctive point, as gravity never seems to end at a point but flows all over affecting all that holds a position in its sphere of influence. The gravity coming from China meets the gravity coming from America at no particular spot but intermingles without distinction.

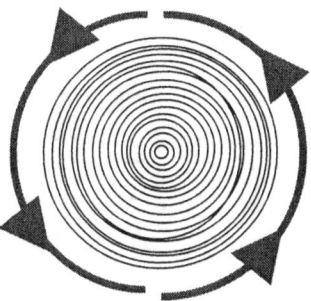

The very centre form an eternal divide that will not allow what is on the one side to present an influence on the other side. It divides spin. It divides direction of spin. It divides all rotation from the outside that one may detect and such divide is there because at one point spin will run to the left coming from the right and just immediately next to that point must run a direction from left to right. It cuts without contributing or participating in movement. It divides without any favour.

By expressing a wish to accomplish time travel such a person wishes to accomplish that material must collide with itself a mean feat if ever there was one. Such a person wishes to have one side collide with the other side as he stops and reverse time while the rest of time is motoring on.

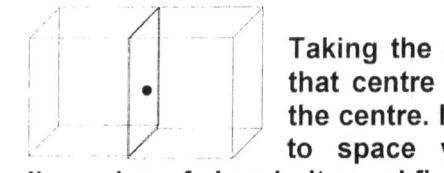 **Taking the outlook from the point the sphere is holding from that centre out into space there are ten points connecting to the centre. In that are the dimensions of singularity connecting to space where five connects to space in the second dimension of singularity, and five connect in the third dimension of singularity**

That is singularity not having a dimension of space and not having a dimension of time, or a radius connecting the rotating distance to Π. Every rotating object holds a centre from where the rest of the rotating direction will differ at any and all given points. Not one point is exactly the same, but in the very middle, the centre no one can draw, measure or see is a point not in motion.

In the centre runs an axis line that forms the division of rotation. No one human will
 **ever be able to indicate the precise line, but such a line must exist because of our logic telling us about such a line. In the centre one will always find one more line smaller than the outside but forever also always bigger as it is towards the inside.**

The sphere holds six sides in relation to form as unit and as does all other shapes and forms. All forms have to have at least six sides indicating different exposures to the Universe. But with gravity having a free choice, gravity always chooses the sphere. As I shall prove later on gravity is the strongest where the form produces the least evenly distributed space.

The first condition for gravity is even-handedness through out the sphere holding the applying gravity and the second is to have most or the strongest gravity located where the space is least. That gravity then has a position in the very centre of the sphere and from that centre the gravity produces all the edges or borders that the sphere consist of. In the case of the sphere this factor makes the sphere much more dominating than any other form does. From the centre point controlling all sides is gravity and with gravity applying control the sphere has seven sides to the square in any other possible form having at least six sides.

The cube can come in whatever form there may be but the sphere adheres to precise measure and behind this principle is all that forms the Universe. The cube has six sides connected loosely and can change form just by changing the relevancy between one side (or more) in relation to the distance brought about by the other sides. The sphere being a complex circle stands related where the sides has to apply precise measure in equality. This becomes a law because in the precise middle one will find the strongest gravity as that gravity holds the object in form and true to form. If there is even gravity spread in all directions the form must be a sphere and the sphere insist on seven points relating to sides or borders.

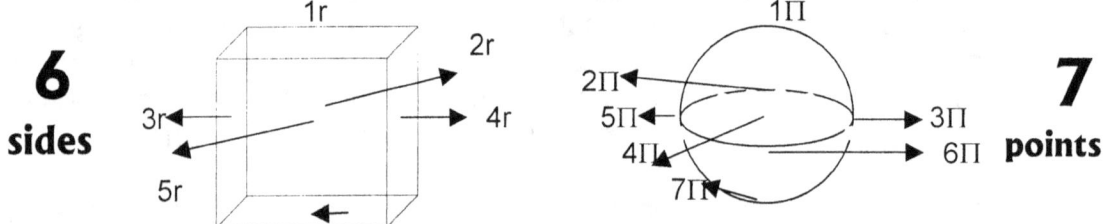

In the sphere there are no radius but only the extending of Π from the centre Π in six opposing directions relating to one another by the square but remaining Π because of the unity the matter holds in relating to space. It is not possible to draw a precise line that would form a precise ring and not cut some atoms in parts. Because there will always be an atom disallowing the precise positioning of the circle the circle continues on a solid basis holding Π as a positional reference and not r. In every sphere there then are the seven Π relating in precise dimensional and positional equality forming equilibrium to the centre Π as well as to one another by 90^0 and 180^0 implicating the dimensional positioning. Therefore the sphere holds 7^{Π} and the cube holds 6 X r^2. Where space comes into contact with the sphere the cube loses one of the six dimensions it has to the more dominating seven dimension of the sphere whereby the seven dimension in equilibrium will dominate the six dimensions loosely connected by r bringing about that the cube then has 5 sides to the seven of the cube. Because the space surrounding the sphere takes on the shape of the sphere and not the other way round where the sphere resolves in accepting the form of the cube, one may presume the form of the sphere is the most dominant of the two choices.

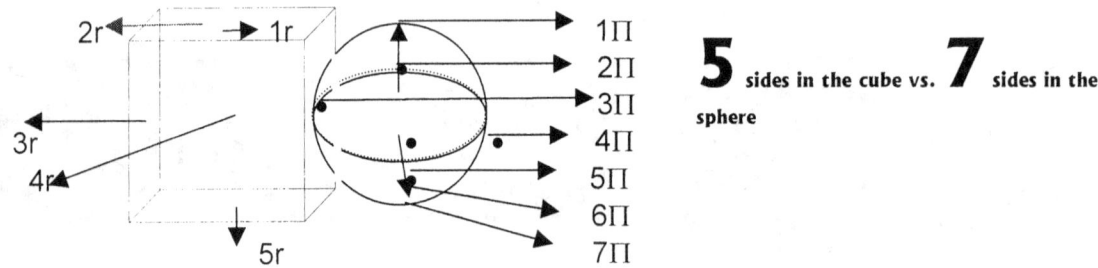

5 sides in the cube vs. **7** sides in the sphere

The sphere is a multitude of innumerable circles that forms one unit with all the innumerable circles that compile a spere all put together. The circle is a constellation of Π where every Π flow from one into another and such flowing varies the number arrangement of r. In order to measure the surface of a sphere the radius carries the torch by going square. It is the radius, which is just another line that come from the outside and run all the way inside to bring the value of the line into the circle.

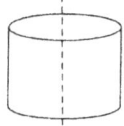 **One must remember that the radius is half the diameter and where the diameter ends on the edges that form, the radius reduces by half. In a circle, the radius that initiates the circle reduces by half as it goes into the circle, which is a single part of a combining sphere. The calculation of such a circle is $\Pi \times r^2$.**

The radius r runs from the circle outwards, from a circle centre point towards Π, the value of the circle. Or we can take it into the circle and have the 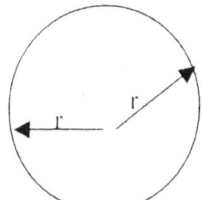 **same results. In the centre of the circle, there is a point where the radius starts. It runs outwards from that point in all directions towards the circle Π. Technically, there then has to be a point where r is zero, an absolute zero. However, the circle therefore remains Π. The circle does not disappear; it remains there for all to see. It is only the radius that the exercise removes. This leaves a situation where at one point we must find the radius becoming r^0 where the radius did all the dividing and cannot divide any more.**

$$\frac{\Pi r^2}{r^2} = \Pi$$

If one removes the radius from the circle, the circle remains, only holding the value of Π. By removing the value of r, Π becomes singularity with no place to be. Singularity is the place where there is no space to be in place. However, Π remains because once r receives the slightest of space Π will find space. Then the circle will grow to Πr^2 and r would determine the space. Without space, there is no r but there is a circle with the value of Π. Singularity is in every single rotating object, be it the proton or the Universe. This situation is part of any and all circles and is therefore part of any and all spheres. The line will end at a point where the line starts going in the opposing direction and that value is indicated by the use of $r^0 = 1^0 = 1^1$

To that end the shortest possible line (hypothetically) must be so short it must have **an initial and ultimate point sharing the same spot.** Any theoretical line being the shortest possible line cannot have the line holding the initial starting point at point zero and advance from there. If it used zero as a start, the zero part would not count, because the line will only start at a point past zero where the line then will start. Zero ultimately means not existing and then that point, as a start does not exist. At one point the reducing attempt of the line would start making the use of mathematics seem silly. The reducing would seem tedious and leading nowhere. But as sturdy as mathematicians can be they would carry on (or so I am made to believe...). Then when the man doing the calculations gets carried off in a straight jacket, while the man is making funny noises, when he is totally cracked mentally, the calculations can still go on and on and on and...and that is where sanity prevail and someone says "drop the affair". It is at that point I would have loved to see Einstein carry on counting stars in so many galactica in his attempt to determine the critical density joke. That is not where we get into infinity. That is where man's brain gets blistered but infinity is still far off. As any one can see the Universe is far beyond some insignificant and senseless formulae invented to impress Academics while others are kept busy and free from boredom, but in the real Universe the attempt is not worth the thought it takes to disregard the attempt.

When the line **has a beginning and an end at the very same spot** and it wishes to extend the position as to further the possibility it has, which direction should it favour. Extending the line in any one direction will favour one direction without any clear reason not extending in other directions. The only mathematically sensible option about extending will be in all directions equally in order to give a meaningful non-bias flow of mathematical equilibrium. That is where one would have to go look for the beginning of the Universe. The Universe is a about lines connecting but where does that which does the connecting end. The Universe used form to this point in development, but then at some point the line came and established the presence it still has. The first form was moving from $\Pi^0 \Rightarrow \Pi$. Again I wish to press the issue, at that stage form was in use and not mathematics. The Universe was just simply too big to measure. If radius did apply, one could use r and r2 but since only Π was in use there was no radius to be used.

There is forever one circle leading to the next circle, which is followed by the following circles. Where the light does not reflect the image that is there, we will still find a concentration of circles leading another and another and another. The end is eventually endless.

When working with concrete and heavy metalled solid objects r would show as a crack distinctly parting solid structures, while Π indicate a continuous flow of solidness giving the material an overall and continuous structural strength, yet engineering never recognised this difference. By confirming Π the circle dot ploys singularity in all components and therefore proves to be a much stronger support as building choice than other shapes.

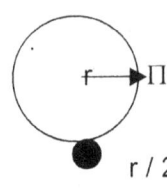

$r/2$

$r/2$

$r/2$

When **the circle reduces, the value** located to **r** will become implicated because **r determines specific size. Not so** in the **case of Π, because** Π in the true sense only **indicate that the circle is a square without corners** and therefore Π **dictates form and not size**. By r**educing size** only **r comes into contest** and will point to such reduction. By **reducing** the circle **radius r by half continuously** will lead to an **infinite small circle but Π will remain because the circle as a form remains** even being infinitely small. By **reducing r indefinitely** to the tune of half each time, **r** would become infinitely small, beyond human calculating means, however as mentioned in the case of the smallest dot holding one spot, **r** would become insignificant beyond human comprehension even, but never reaching zero and still Π would remain intact and dictating form.

The spot forms a full circle, but the line running through the circle is forever present because that is the future radius of the circle that will one day develop the circle, which is equal to the present diameter. The fact of the presence of such a possible line in such a possible circle dividing the possible circle into two parts makes the centre line equal to the half circle. The line forms the half circle but not only that the line presents the half circle as much as the line is the half circle. The line then is 180^0 and the half circle is 180^0 because in singularity the two factors are the same. The same value is of course $\Pi^0 = 1$.

In this half circle of the future, which is no half circle as yet because of a lack of space there are three future points indicating the space less ness that will go on to become space filled with something. On top of such a circle to form must be a marker indicating an awaiting boundary or future border and at the bottom of the future circle there also must be a similar marker that is no marker as yet. Between the two possible points that are not there yet is a future line running that is not there yet. Then indicating the possibility of a position to come, that will bring about the half circle being a future distance apart from the future line indicating a diameter that will one day be there. A third such a marker must be established for the future. That forms a triangle with two more sides being connected by either a line being one or half pi being one. From singularity comes about that the line is the same as the half circle is the same as the triangle and all has one value being 180^0. From this come the most basic principles in as much as forming the ground rules of the law of Pythagoras.

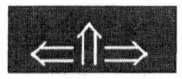 The one form we know there was and was valid at the time was the straight line

 The next form we know there was and was valid at the time was the half circle

 The other form we know there was and was valid at the time was the triangle.

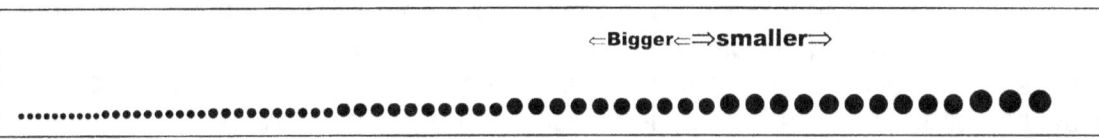

It eventually gets so small we humans can fit into it... remember we are seeing the reverse of the truth. It was never so big that it contained nothing because all that and we came afterwards being smaller that the dot filled the dot.

Everything at the time that was outside singularity and was in form at that time was equal. Think how big they were. They filled larger parts of the Universe than our brains can cover by thought. They formed the holding tanks that still hold us in the massive Universe. They were at the time too small to have size, but since then they grew into structures that are too big to have size. Those dots still are bigger than mathematics can apply because there still is no measure quantifiable mathematics can reach. Just because they compare with what we seem to preserve as small in cosmic relation they are too big and too large for us to comprehend. Even if they were immeasurably many they filled an immeasurable Universe in the same way they still fill the immeasurable Universe and we are so small we and our surroundings are measurable and quantifiable. They were so enormous there were no relevancies applying to compensate for distinguishing. Distinguishing only followed later when size started to matter. That meant the Coanda effect was in place without the Roche or the Titius Bode law. It was the start of the relevancy principle from which the atom later came and which is the result of the Kepler expression.

$$50 + 50 = 100 = \sqrt{100} = 10$$

1^2

$7^2 \qquad 7^2$

7 to and 10
spinning
7 through to 10

Space is already in the square but through time ce goes
from one point to another point in the square of motion.

It eventually gets so small we humans can fit into it... remember we are seeing the reverse of the truth. It was never so big that it contained nothing because all that and we came afterwards being smaller than the dot that filled the dot.

With no line possible there had to be another dot that formed since the Universe has many dots that formed lines. But let us not to get confused and lost in the range of possible diversions but let us stick to two dots. One dot was next to the dot next to the dot, but as I said, we stick to one dot next to the second dot Π is the first step where gravity began with. That leaves us with a huge problem in as much as when $r = 0$ then $r^0 = 0$ and 0 dividing any value will leave 0 as the answer. If the particles were inseparable at the start it must bring about that gravity would not be forming since the distance will not permit any dividing. By allowing the distance separating the particles to be zero, the particles melt into a unit. Again this is Mathematics and not my incoherency as some Academics dismissed my work. Let me run through the argument one more time because I have been insulted by Academics in the past telling me I am bending mathematic rules with my applying double values to try and produce some argument. The two particles formed by an inseparable unit separated by a sharing of a spot. We know that at least two spots formed because there are many more than just two that remained to become part of the visual Universe. Let us name the spots because that is what humans do best if they do not know what to do with what they have to do. Let us call the on dot and the other one spot next to dot we then call dot two. Between dot and dot two there were nothing because dot and dot two were inseparable. By they're being inseparable we would naturally be inclined to think that the separation value should be nothing or at least zero. But putting zero in that place is a mathematical excluding procedure leaving future mathematics excluded. With m multiplying m_2 and then dividing $\div$ r with zero ($r=0$) such a procedure will leave the lot at zero and with that nothing is going nowhere. That means although we think the space between the two parts are nothing the non-existing space has to be at least one to be a future factor.

At fist $\Pi^0 = \Pi$. Then after a while $\Pi^0 = \Pi^3 / \Pi^2 \Pi$ and gravity comes about forming space-time by motion of form. Being the sphere that formed the 7 holding relevance in the form of the sphere took shape. But the Universe is layer in dimension forming the next layer in dimension forming the following layer in dimension. The Universe was $\Pi = \Pi^3 / \Pi^2$, which is taken from Kepler's formula he received from the cosmos as $k = a^3 / T^2$. Where there is a sphere involved there is a natural tendency to grow by developing the sphere.

Every part of the argument is sound but was never yet used. I repeat once more if my argument reflects on inconsistencies those inconsistencies are not about my work. In order to disprove my argument replace Mass one and Mass two with any number possible, then divide such a number with to the square being zero. If there was no space then the value of the particles had to be one. If there was no space between the particles the particles then had to form a unit. But if there is a mathematical possibility of reducing a line to the single dimension then there had to be a factor representing r as a factor of one. Take $(M_1 \times m_2) / r^2$ and substitute any of the factors with zero and the result coming about has to be zero. The factors in the equation have to have any and all the elements at a value of at least one. Only if r was a factor of one can gravity bring about any mathematical equation developing from this argument. That means the mass on both sides must have a factor of one being a limit, which does not allow such further reduction of r and any further reducing of r beyond the limit will not be tolerated. Only if $r = 1$ then r^2 can be 1 and mass can be apart. Like it or not but believing in the Big Bang must also bring about the accepting that the cosmos moved apart somewhat. The fact that r brought increase in the space separating the mass produces a problem that was solved already. About a century and a half ago Roche found just such a limit. Once again I were confronted by zero becoming growth. There is a huge hole that needs filling when bringing into a relation any forming of an alliance between a cosmos coming from nothing and filling with nothing and a cosmos growing spontaneously through balance shifting prominence. Mathematically the fact of applying nothing as a vale applying in the cosmos is not a strong and convincing argument. The minute one brings in zero as a multiplying factor forming a definite value working into the calculations of the cosmos, growth disappear. If growth was not a factor, the zero factors could be involved with some form of maintaining stability and where then further growth will accept the responsibility of zero

Taking nothing to mathematics zero locates in infinity by abolishing nothing from the universe. The sphere is 7 X Π = 21.991. It is a formula or a recipe for gravity. It shows how a sphere grows into a sphere by form on a continuous basis.

In the circle using $r^2\Pi$ the r has to have distinctive qualities placing it as a factor apart from Π. Where the growth shows no separate distinction but a continuous flow from the precise centre to the precise edge the flow would become in relation with Π depicting the circle and Π replacing r as reference to any point on the circle. By using r distinction in the circle is possible but by using Π there is no distinction possible.

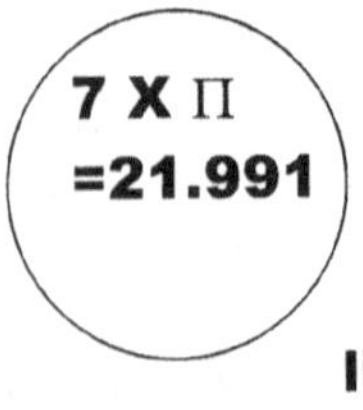

7 X Π =21.991

The sphere is 7 X Π = but from the other view the sphere is singularity relating to ten. That starts the Titius Bode principal which in relation to the Roche limit at the limit of Π use $\Pi^2 / 4$ to form

10

21.991
Singularity
= 1.9991
10

gravity Π^2. However this is a little more involved, as it seems because the one sprouted because of the consequence of the other bringing the one into the Universe as a relative that will sprout to bring about gravity.

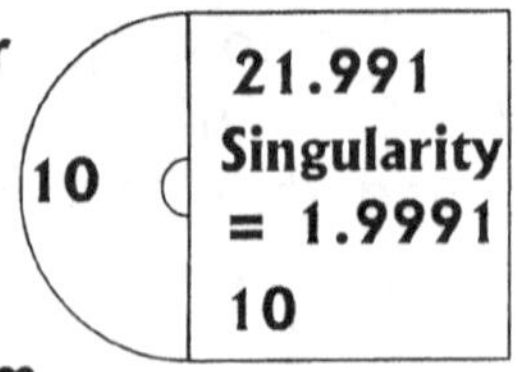

However if the attachment was $7 + 10 + \Pi^2 / 2 = 21.93 / 7 = \Pi$ is the circle that serves as an attachment meant being a sphere and holding as well as sharing singularity. This is what Newton saw, but this is not gravity in the cosmic sense.

7 A sphere being formed by the six ends crossing as it incorporates the centre singularity
+ 10 anther sphere with an identifiable motion keeping the independent singularity apart but within the relevancy of the unit formed.
+ $\Pi^2 / 2$ Singularity by gravity shared by two in one unit.
= 21.93 / 7 still holding the unit to form where the overall containing form will be a cosmic sphere.

The sphere holds many dimensions relating to seven but also by the square of space which is 5+5 = 10. This brings about gravity generated by means of the Titius Bode law and is principle proof of this statement because it indicates an infinite number of numerical positions influenced by quarterly divided sectors around a point holding singularity.

Boys playing games will never realize scientific breakthrough explaining and grown ups do not play with toys. In this little toy played everywhere everyday by almost every one is the answer most brilliant of human Brainpower seek answers about all the cosmic riddles no one seem to understand.

Newton said the rotation delivers no work and therefore the effort of the r0tation results in a zero. Firstly it bring us back to the zero idea where with all the reasoning in the world and all the leniency I allow I cannot find zero as a value being part of mathematics. Let's move back to the circle to try and find the zero Newton saw in the rotation.

$\Pi \times r^2$ = CIRCLE

If you remove r it then is $\Pi \times r^2 / r^2$ = CIRCLE.

You cannot then say $r^2/r^2 = 0$ and therefore $\Pi \times 0 = 0$. That is nonsense. $\Pi r^2/r^2$ will always be $\Pi \times 1$, and that is the eternal circle. When looking at any rotating object, there has to be a point of no rotation and no rotation means "no rotation", not no existence. No rotation means a factor of 1, not zero.

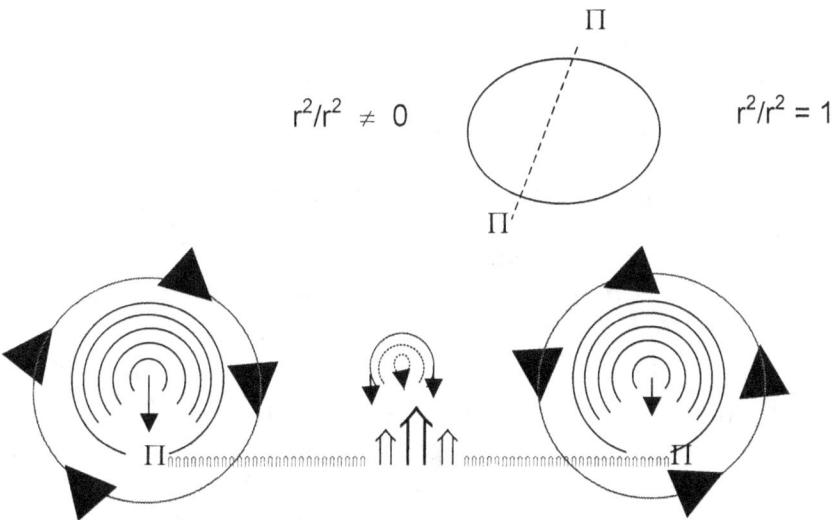

Not only does atomic individual singularity maintain self preservation, but in doing that it also sustain a governing singularity holding structural composition and form within a cluster of matter for example a star. As there is between stars so there are in the same manner a mutual or bonding singularity between atoms in stars, which we see as fusion.

Any object in rotation will have a middle point, a very specific centre point that does not spin. That point once again hypothetical but none the less must be standing still because every line running from that pint in opposing directions are also in opposing directional spin to each other

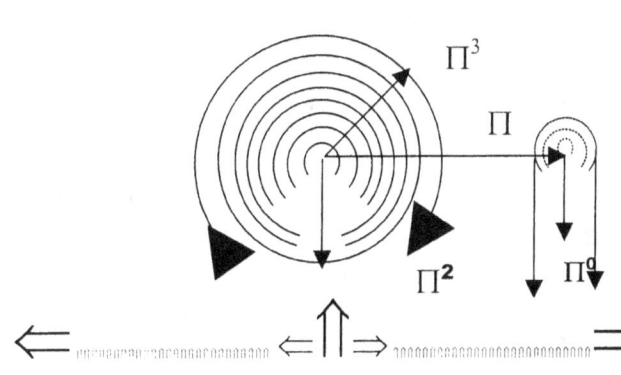

As the stop starts to spin the motion establish the centre line, which activates singularity, which activates space-time that activates gravity at a specific relevancy. Where we locate singularity there is not nothing because gravity cannot come from nothing because only nothing comes from nothing.

After all it is gravity that keeps the top as it is spinning in an upright position while it is spinning because it is gravity that stabilises the cosmos. Moreover, what is actually in progress from the top spinning is the Coanda principle activating gravity and that happens in accordance with Kepler's formula

This means that in the cube at the point of contact between the cube and the sphere the cube experience such a contact point as if the "bottom falls out" of the cube and without a "bottom" to support objects they fall to the sphere as objects does fall to the Earth. Remember that a body "floats" in space, but at one specific point it starts to "fall" to the Earth. That is gravity and it is a dimension change much more than any force. I shall explain this last remark later on. That too is the Lagrangian system with five cosmic structures holding relevancy to the centre structure where the centre structure stands in for seven positions diverting from singularity and the orbiting structures standing in for five positions in space.

In the centre runs an axis line that forms the division of rotation. No one human will ever be able to indicate the precise line, but such a line must exist because of our logic telling us about such a line. In the centre one will always find one more line smaller than the outside but forever also always bigger as it is towards the inside.

From such a point every other point will be opposing any other point not pointing in the direction to which the first point is pointing, whereby it extends the direction it holds. No matter what the point is or where the point leads, such a point holding a specific direction will be unique in the direction it is rotating because at that or any other specific point wherever, it will be directing not in the direction it spins but in the direction flowing from the centre point outwards.

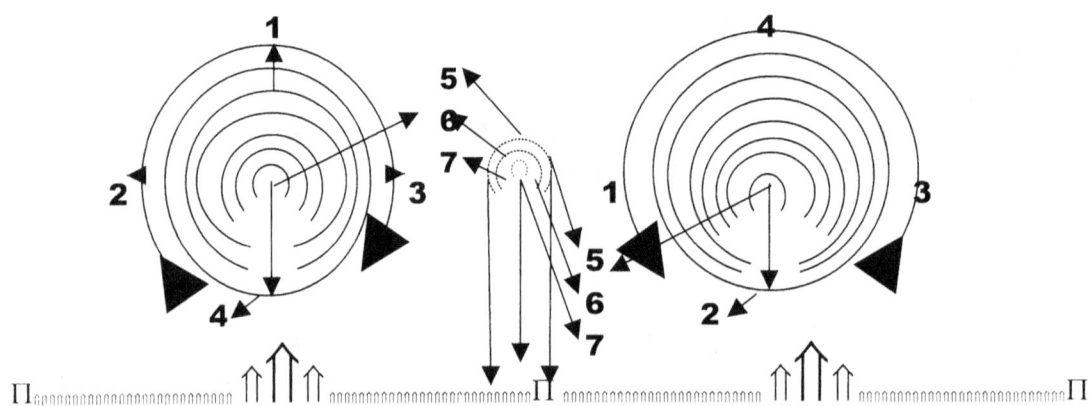

Any point will be it opposing itself within the rotating of 180° changing every aspect of its previous flowing characteristics it previously had or will once again have in 180° from there. While in rotation from the point of an outside observer all may seem static and never changing but to the object in spin every next second will be a diverting from every aspect it was in very second passing, and the direction it held in relation to the direction it held the previous mille, mille second will totally be incompatible with the direction it holds the very next mille, mille second of rotation. That proves no point can be static or constant, all though it may seem that way to outsiders.

In the very centre of the sphere the form of the sphere dictates that the shape will relinquish space as the line run from the outside towards the very centre. With this natural state of affairs the sphere are naturally inclined to dismiss all space that it can form in the form as the sphere holds space inside and the form will finally be without dimension. All that I attribute to the line shrinking by reducing actually takes pace in every sphere as the diameter reduces to the centre. In the centre where the radius line goes single the form relinquish the three dimensional form it has inside. Being without dimension in the very centre means that at a point in the extreme centre of all spheres there are a point that holds singularity because this point with no space has a mathematical position although it is invisible since there is no sides to such a point to give that point any dimensions. The shape of the sphere is calculated by using the formula $4\Pi\,(r^3)\,/\,3$. By reducing r to a point where r is r^0 singularity steps in because only the form remains as Π. Going even further we find that there then comes a point where Π goes singular Π^0. At that point absolute singularity is present but so is absolute gravity present at that point. When holding the strength of the shape of the sphere in mind as well as taking into account that all cosmos objects of importance is in the form of planets or stars and they are all in the form of a sphere, we therefore may contemplate that it is where gravity originate. We now only have to find the reason why gravity will hold a base in a space less ness as Einstein predicted. It is clear to be seen that gravity is in the centre of the sphere controlling from the centre everything that is outside the space less centre. We can reason with confidence that gravity is the strongest where space is the least. We can further reason that it is gravity that is holding the sphere in true form and since the sphere allow gravity the best working opportunity, gravity can form the sphere in as strong a shape and form as the sphere seems to have. From every point on the surface of the sphere is where that point connects with the other side of the surface of the sphere by a line that runs through the space less ness of such a centre of the sphere. Such a line also connect by an angle of 180^0 as well as 90^0 to six other lines running from top to bottom, right to left, and back to front, where all join and cross in the centre of the sphere. There are therefore six lines crossing and connecting by a centre from any given point on the surface of the sphere. Such points connects in total six surface points on each side of the sphere while they all support one another through the space less centre. In that absolute space less ness in the centre holding singularity we find gravity supporting and controlling all space within the sphere as well as space connected to the sphere. That is where gravity control and guide the space, which falls in the parameters as well as under the influence of the form of the sphere. In the gravity centre space goes singular meaning space becomes space less or flat.

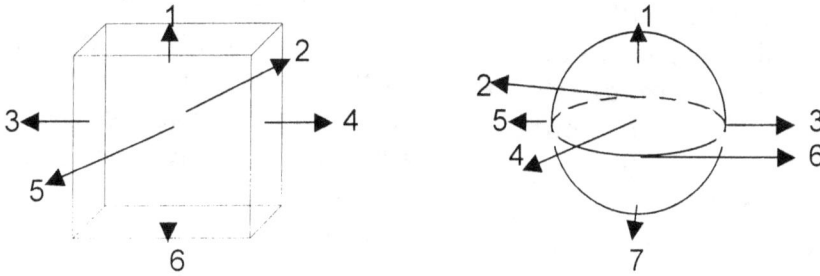

Also it is true that the entire form that is the sphere is controlled from a centre within the sphere. That centre holds the sphere in form and shape. Therefore the strong form is dictated from that space fewer centres where there is no space and no form left. The natural inclining is in the form of the sphere. It is part of the roundness that the overall shape of the sphere represents and this structural strength is carrying down to the very centre. Because the circle is forever reducing that reducing which is inherently part of the form of the sphere becomes a tool in distorting of space in the sphere and is eventually removing all forms of space from within the centre of the sphere. The very centre ends up as having no space because of the reducing that continuous down to become the space less inner centre. The all roundness is the ingredient that forms the backbone of the absolute strength that the sphere has and that is the component that the sphere is so famous for. The form the sphere has allows the sphere to have a control that is coming from the centre deep inside the sphere where the space vanishes and being without space seems to keep the entire structure rigged. The strength of the sphere comes from the centre of the sphere, which is inherent of the shape. That is why the sphere has such and the fact that all connecting sides refer to a centre brings the strength that the shape has. How does it work in its most basic analyses?

It is from the layout that the sphere uses as a natural form that we are able to locate singularity. In the case of the sphere the material naturally reduces by measure of the radius becoming smaller to a point where the radius is r^0. At that point the line that will form the radius has gone single dimensional r^0 and that is equal to 1^0, which is singularity.

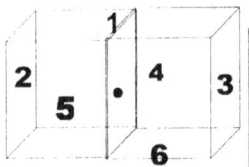

There is one more point in the sphere in the centre forming an addition in the sphere. That point holds gravity secure.

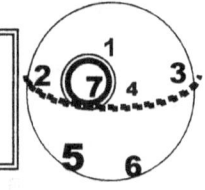

The cube has sides and the sides form a rather weak and flat surface that connects four corners. The flat surface produces a rather indifferent contact point with no special features on the surface. The corners connect to other sets of corners and those corners form a weak structure without any direct support coming from the other five sides. Without material to fill the body of the cube the cube has no direct connecting between any of the sides other than corners connecting at the edges of the sides.

Taking the vantage from the point the sphere is holding from the centre out into space there are ten points connecting to the centre. In that are the dimensions of singularity connecting to space where five connects to space in the second dimension of singularity, and five connects in the third dimension of singularity. On the other hand, the cube does show a very different characteristic, which involves only six sides (at least) connected.

In the very centre of the sphere the form dictates that the shape will relinquish all grounds in space that it can hold and the form will finally be without dimension. Being without dimension means that at a point in the extreme centre of all spheres there is a point that holds singularity because this point with no space has a mathematical position although it is invisible since there are no sides to such a point to give that point any dimensions. When holding the strength of the shape of the sphere in mind as well as taking into account that all cosmos objects of importance are in the form of planets or stars and they are all using the form of a sphere, we therefore may contemplate that it is where gravity originate. We now only have to find the reason why gravity will hold a base in a space less ness as

Einstein predicted. It is clear to be seen that gravity is in the centre of the sphere controlling from the centre everything that is outside the space less centre. We can reason with confidence that gravity is the strongest where space is the least. We can further reason that it is gravity that is holding the sphere in true form and since the sphere allows gravity the best working opportunity, gravity can form the sphere in as strong a shape and form as the sphere seems to have. From every point on the surface of the sphere is where that point connects with the other side of the surface of the sphere. All other possible points connect by a line that runs through the space less ness of such a centre of the sphere. Such a line also connects by an angle of 180^0 as well as 90^0 to six other lines running from top to bottom, right to left, and back to front, where all join and cross in the centre of the sphere. There are therefore always no less than six lines crossing and connecting by a centre from any given point on the surface of the sphere. Such points connects in total six surface points on each side of the sphere while they all support one another through the space less centre. In that absolute space less ness in the centre holding singularity we find gravity supporting and controlling all space within the sphere as well as space connected to the sphere. That is where gravity control and guide the space, which falls in the parameters as well as under the influence of the form of the sphere. In the gravity centre space goes singular meaning space becomes space less or flat. That is where Einstein's Universe goes flat because that is where gravity is at its strongest. However my bringing up this statement brings me directly to the point where I get very confrontational about how the brilliant mathematicians treat those they suspect are less inclined to think.

By examining the form of the sphere, we find that there are 6 points on the surface of the sphere that is holding the form at a specific and equal distance from the centre. Lines run from the centre into space at 90° and 180° angles of each other from six opposing sides. There then are six lines at 90° and 180° connecting to the centre from six points on the outside edge of the sphere. As a result of the basic shape that a sphere has, there is a spot in the extreme inner centre of the sphere where the lines in 90° relevance cross each other and others connect by 180°. There is also at that point a spot where all space relinquishes a position and only singularity 1^0 as form remains. At such a point we find the measure of the sphere being Πr^0 with $r^0 = 1^0$. That is where the line that represents the radius as a line disappears, as it becomes singularity r^0. After more reducing continue we get to such a point where we find only Π^0 left. At that extreme point is where space in all form disappears, as the circle providing the sphere the form the sphere has, removes all possible form by going into singularity $\Pi^0 = 1^0$.

Then in that area all form of any possible space disappeared leaving only the dimensions of singularity 1^0. I cannot delve deeper into the argument. However, from such a point there runs lines that connect to space on the outside where six points on the outside points connect to the space less point in the inside. In this book I take this argument much further but for now I leave the argument at that. Those lines carry the structural strength the sphere has. Contact with one point has support of six other points across the whole structure where the other six support every one of the six by singularity and the support runs through the entire sphere including the middle. Where there is no space, there must be singularity 1^0 just because the space filled with material removes zero and only material filled space is present. That means material fills the lot although in singularity 1^0. If zero was a factor where all space finally halted in zero as the value, then zero would be able to remove the space from the centre and such removing would continue to remove the space until all space was removed. It will finally abolish all space in the sphere and it would remove the sphere. Zero removes all possibilities of anything coming about. Since the sphere is there, a zero factor in the centre cannot be present. Only infinity can be a factor from where space may grow because infinity can extend and grow into and up to eternity.

The implication of this is that following the line down to the centre of the sphere we located the centre of the Universe. That is where gravity is. There is a lot more to that but be patient, we are getting there. In every centre we find a point, which is in truth not there but is the mainstay of all that is within the sphere. The mathematical value of such a point is $\Pi^0 r^0 = 1^0$ and 1^0 is singularity. That is the point where the Universe started and that is where the Universe will finally end. That is the Universe without space-time. That is $k^0 = a^3 / T^2 k$ which proves the Universe is without doubt a sphere...and we just located the centre of the Universe!

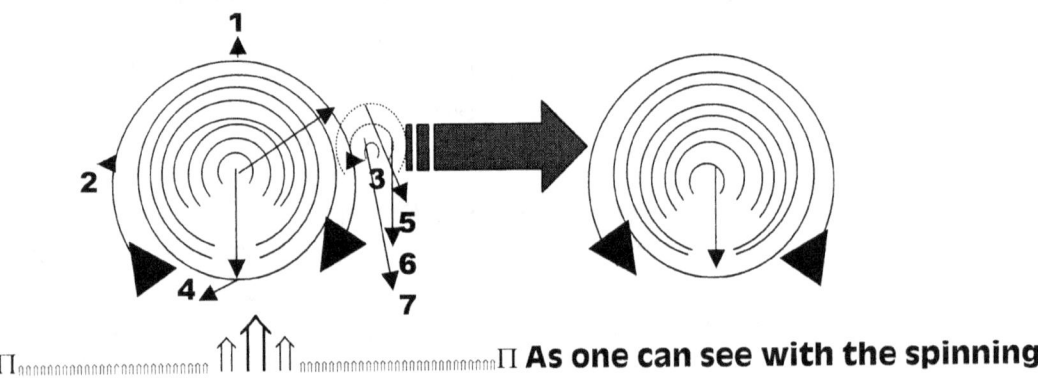

Πⲛⲛⲛⲛⲛⲛⲛⲛⲛⲛⲛⲛⲛⲛ ⇧⇧⇧ ⲛⲛⲛⲛⲛⲛⲛⲛⲛⲛⲛⲛⲛⲛⲛΠ **As one can see with the spinning top delivering the Coanda principle, every point overheating can spawn space-time by centralising singularity**

One can see from the top that singularity is established wherever spin occur. The motion generates a position of seven in relation to ten and singularity manifests as 1.9991 as is explained elsewhere. That means any point formed by the sphere spinning can and does start a centre in which no motion holds no space and of which motion surrounds such a point by forming space. Although everything at the time was in the form as a multiple circle, which results in a sphere, the sphere was not the only form present. This too has to do with singularity interpretations. We see a cube, as we know the cube but at first when form came about the cube were not yet a form.

While the one sphere forms on this spot where the dominating sphere secures an edge the dot may be reserved as an edge marker to the dominating sphere. To the forming sphere in progress of emerging heat gathers at that point because the rotation is a result of duplicating and duplicating is the tendency of naturally growing in space-time **$k = a^3 / T^2$**. In order to find duplicating coming about there has to be heat in order to duplicate what will form heat. The duplicating process is a process of one factor going softer or less solid and therefore more dynamic than the other. To have singularity is to have gravity but to have gravity there has to be a point of motion and a point of sturdiness. The point of sturdy may be in the centre of singularity, but then the solid must be motion. However even today it still apply: what moves forms liquid in the presence of a solid and at that point singularity presented the solid therefore what we might think of as solid was the liquid because it moved around the solid. Where the one factor is duplicating the other factor is compressing $k^{-1} = T^2 / a^3$

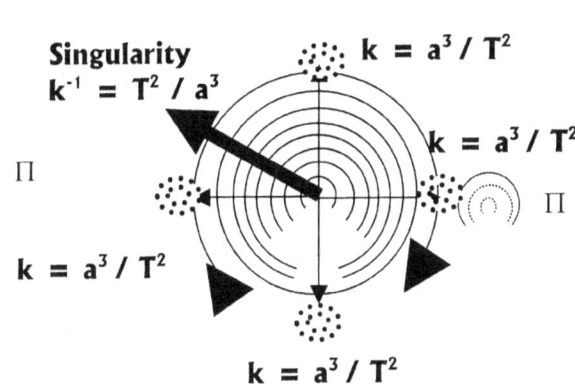

Singularity
$k^{-1} = T^2 / a^3$

$k = a^3 / T^2$

Π

$k = a^3 / T^2$

$k = a^3 / T^2$

Π

$k = a^3 / T^2$

The points duplicating is four moving around a centre by the square of gravity. The motion is the sources of heating because the heat is bringing about the movement. The heat growth therefore provides the action because the action is what energises the points to provide the motion. The motion is purely is space-time duplicating and the duplicating is feeding heat to the centre from the four points overheating thus the points that shows expanding.

But also the duplication leads to the spawning of one point of singularity that provides the installing of the next centre for the next sphere.

Because of the principal in which the Coanda works the motion will centralise a new sphere and by appointing six position around the centre three points will not move while four will move about the three points forming the centre line. The result is that the four points by duplication will reserve the point moving as the next point in singularity because of $k = a^3 / T^2$ singularity will be a natural result of the motion. Then that point will secure a position $k^{-1} = T^2 / a^3$ which will secure six points about such a centre. The centre will bring about four points spinning around three points holding a line singularity. The line in singularity will stand in relevance the contacting factor $k^{-1} = T^2 / a^3$ and the duplicating by expanding points will be four and serve the relevancy by contributing $k = a^3 / T^2$ as space-time only in form. From this the rest of the Universe burst into the next phase of Creation.

The gravity is in relation to the spin, which is in relation to the four points spinning which are $\Pi^2 / 2$ and that is the Roche limit. It is the dividing of singularity sharing space-time just as we on Earth share singularity by division between the Earth and us others that is not part of the Earth. The total that forms from the point that spawns is seven plus five plus pi square in division of four totalling twenty one that stands related to the first seven and once again another sphere formed. However this is an eternal relevancy that can never break.

Any object in rotation will have a middle point, a very specific centre point that does not spin. That point once again hypothetical but none the less must be standing still because every line running from that point in opposing directions are also in opposing directional spin to each other. Although the points had the same characteristics only seconds before, they oppose the characteristics it had just before and just after the very second in which they are and to which they relate by similar points also in rotation. Due to the spinning nature of such a point with all surrounding the point very varying second, the value of such a point can only be Π because of its constant changing.

The sphere has seven points. The cube without truly being a cube but is just in consideration of having a cube in form holds five points to singularity. In the centre runs singularity to the value of Π^0, which means that which surround Π^0, holds a position of Π^2

The spinning sphere activates the seven points, which places gravity in relation to a centre. Outside the centre there are five sides by dimension. The sphere has seven points of which four is spinning. The four spinning stands related to the gravity of spin, which are Π^2.

Using r would specifically oppose another r from every angle. From such a point every other point will be opposing any other point not pointing in the direction to which the first point is pointing, whereby it extends the direction it holds. No matter what the point is or where the point leads, such a point holding a specific direction will be unique in the direction it is rotating because at that or any other specific point wherever, it will be directing not in the direction it spins but in the direction flowing from the centre point outwards. Any point will be it opposing itself within the rotating of 180^0 changing every aspect of its previous flowing characteristics it previously had or will once again have in 180^0 from there. While in rotation from the point of an outside observer all may seem static and never changing but to the object in spin every next second will be a diverting from every aspect it was in very second passing, and the direction it held in relation to the direction it held the previous mille, mille second will totally be incompatible with the direction it holds the very next mille, mille second of rotation. That proves no point can be static or constant, all though it may seem that way to outsiders. Although matter is matter, matter can also be anti-matter at the same time.

At this stage time was still eternity being interrupted by infinity. To say the Universe is or was 13.5 X 10^9 years old is shear Newtonian thinking. Was it 13.5

$X 10^9$ years and how many days in the year of our Lord and what about all the years that passed since this date was revised? Time was flowing according to interruptions in eternity changing from what was to what is to what will be. Time is a norm that comes as things in the Universe change about things that are places around and scattered throughout the Universe. We may presume time at this point somewhere became a factor since sphere sprouted from points on sphere edges and differentiation in development came in place. Considering the role that the Roche limit played one can see how points in singularity grew from contraction and secured ever stronger centres by divulging hear points within the realm of singularity control. When a point in form developed at a position that was close than the original Π^0 to Π, the singularity in control took control.

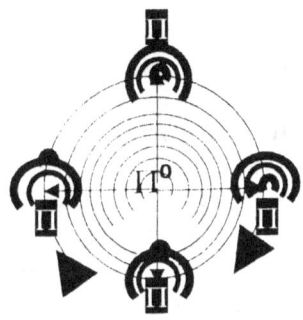

With everyone of the four rotating points duplicating the value of Π in relation to the centre Π^0 at a measure of $\Pi / 2$ and where Π^2 is the establishing of Π by the motion thereof therefore $\Pi^2 / 4$ became a limit in relation to the developing centre. One has to remember that the star of the present takes characteristics of the form from the era before space was a factor.

Consider what happens to a star that developed closed than the Roche limit of Π to $\Pi^2 / 4$ would allow, it is easy to see how the singularity centred grew by concentrating the heat the points in singularity brought about

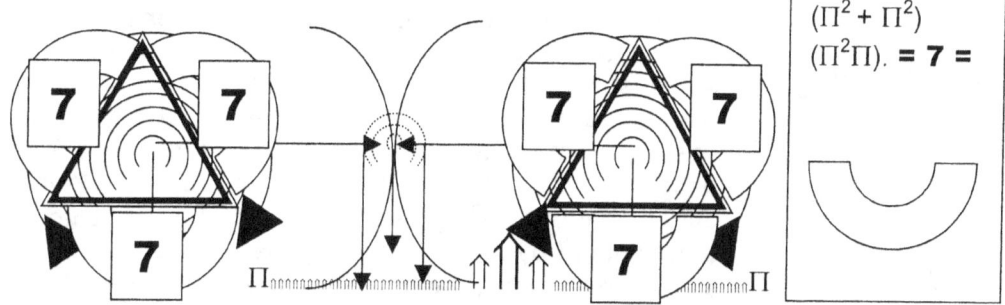

$$(\Pi^2 + \Pi^2) (\Pi^2\Pi). = 7 =$$

$(\Pi^2 + \Pi^2) (\Pi^2\Pi)$ 3. = 1836, which after wards the atom was about and the Big Bang could then proceed.

While very line is circling bringing about time in space to the value of Π repeating Π to form Π^2 at the same time Π is extending in one specific centre to the value of Π^0 and only the spin value keeps Π not becoming r The spin keeps the immovability from becoming Π and maintaining Π^2 by performing duplication, but with any slightest reduction in spin reducing Π^2 to $\Pi^{2/4}$, Π will start extending and as one can see from the behaviour shown in the Roche limit, the heat will be concentrated at the centre and the singularity in the centre will grow four time in concentration. Only at points exceeding Π in diameter was time as Π^2 able to retain form and also grow. From that space slowly developed because at Π could Π^2 bring about a form which provided motion. In the centre there developed Π^2 and Π^2 kept all form at a safe distance of Π to bring about the needed solid immovable centre with the form $\Pi^2\Pi$ about the double Π could $\Pi^2 + \Pi^2$. That secured the makings of the atom by applying the Coanda principle of enticing gravity in the centre of motion, which then provides space-time by measure of $(\Pi^2 + \Pi^2)(\Pi^2\Pi)$. This totalled seven in dots and with three of those seven circling singularity at 1.9991 the atom came about. **I again wish to repeat the centre in the sphere and the centre of the sphere because in this is where the realisation comes from how the Cosmos started.** If there were gravity at the very first instant then there was a sphere at the very first instant because gravity can only be in the sphere because of what the sphere represents.

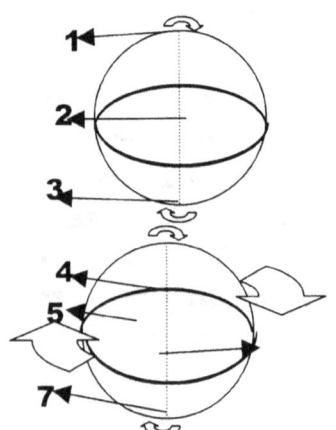

By not having motion the lines also have no space as the space extends from the line serving the three points to the outside. Where there is no motion, there is no space and where there is little motion there is little space. The only space the line may relate to can be a point that is on the border of the sphere that is crossing singularity and connecting the two edges on either side of the sphere that is forming the sphere. That means the line from one point holding singularity to another point holding singularity that line will cross the centre line which gives the line in singularity valid space-time to control. Singularity does not have the ability of motion therefore singularity does not hold space. Singularity is also eternally indifferent to motion and motion can excite singularity but singularity cannot be shifted by motion.

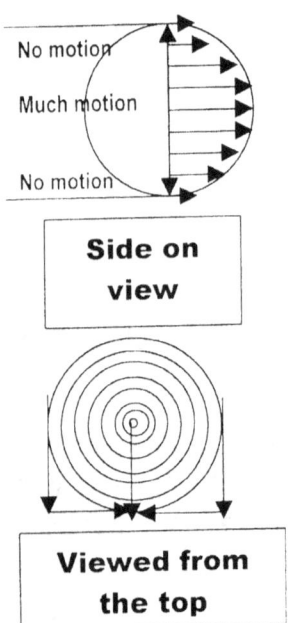

Side on view

Viewed from the top

The spinning releases Π from Π^0. In any sphere Π^0 is part of the allocated centre and is part of the construction. As a sphere the centre however does carry singularity and all aspects in such a centre of any round object does comply with singularity. Without motion the singularity however has no direct influence on the individuality of the top except by the motion of all the atoms producing such an allocated centre Π^0 in the form of keeping the structure independent by form. However when motion comes about a line forms within the centre removing Π from Π^0 and that line that forms in conjunction with the rotation forms a driving ability that keeps the top erect as long as the motion applies. The other factor that establishes a presence is the time factor, which the motion facilitates. By spinning the time produce a duplication of material and this duplication we see as movement. In the movement time forms a continuing of the past location of material through the present location of material into the future location of material. It is the inner three points as well as the outer time carrying three points is that which the top loses when the top loses its initiative in the form of motion. By losing the inner three points as well as the outer three point the top loses an individuality which with such individuality the top receives an independence as a cosmic object. The motion allows the top a cosmic birth or removes the top as a cosmic independent and free object.

In this one can clearly see that it is the motion that sets the top free and independent from the gravity of the Earth. By motion the top generates individual gravity that allows the top an individual gravity and that motion frees the top from the gravity by which the Earth restrains the top. The motion gives the top independence that the top immediately loses when the motion subsides. This fact is the utmost important issue of all physics in the Universe.

The top is one of (perhaps) the easiest and most common examples one can find to demonstrate the cosmic generating of gravity. By spinning a "force" which is no "force" keeps the top erect and it is only motion that accomplishes the act of gravity.

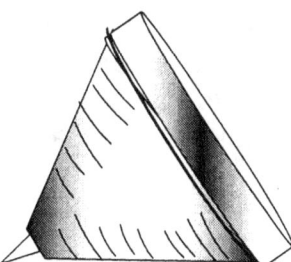

The top lying still holds the same singularity principle that the sphere holds because if the shape the top has. The roundness protects singularity at a seventh position deep inside. However the top is a dot Π or even going down to a spot Π^0 and is only by the form the material has which puts singularity in place.

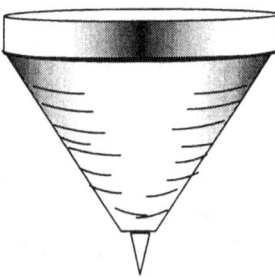

According to Newton it takes no effort $\dfrac{dJ}{dt} = 0$ to get the top from where the top was motionless to where the top is spinning. I say this on the work that Newton suggested comes about from the effort it takes the top to circle.

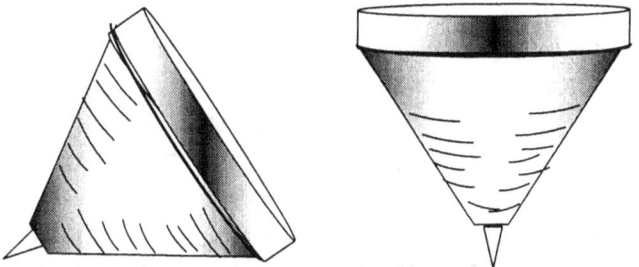

From a position where the top is lying down being collapsed the top generates a position by spinning and the motion puts the top erect. It takes motion and not nothing to establish such an independent and secure position. It puts the top in the centre a newly established and independent Universe as the top then finds courage to fight the gravity of the Earth up to the last "breath" is fought.

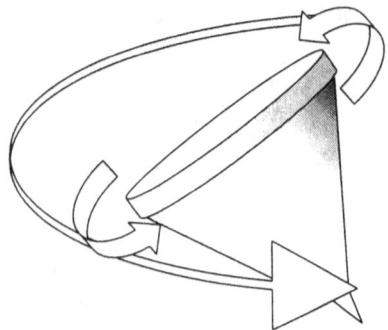 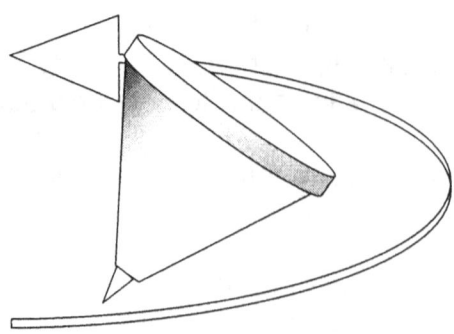

In the motion a line comes to life running in the centre of the top. This line is not just another line but can focus the top to spin upright and erect. The line was not there when the top was on its side. By motion the line can concentrate an effort that will unleash such dependence to the top that the top will come into a position where the top has the tenacity to take on the Earth gravity in a struggle for life and supremacy.

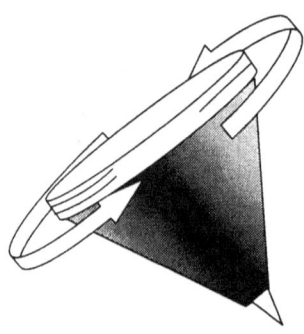 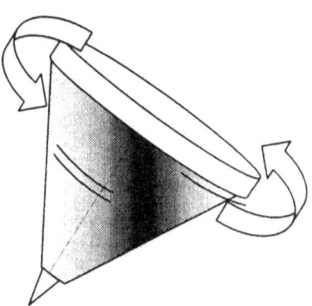

From whichever angle one looks at the top, the top seems possessed and I can even be slightly forgiving towards Newton for calling it a force because although not a force the stance the top takes when spinning upright leaves on with an impression of forcefulness being part of the situation. One must not see a force but one should see the manipulating qualities of life extending to the top and by life's ability to manipulate space-time and control motion in space-time with space-time, the throwing of the top is as little a cosmic event as the apple Newton saw falling from the tree. In every event, the top as well as the apple the drive was life controlling events and as far as there is proof there is no possibility of such an event taking place anywhere in the Universe by something as small as the top or the apple. In the case of stars such development does start the star on a course of independence and it sets the star development on the rode of becoming independent from the galactica. The drive generates gravity but the driving that allow such rotation is inspired by the accumulating spin of the entirety of all the atoms in motion within the young developing star. The heat blanket in which the star cradles has a lot of influence but it is the atoms becoming a driving factor that inspires the motion and such inspiration allows the top the initiative to form independent gravity as a star. In the case of the top the spin is completely cosmic unnatural. If you start to imagine about life in the Universe you may just as well start believing in ghosts fairies and all other fantasy creatures. Science must decide whether they wish to speculate about the life's abundance and being a dime a case, found all over and everywhere you look throughout the Universe, but in such an event distance their fantasies from science and reality, or stick to science in reality and believe only in facts as science presents facts.

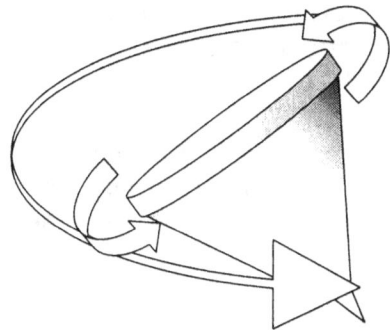
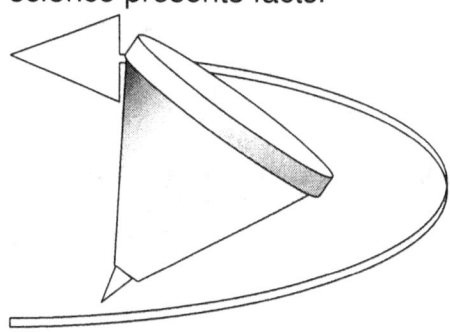

Let's consider what are facts with the top spinning as the top does. This is no fantasy or life coming from same imaginary source but it is a cosmic reality which life found a way to manipulate.

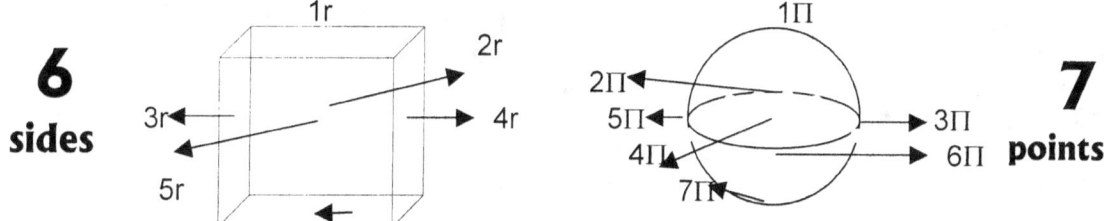

The effort it takes the top to spin gives the top a distinction of extreme significance. The top is promoted by the motion initiative to that of a star in motion because it charges singularity into existence where singularity then controls space-time.

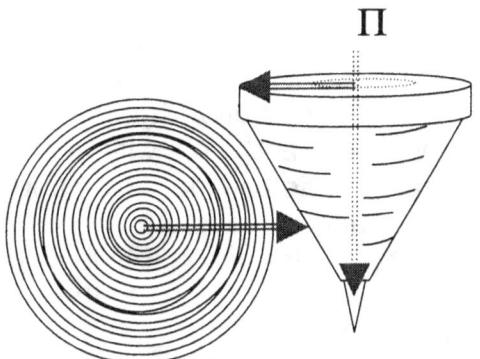

Singularity is a mathematical point hidden in every sphere. Singularity is very much inactive in every sphere but to keep a centre of structural bonding in every sphere. Something happens to the top in having a singularity just like any other sphere has to a point that takes charge with all the cosmic dynamics the sphere may show.

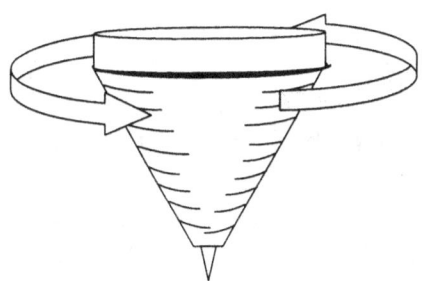

The top is charged with an energy, which not only takes charge of the top as well as the body of the top, but also the immediate space surrounding the body of the top. When a person with skill manages to put a high degree of spin into the motion of the top the top then spins in the surrounding space so vigorously the top stars to whistle vigorously

The linear remains linear because the linear redirects its intentional direction because of the rotational change that the linear motion always ends up doing. The line forms an eventual circle because the linear line must constantly entertain the centre.

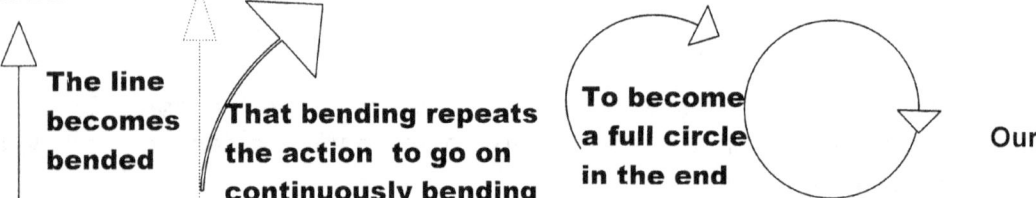

The line becomes bended **That bending repeats the action to go on continuously bending** **To become a full circle in the end** Our

gravitational   falling to the Earth is a result of a **circle going straight and forcing us straight down to an everlasting directional alternating circle, we have as we spin with the Earth as we spin around the sun. As we fall straight down we, change direction while we are falling straight down because that point we are heading to what we are falling to is changing too. From the centre of the axis, everything seems neutral. The axis does not spin at all, because the axis brings about spinning motion changing eternally. That is in nature and not man-made motion.**

As the top comes to motion the top finds the characteristics that the top shows very indicative of the characteristics that all moving objects in the cosmos show. The top spins in a straight line that bends by a 7^0 inclination.

Apparently an idea concerning the subject of gravity Einstein came to was about him falling off a multi story building and the gravity mass that he then would experience. Remember, this was long before flying and parachuting with free fall acrobats showing on TV how a man and a car with a man in the car can descend five or six kilometres while falling side by side. This happened while Einstein was still being a patent clerk in his younger days. Apparently Einstein was looking out a window of the multi story patent office, when Einstein suddenly realised that had he, Einstein fall out of the window from the roof to the ground of the patent office where he was working at the time, then he (Einstein) would feel as if he was weightless during the time of his fall. By falling with him, those articles would feel equally weightless should they accompany his fall down as being part of the falling process in his imagination. As the objects were travelling alongside Einstein down the building to the ground the lot would travel at the same speed from the top to the bottom of the building. Then I went one step further by supposing the Einstein group's falling was real and no imaginary thoughts were set in the fall, then what was the imaginary factor then? Let's pretend Einstein did fall with his pen, his chair and his desk and Einstein was not imagining his fall. Einstein as a human being can imagine but his falling companions can't. If Einstein was imagining his weightlessness, it might be psychological, but in the case of the other travelling companions it was not possible to imagine anything. There is an immense difference in size between the falling companions and that notwithstanding they travelled the same speed while descending. If they travelled the same speed as Galileo proved and they all hit the Earth the same time, which then indicated that their weight and mass, that which gravity used to drive and what propelled them downwards and that which was causing the drawing of what the mass was instigating to allow the motion of fall to commence, was equal. Kepler found space a^3 being equal to the motion thereof T^2 in relevancy to a centre point **k**. Kepler found space had to move.

When reading this that evening so many years ago, I came to realise that Einstein could only feel weightless if it was true that he (Einstein) was weightless. He could not feel as if as if was part of his imagination because he was truly falling, and in truly falling the falling was then without his imagination doing the pretending. Einstein had to feel his weightlessness as a cosmic fact in the true sense because if he was truly falling, then the part, which was the falling experience, was what he was experiencing in reality by three dimensions with one dimension in time. If Einstein was experiencing weightless ness, it would be because he was weightless while falling, then Einstein would not imagine the weightless ness because Einstein was truly falling, thus carrying out his cosmic state he was in. His body being in motion ($a^3 = T^2k$) was at that moment truly weightless while experiencing unrestricted gravitational motion. Einstein, the pen, and the chair had the same weight since they were all weighing the same in falling. If there were any mass differences there had to be speed differentiation for the force of the one would generate more motion than the force of the other onto the different mass components but since there is not mass discrepancy amongst the falling while falling, the lot is having the same state of weightless ness, they adopt the same speed in the fall. After all it supposedly is the mass that is doing the pulling and more mass does more pulling…except if the mass is not doing the pulling in the first place. All four items including Einstein, would be equally weightless during the falling…that was what Galileo found because objects of different size and different mass travel at an equal pace (distance over time or space moving divided by time flowing while the object changes position in relation to the Earth ($a^3 = T^2k$)) while descending. The bigger objects do not fall quicker than a smaller object and that can only be attributed to one fact; it can only be true if the four weighed the same while falling and no one weighed anything while falling. That means the gravity applied while time flow in relation to the space that was applying the motion, which was what gravity is $k = a^3 / T^2$ according to Kepler. The single line falling is represented by the factor **k** being the relevance of space a^3 that was relocating its cosmic position while all that was happening in relation to the motion of the Earth T^2, which was in relation to the Earth spinning around the Sun and that rotation gives us our time T^2. While in motion the four different objects weighed the same since they travelled at equal speed downwards. By standing still the objects had mass differences and when they were in motion they weighed the same. When the motion became frustrated by being blocked by another space that was also filled with material and that was holding the spot too where the motion was directed, they then had different weight. The pushing resulted from the bodies striving to remain independent. The two objects were in a fight to claim the position each desired, and that was to fill the centre of the Universe. Being ($a^3 = T^2k$) was being in the centre of the Universe because the centre of the Universe was $k^0 = a^3 / T^2k$. Then one may conclude that gravity is motion of space and mass is the restricting of the motion of space. Having mass does not bring about gravity but it does restrict gravity's motion, which is what brings about the mass and weight. Gravity produces mass but mass does not produce gravity or in fact mass produce weight but mass is not responsible for the intended motion. The intent on moving while being blocked by another object is frustrating the motion of gravity in both cases and the higher the frustration on motion is, the more mass there is coming the way of the bigger object who then has the greater desire to move. The reason why it has the desire to move and why space is equal to the moving in time of the space in relevance to the centre of the Universe (which at that point might be the Earth or be the sun) is what the have the effort to explains. Mass is the restraining of motion and gravity is material moving about by committing gravity. Mass only comes into the application thereof when two objects filled with space moves into a position where both want to claim the very position in space the other occupies.

It is the motion and the independence they show to hold onto their individuality as independent cosmic structures that prevent them the sharing of space which in turn prevent further motion that causes mass. Gravity is in essence where mass is present, still in a tendency to commit motion but is then in the frustration of motion and gravity at such a point is the commitment to move once the blocking of space is relinquished. Because the one object that has more "mass" would put in a more assertive effort to move in relation to a smaller object and the effort to move will constitute to a greater resisting effort by the blocking objects in a fight not to relinquish its position on the space both object claim that the tendency to move and the tendency to block the movement will bring the effect of greater or smaller mass being present during the effort and in line of resisting the effort. However while any space is in motion, the gravity of motion is equal to all and puts everything on an equal basis. Therefore there are no big and small and the big Sun does not pull the small Earth closer. Mass is when the motion is prevented that a differentiation in motion effort becomes part of the picture.

Do not be fooled by the seemingly innocent explanation that space is the motion thereof which is what gravity produces because of all things the cosmos creates, motion of space through time is the utmost complex manoeuvre and without bringing a restraining of mathematics into science, it is so complex there is no viable explaining in physics about how the cosmos produces the act of motion of space in time. In order to get every atom to spin as every atom follows the lead of the atom in front. This gives direction to follow to the atom just behind while giving coherency to the structure. By following the one in front and being followed by the one behind the lot of atoms are holding as an individual unit times the units there are going around in the entire Universe. The measure of this complexity is beyond what the human mind can absorb. While the atom in front is vacating space to fill the space of the atom in front is vacating at that instant, the atom behind is filling the space that the atom in front has vacated in order to vacate and relinquish the previous position in favour of the following position to honour the direction gravity is insisting upon. Times that with every atom there is in the Universe and one may grasp the significance of the calculation. Removing material from space by filling material into a position of new space sounds simple because the complexity has never been realised. I am in the hope that in this matter I will be able to will reveal what the factors are in understanding the commitment of material to move through time. This was all a result of understanding the dynamics of Einstein's arguing about gravity and mass. Then I kept this information in mind and it helped me to further realise gravity is motion differentiation between objects.

It is the independent motion providing a different speed while sharing a common centre of attracting that allows a discrepancy to establish mass under specific conditions applying between the two in relevancy. While falling the gravity applies as moving of space that is putting time in relation to the distance travelled. That means there is a speed relevancy between particles in motion and synchronised motion, which would bring about equal orbit around a shared centre. That is the result of gravity functioning. While the object falls, the motion confirms gravity. When motion ends mass sets in and becomes the constraining of the object preventing further motion. The motion is still there but now it is reduced to a tendency to move thus establishing the object mass as the limiting of further motion. Preventing the motion by implementing mass is the resting of objects against each other by resisting the motion to continue, which then is where the mass takes the place of the motion. Where a confronting of objects restricts gravity, the action then implements an introducing of the mass as a substituting factor to motion that then replaces motion as substitute to the motion that would be and the mass is providing the tendency of gravity being the motion of space.

However mass then restricts motion and becomes motion in a tendency to apply motion. While falling, gravity applies and motion neutralizes size, mass or weight. Mass counters motion being when the Earth restrains further motion of the falling object and the moving object is stopped from further movement where mass is then preventing or hindering gravity. This is the result of objects claiming an individual and personal claim to space occupied in a dual or in fighting for their individuality and independence of each other while wanting to be in the centre of the Universe. While falling or moving there is no opposition to the body being independent. When the motion seizes, the falling object remains individual and still tends to move while Earth individuality resists further movement of the falling body's movement. Further movement is disallowed as other material fills space a that falling body wants to lay claim to. The only manner to remain independent by the falling object will be to relinquish to motion in the securing of mass as a substitute to motion where it then finally comes to rest. Mass then sets in not causing the motion but substituting the motion and from that motion restriction becomes resistance that becomes mass. While falling the object is experiencing gravity because the object is in gravity but when on the soil the object experience mass which is the restricting of gravity or motion by other space filled with material.

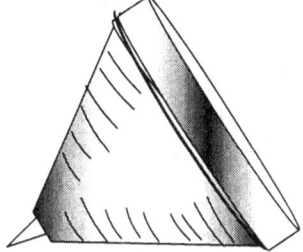 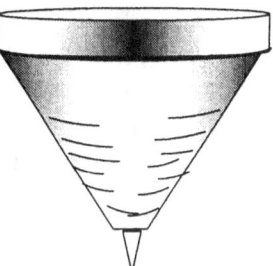

Looking at the top spinning and not spinning, brings a question to mind: why would the motion of the top beat the gravity of the top and that of the Earth hands down when the top is spinning. Surely the mass is in effect while the top is spinning just as much as the mass is in effect when it is not spinning and yet when it is spinning the pulling subsides to give way to free the top from the mass restriction and charge top with much excitement. The excitement is so much it seems to relieve the top of the pulling there is between the top and the Earth. That even strengthened my suspicions more about the fact that gravity is motion. By spinning the top finds additional motion on the side of the top that brings the side of the top in a more favourite position than was the case before the spinning commenced. The top finds additional gravity in the spinning and the gravity in addition has to bring reconsideration to the position the balance in gravity sets in margins. The top secures a better margin or stretches the parameters of location because the top inherits the Motion of the Earth and in addition to the gravity motion that the Earth provide, the two secures more gravity motion. With more gravity motion in addition to the Earth's gravity motion the top has to have more gravity motion than what the Earth has. That will allow the top the spin while facing such enormous disadvantage there is on the side of the top in mass difference when considering the size disproportions. Let's face it, if it was about mass pulling the top in relation to the top pulling the Earth, then the top had no chance ever to move by the very slimmest of chances there may ever be.

However if it is the gravity motion that the top inherits and with the aid of what motion life can add in addition to the motion already the which is the motion the Earth already contributes and considering the mass the Earth provides then it will not require that much a bigger effort to get the top going. Singularity charges motion by instigating motion without ever moving. Coming from a spot to a dot and then producing a line running from dot to dot though a spot signify the birth of the cosmos. A line holding time by the square and by the square of ninety degrees announce a birth in the Universe of a birth of the Universe. That which was not there suddenly is there by not being there. That which was undetectable suddenly is detectable by being undetectable. It is not my forte to write riddles but in this case there is a Universe within a Universe, which is not in the Universe and does control the Universe from a point no one may ever locate inside the Universe. The Universe is built up by innumerable dots and each dot is charged with being the Universe while being in a representative position since it is not in the Universe.

Space-time is a four dimensional position of the Universe where the position of an object is specified by three coordinates in space and one position in time. According to the theory of special relativity there is no absolute time, which can be measured independently from the observer, so events that are simultaneous as seen from one observer occur at different times when seen from a different place. Time must therefore be measured in a relative manner as are positions in three-dimensional Euclidean space, and this is achieved through the concept of space-time. The trajectory of an object in space-time is called world line. General relativity relates to curvature of space-time to the positions and motions of particles of matter.

In view of the definition of space-time I wish to elaborate on my view of singularity and my deriving of space-time from the likeliness that singularity may produce space-time. In the past singularity was mentioned in the manner one would speak of a ghost hiding in a haunted Black Hole. Let's put singularity in the clear.

Singularity is within every sphere due to the natural shape or form the sphere is committed to.

Positioning, finding, allocating and valuing

Singularity

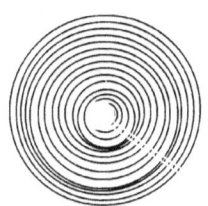

In the **precise middle** of all **objects in rotation** is a precise centre dividing the object in sectors that will **start the spinning initiation** from that centre point. Thus, the spinning object **will have a middle point**, a very specific **centre point that does not spin** and only holds Π as a specific value because no radius can apply. But also the one value such a line **cannot have is zero** because the line **is there and holds contact** to the rest of the material bringing about that **zero does not start any** line and therefore the **value of the line must be infinite**, just as described in **accordance** and by **the definition of singularity**

As I am introducing a very new idea, I whish to explain in more detail what I try to convey.

Move the rotating line progressively to the middle by reducing the length the line have from the edge to the middle. At one point all further reducing ends.

> As the rotating direction moves inwards, the rings will become smaller and smaller.

That point albeit **hypothetical, is also as much a reality none the less and is** placed where that point **must be standing still** because every line **running from that point** in **opposing directions** are also **in opposing directional spin the other or opposing side.**

In considering the spinning motion in the fraction of time in the detailed instant every aspect of rotation will turn in every instant of change in time. Although the points had the same characteristics only one instant before, they oppose the characteristics it had just before and just after the very instant in which they are and to which they relate by similar points also in rotation. The fact of the graph proves my point in quarterly opposing dimensions and values,

There must come a point where the ring is infinitely small, where it can reduce no more, where it reached its ultra limit, but at that point it cannot be zero, because the point is there. Understanding all the following is connected intimately and all conditionally to the fact of accepting that all individual particles in the universe use motion and therefore spin.

The definition of space-time is as follows:

According to Einstein singularity is a mathematical reality within the Black Hole but much more so in every sphere. Einstein may be the first to name it and Galileo (unwittingly) may have been the first to define it as Kepler was the first to formulate singularity, but in mathematical terms singularity is the most basic principle. At this point I wish to establish a fact that seems lost in all other grandeurs of cosmology. When tracing the radius down into the sphere the radius stars where all lines start and a straight line cannot begin at zero or nil it can only start at infinity. Such a statement will hardly seem appropriate but the relevancy of this fact has no limits. If gravity is motion then motion starts with a line. Let us follow the line as motion abides by the rules of the line.

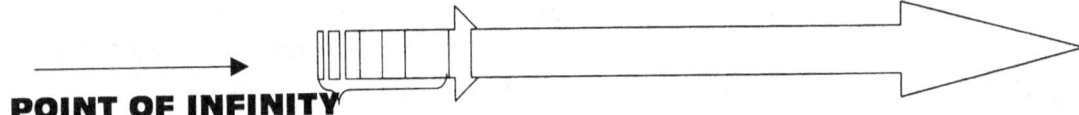

POINT OF INFINITY

If the line started at zero there was no line to start because zero multiplied by whatever results in zero as the answer. That must also be the cosmic starting point. Einstein introduced such a point and named that point singularity. When looking at the cosmos from whichever angle all indications lead to the fact that the whole cosmos is in motion in its entirety. It is forever spinning and it is going too as much as it is coming from. Everything is on the move and always encircling something of greater importance. A top can spin but the parameters of its spin are limiting the motion it can apply. By not spinning the top is still spinning as the Earth is doing the spinning on its behalf.

When spinning too fast the top fights something because the alignment keeping it upright starts to tarnish. The same apply when spinning too slowly but that makes sense. It is the fact that the same affect comes about when spinning too slow that triggers the questions. Why would the top stand upright by spinning. It must be because singularity charges the top into a cosmic independent reality.

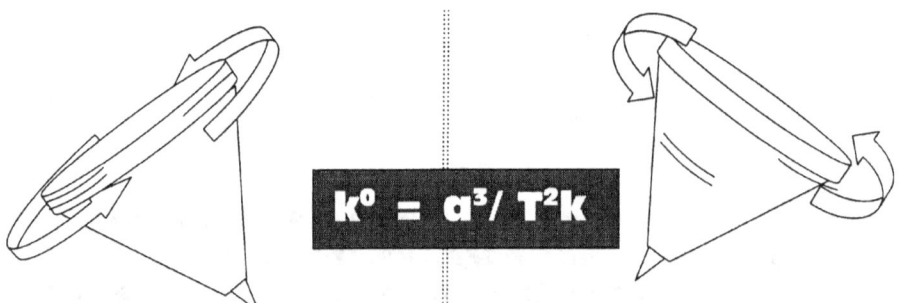

$$k^0 = a^3 / T^2 k$$

The spinning top is all the evidence any one needs to come to such a conclusion. I know probably as much as any graduate about cosmology but lack certificates to prove my knowledge. I am not part of established science. In my developing of knowledge accumulation I came to some conclusions about cosmology that are unique and divert somewhat to drastic form the accepted norm. Most of the work I see the same way as the norm does but in a reverse. Allow me a short explanation

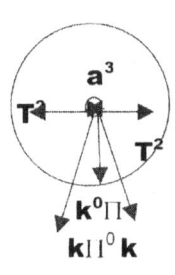

In dimensional terms, which I explain later on the value of **2k** relates to T^2. That relation extends to the next value where T^2 relates to **k**, which relates to T^2. The first space in the circle will then be T^2 **k**. From the centre being in infinity, one can realise by applying mental power the single dimension factor not seen but present all the same. Extending that into the 3D comes six **k** and any one of the six will further extend to form a seventh point as T^2 All this is a multiplying of $k^0 = a^3 / (T^2 k) = 7$

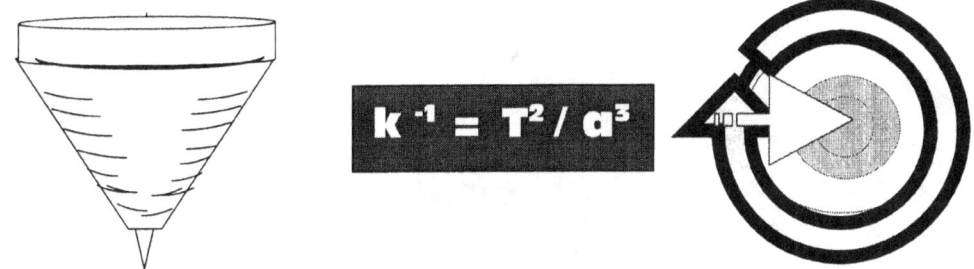

$$k^{-1} = T^2 / a^3$$

The motion established by singularity results in the implicating of the Coanda effect as much as the motion establish the Coanda effect. The spin realises the space limit while the space limits attaches the motion onto the space in the time within the time.

We have to be clear about what we think of when we think of the Universe. Most people think of a picture recalling the black night sky when thinking of the Universe and that thought is most incorrect. Einstein was most correct when he declared the Universe was going flat where gravity is at its utmost, but the concern we should have is not with the mathematics being valid or not but with the vision about the Universe being what we think of and where we place the Universe. The Universe is in the centre of what is spinning and the biggest single particle that is spinning in total independence of the rest of what forms a total Universe is the atom. The atom spins and by the motion the atom evokes the Universe forming what must be the group effort of all the atoms then spin by the motion the atom renders the rest of the larger Universe. The Universe is the part that allows the rest of what the Universe establishes to spin. What spin you may ask. Kepler said it without saying it: $k^0 = a^3 / T^2 k$ and not even Einstein with his super human mathematical skills could say it better or more accurately.

The motion established by singularity results in the implicating of the Coanda effect as much as the motion establishes the Coanda effect. The spin realises the space limit while the space limit attaches the motion to the space in the time within the time.

With the top spinning the Coanda effect steps in and do justice to Kepler's formula.

Time is always a displacement of space in relation to the implication of singularity, and comes about between two points in space relating to the centre of singularity as positioned by **k**, either to the value of **k** or to k^0.

With the top spinning the Coanda effect steps in and do justice to Kepler's formula.

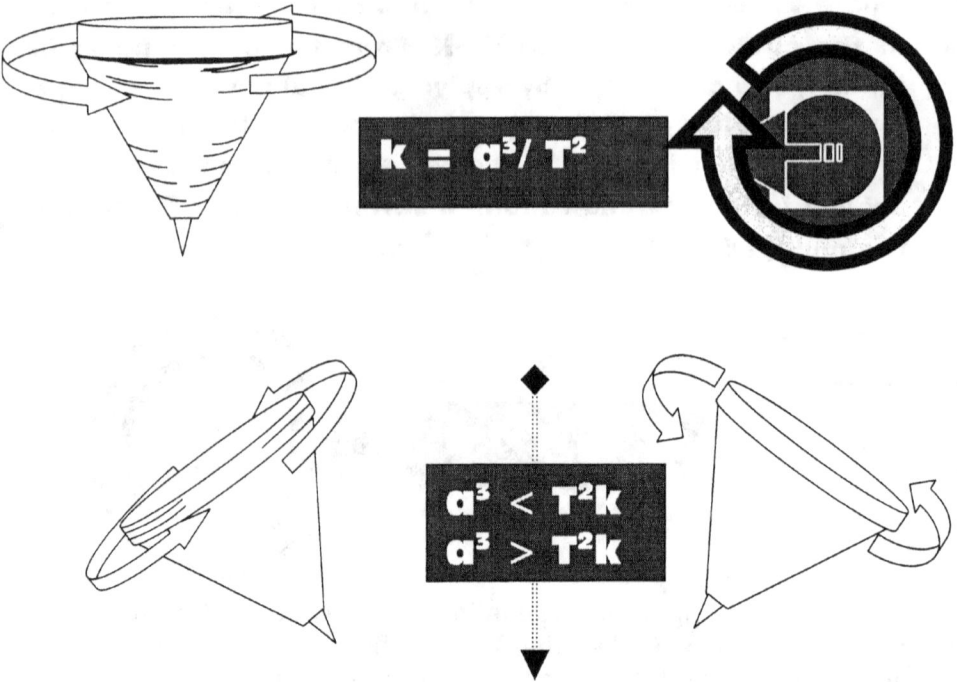

$$k = a^3 / T^2$$

$$a^3 < T^2k$$
$$a^3 > T^2k$$

Time is always a displacement of space in relation to the implication of singularity, and comes about between two points in space relating to the centre of singularity as positioned by **k**, either too the value of **k** or too **k**^{0}.

When one takes Kepler's equation into consideration the whole process of motion starts to make sense. It is motion that keeps the top erect and that was accepted from the time of Newton. Now however, by close scrutiny as well as considering Kepler's equation where the statement emphatically reads that the space **a**^{3} is equal= to the motion of the space in relation to a very specific centre. But this relation works both ways and not only from one side. It has nothing to do with pulling because if the two were pulling the top had no chance of any motion. The top uses the motion of the Earth to the advantage of the top and then on top of the Earth's motion, the top applies individual initiative and claims even more independence than the structural independence it had before. If it were merely gravity of mass and nothing more, then the gravity disappearance the Earth produces would be such an imbalance that the top would never stand a chance of committing individual motion. However should my view be correct about motion and mass being the frustration of motion hindering the motion of gravity, then yes, the top by motion free the distorting of the Earth mass, which is a frustration to the top gravity motion and that would enable the top to spin even with such slight energy applied. Merely taking into account that the top has to overcome the considerable mass f the Earth brings the Mass theory into dispute because how can the top with such slender energy find freedom from the enormity of the Earth mass.

Every round object has a point establishing a very centre, a middle dividing one side from the other. That division determines the space from one side away from the other side. At one point there must be a point that does not fall on either side of the divide. Such a point will still be a circle, because from that side the circle divides into two sectors.

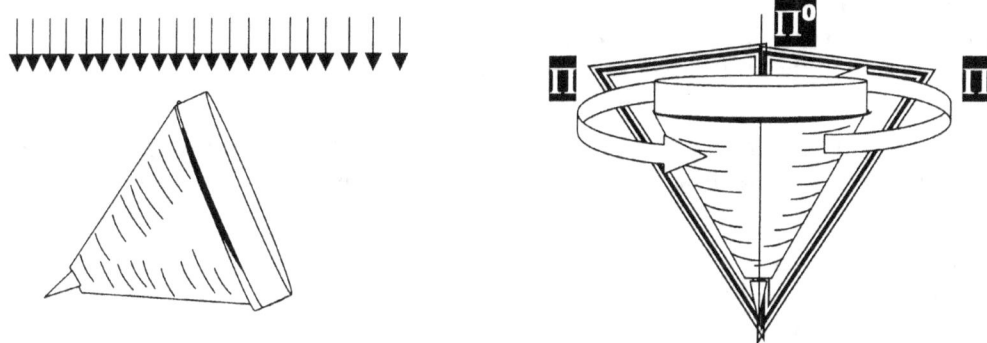

In every spinning object there is a point of infinity, a point that does not turn because it holds the dividing spin. However when such a point becomes a line that cannot spin a new Universe is born in the midst of many others. At the birth that point diverts space outwards and from that point the spin is either clockwise or anti clockwise in all directions. As I pointed out no line can start at zero because then there is no line and no rotating point can start at zero because then there is no rotation. Calculating a square involves two aspects that we think of as sides.

There is a Universe in differences between the top lying down without any individual motion and ostentatiously independent, self assured spinning top that even produce a sound to match the occasion. While without motion the top submits to the contraction lines running as the straight line holding half the value of the square being 180°. The top seems dead as it surrendered its long-term position and would eventually succumb to the Earth's gravity by relinquishing the structural independence it has. Then the motion brings life into the top and gives the top reasons to fight the Earth by fighting for independence. The top just became independent by the motion it received from the combined efforts of all the independent atoms forming the structure of the top.

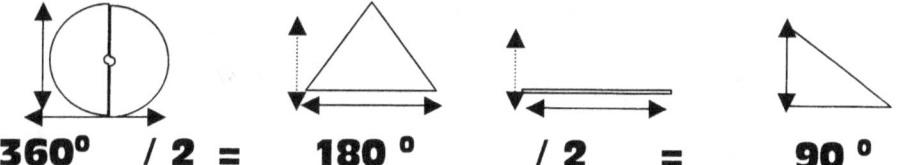

360°   / 2 =   180°   / 2   =   90°

The circle is a square holding a round shape, as the straight line is a square holding one side to infinity. Calculating a circle involves two aspects where the one is either the radius or the diameter that is double the radius. The other is the factor Π

Because gravity work both ways and not singularly in one direction as the Newtonian myth would have us believe, there is the interaction in the neutron position between the total of material in relation to time formed in space as space and time formed in space in relation to the total of material.

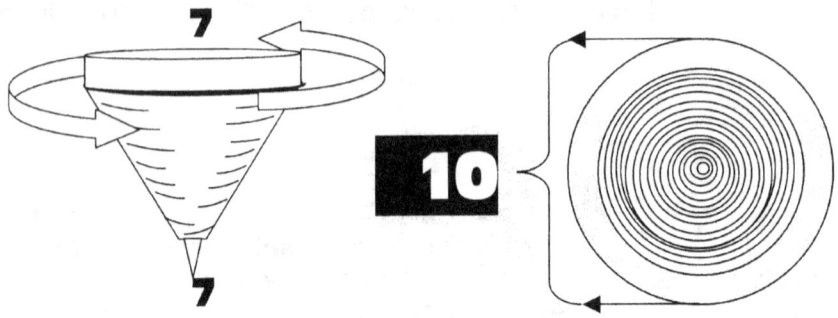

With everything in a cube or a circle or a potential of the two, brings about the implication of eternity in a form of singularity or the point of creation. Removing the radius of a circle does not remove the circle, because the circle is there, securing the ring. If the line (or imaginary line if you wish) holding the value of Π^0 = 1there has to be a point where the circle is no longer in infinity but claims existing outside the imaginary. At that point the radius may be lightly more than infinity, but to all calculating purposes it still remain as infinity.

The spin was going on for eternity because the spin does not apply, it has a value of infinity and infinity at the time was combined with eternity.

Having edges where Π^0 duplicate to present the edges singularity lost the value of Π^0 to the value of Π^1 with the same value singularity had being Π^1 to the one side and Π^1 to the other side, Π^0 must be the point splitting singularity into two parts of eternity, the eternal value of the first dimension outside eternity. It was the square of Π^1 being Π^{1+1}. That was the first dimension outside singularity Π^0 where singularity has a value of Π^1 in the form of $\Pi^{1+1=2}$. The first claim to space had a value of Π^2. This applied to both sides of the claim to space outside singularity, and the double proton became the dominant factor on matter.

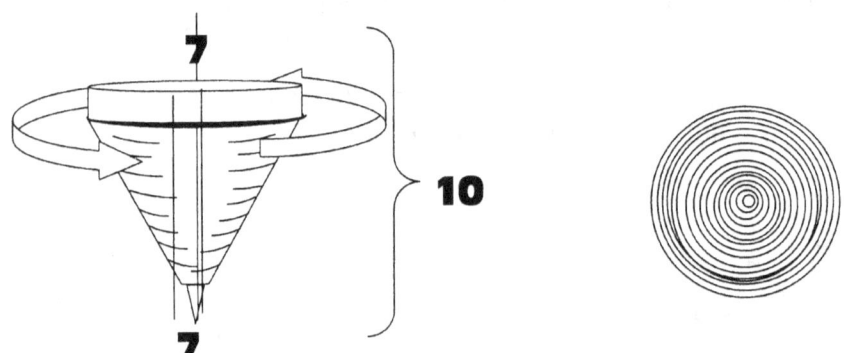

The top has space but there is a space in which the top spins that covers the time part of space. That is the time part Einstein identified (1) as coming from (2) being at and (3) going to and the position holding singularity is represented by the entire body that holds all the space of the spinning top. As soon as the spinning of the top commences, the time aspect releases space in which the top spins from the space holding the time of the Earth and the rest of the Earth within that time. It is this space in time that becomes so hot when the aircraft is speeding because the motion takes the time back to what the time was when time was nearer to the Big Bang.

By receiving space, singularity received a value outside eternity as Π^0 received edges. Granted the fact that the edges were so small there still was no r to present a circle.

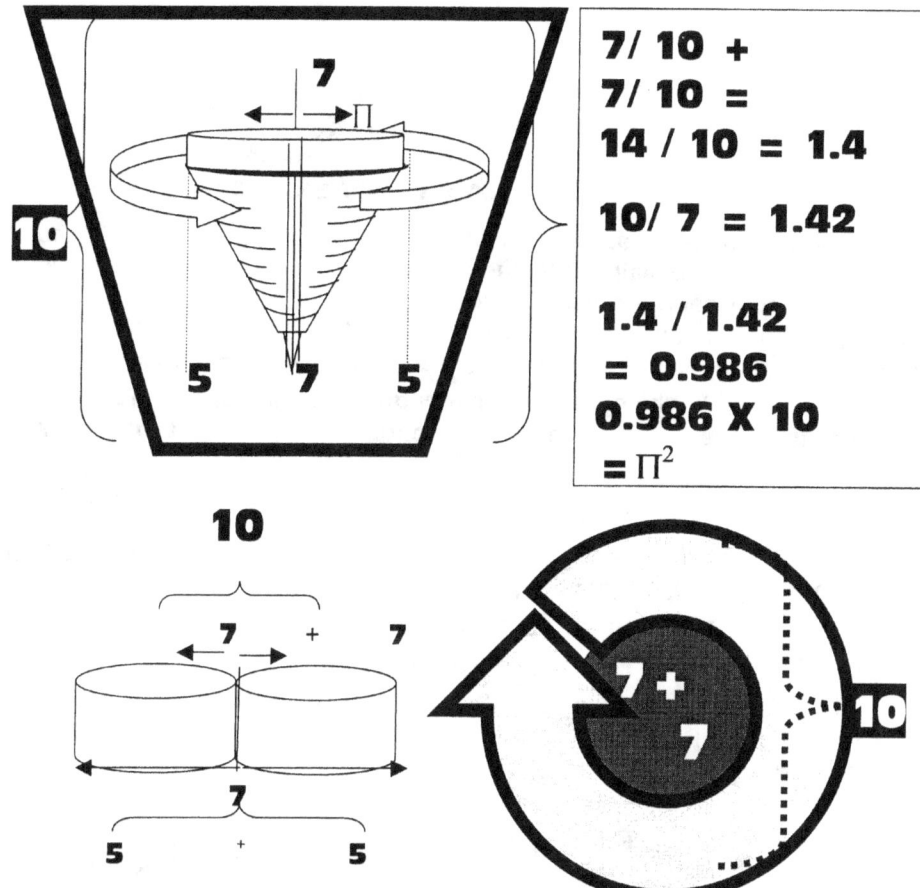

$$7/10 +$$
$$7/10 =$$
$$14/10 = 1.4$$
$$10/7 = 1.42$$

$$1.4 / 1.42$$
$$= 0.986$$
$$0.986 \times 10$$
$$= \Pi^2$$

Taken from the point of rotation the two sides are in opposition to each other in every aspect that they may contain and with all that they hold.

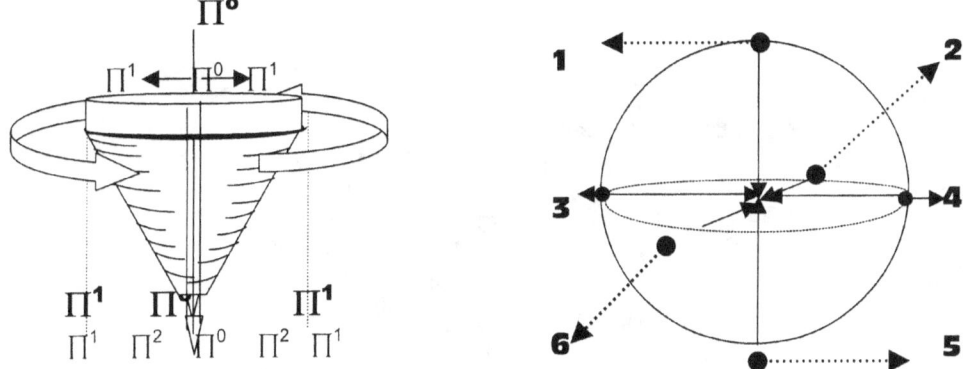

With Π^0 little more than a figment of the imagination there is actually to values of Π^1 facing each other in a relation combining Π^1 to hold the value of $\Pi^{1 +}$ $^{1 = 2} = \Pi^2$ and with two sides being the very same but opposing each other there will therefore also be Π^2 to every side that holds Π^1.

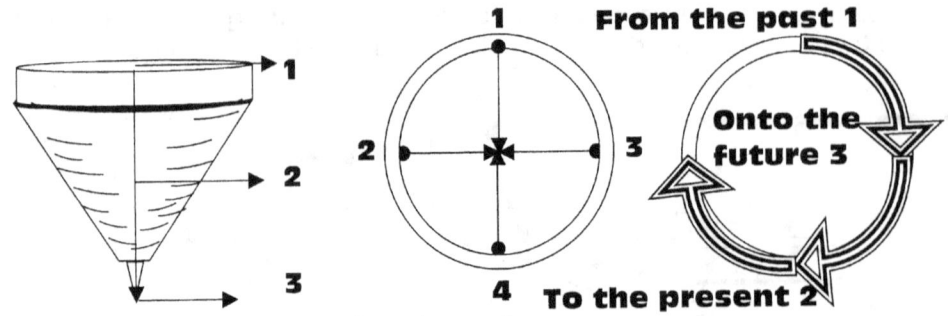

Using such logic makes science appear foolish. There is just no rational in the time verses events that can explain facts without. Since the time of Newton, the arguments tarnished from being brilliant to clever to fair too poor and a hundred years ago to the point of being stupid. That is what Kepler's formula is all about? That is what Kepler indicated with his formula $a^3 = T^2 k$. The space of an object (a^3) is equal to the time (T^2), which it is in, in every given instant (k). If the space becomes smaller, the time duration becomes longer every instant of time's progress.

In they're using of such logic makes science appear foolish. Since the time of Newton, the arguments made by those in the time of Newton tarnished from being brilliant to clever to fair too poor and a hundred years ago it reached the point of being stupid. That is what Kepler's formula is all about? That is what Kepler indicated with his formula $a^3 = T^2 k$. The space of an object (a^3) is equal to the time (T^2), which it is in, in every given instant (k). If the space becomes smaller, the time duration becomes longer every instant of time's progress.

Motion creates space Π^2

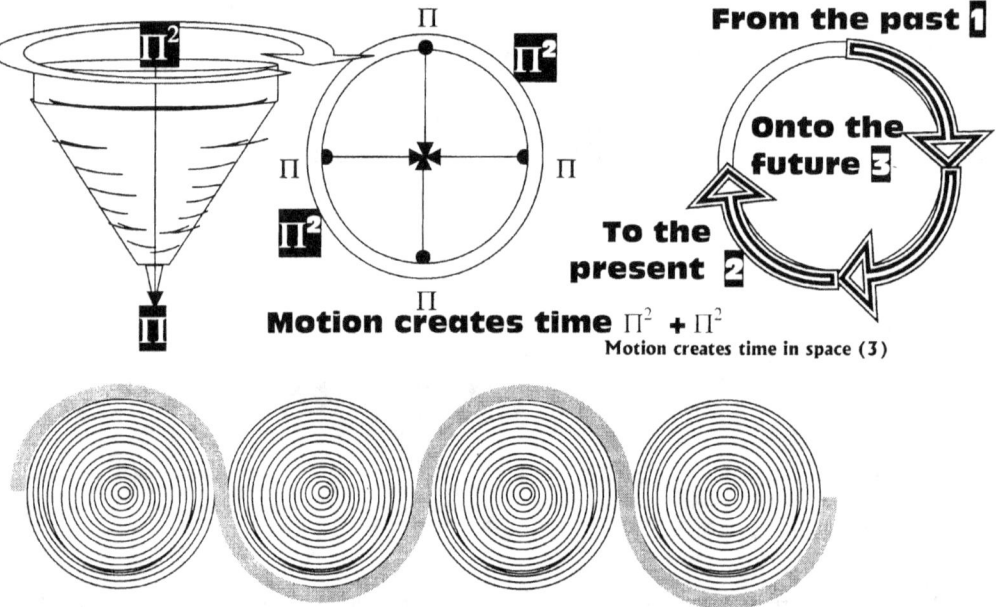

Motion creates time in space (3)

$R^3 / T^2 = k$ or $R^3 / T^2 = k^0$. With this fact established we then must return to the value as indicated by singularity being Π. In this we find that $\Pi^3 / \Pi^2 = \Pi$, weather k is Π or Π^0. This brings about the value relating to space-time relevancies as a formula consisting of $\Pi^3 / \Pi^2 = \Pi$ in various forms and relations. One also must keep in mind that there are ALWAYS four sides relating to the Universe from any point holding singularity, and since every point in the Universe contains singularity in what ever form, very spot in the Universe comprises of four points initially extending to the next spot by means of $\Pi^2/4$, which we know as the Roche factor.

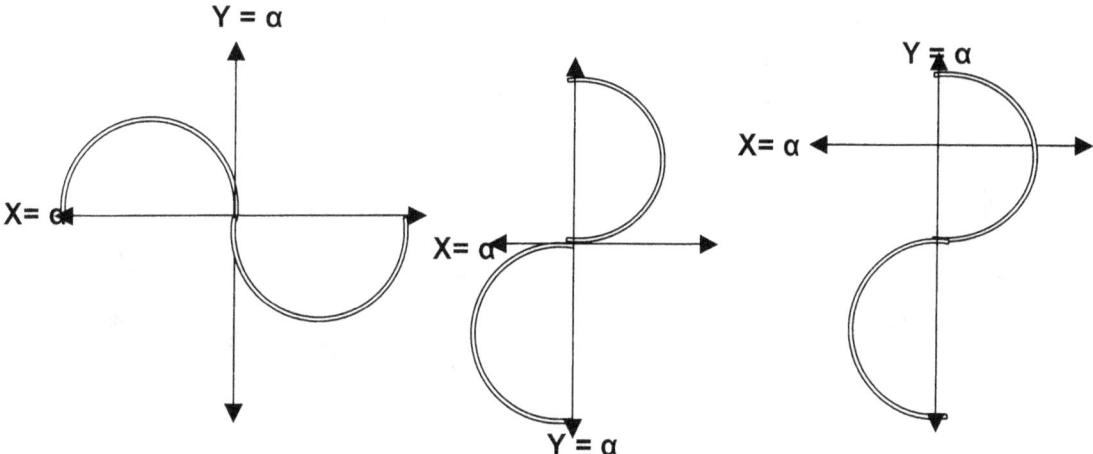

From the graph one can establish the link in the circle's rotation around a conforming unit being singularity. Saying that one therefore has to admit that the smallest spot has to hold space because the most insignificant dot can transmit light and being able to accomplish that, one must accept it to carry a value of something. If that spot had the value of nothing, it means that spot was not there to begin with. If the graph connected by zero that would mean there is no connection at all. With no connection the graph would be a mathematical tool with no value or use in any way.

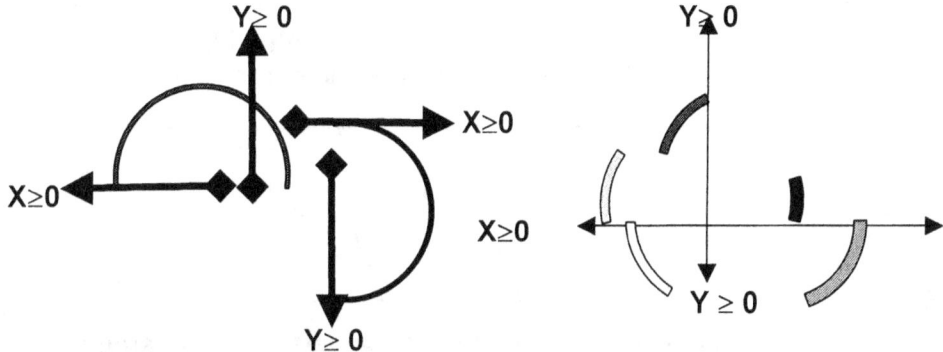

The graph with no connation between points because zero or nothing connects the points will render the use there of in mathematical terms quite obsolete.

Holding space-time one should return to the original formula indicating space-time in as much as $a^3 = T^2 k$ where a = R and T = T. Being time it has to alternate positions and that can therefore only apply to **k** where **k** will indicate a relation to the space-time in question or the relevancy to singularity being $k^0 = 1$. By receiving k on top of the already $k^0 = 1$ that is in place the top becomes an atom by erecting the line of singularity from $k^0 = 1$ to $k^0 = a^3 / T^2 k$

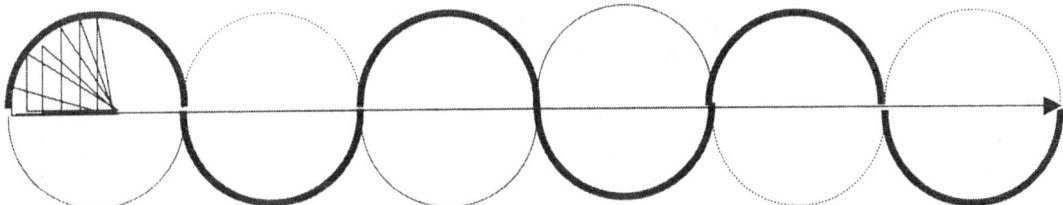

It started with a dot, because that is the only form, size and dimension mathematical logic will allow our brain to accept. From the one dot had to come a second dot and a third dot. The dynamics of such a dot is smaller than we can understand because such a dot is in negative relation to what we see Π to be, and the deeper we delve in finding the smallest fragment where space started, in the spot where time is still eternal as much as we can accept eternity to be.

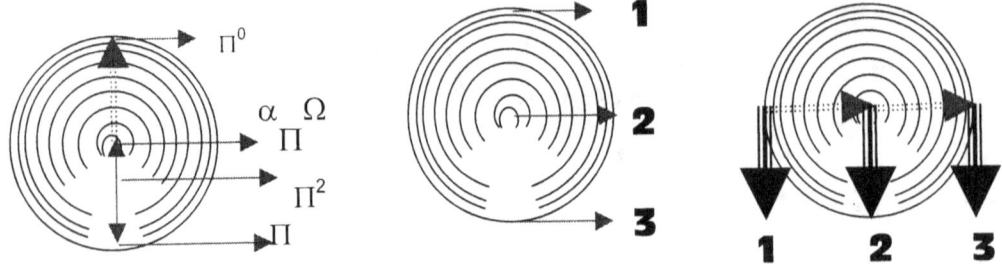

The reason why we should first locate the spot is because we can only work from that point forward. By working forward we have to work backwards to locate where we are heading. The cosmos started at a point and where such a point is, we will find the Universe. Every one knows where the Universe is, because we can see where the Universe is, but if we can see where the Universe is, then we should find the centre of the Universe in that spot. Einstein theoretically positioned the point of beginning at a place he indicated where singularity should be.

When realising the error of science in accepting a value as zero to be legitimate in mathematics, one can establish from that that the circle does not employ zero as a value after the completion of one rotation therefore $F = G\,(M_1 \times M_2)/r^2$ is invalid, one has to return to Kepler's $a^3 = T^2\,k$ and establish a value from that.

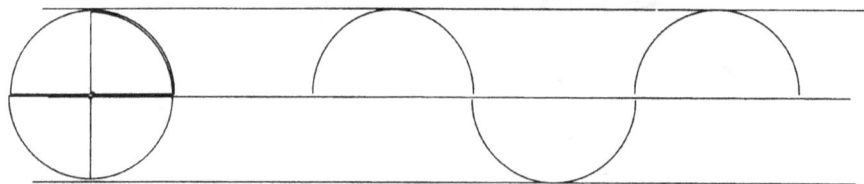

From the graph one can establish the link in the circle's rotation around a conforming unit being singularity.

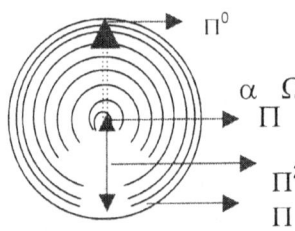

The motion establishes a centre and the curve builds up after which it once again breaks down. It builds to a climax and removes that which it built up until it becomes a line once more. The building up of and the breaking down of is a cosmic reality and that develops power driving many machines. Newton's disregarding of that makes his views forming part of the dark ages. In the motion of the circle is three relations where the one is a motion in four points building around a complete circle while there is one establishing points which relates every time with a centre position Then there is a third having the top separated by the bottom by a line running parallel with the developing.

With the cosmos the size it is and space so large compared to our smallness we have no chance in finding the centre of the Universe. The Universe started where singularity is and singularity is the sure indicator of the Universe. With all spinning objects holding singularity we then have located singularity in as much as finding the centre of the Universe. The Universe started with a dot forming. That answer arrive from taking mathematics back to a point of being the smallest possible position, far smaller than we may be able to calculate form. The ten dimensions I named the atomic relevancy is also showing the double value of singularity as singularity extends into as well as beyond space. The atomic relevancy is $(\Pi^2+\Pi^2)(\Pi^2 \times \Pi \times 3) = 1836$ that is the mass relation between the electron (3) and the proton. Proton = $(\Pi^2+\Pi^2)$ Neutron $=\Pi^2\Pi$. The atomic relevancy holds the dynamics of singularity control. In the ratio and dimensions we find in the atom, all space-time derives from the atom, whatever the atom is. Our instincts, our logic and our calculating process all indicate that the sphere holds a centre point from where six evenly positioned point's position matter to be. Using The formula $F=G$ $(M_1.m_2)/r^2$ it indicates to a force pulling objects closer, where each force is coming from each centre point the body in question has. The contraction must commit the two bodies towards a point in each case being spot on in the middle, not withstanding what direction the force is applying, the body will draw to the centre. If the Universe spins around a centre point holding singularity, and singularity confirms the centre of the Universe, then every particle holds the centre of the Universe making the number of universal centres immeasurable many, and every atom and sub atom particle presented outside the atom in smaller bits, are all not pieces of the Universe but they are a Universe surrounded by many Universes. If every atomic particle no matter how small is holding the centre of the Universe, then the gravity is coming about from that point because that is where the gravity applying in the Universe is applying contraction. If the Universe did start from one single point and time, matter and space flowed from that point, then that point must have a relative connecting base because such a point holding singularity must be eternal as space, matter and time link eternal. There therefore must be one point linking the entire Universe when regarding the fact of singularity. Then according to the theory off relativity there has to be one exact point holding time in relevance notwithstanding the fact that time departs from that position and relate differently to all space-time away from such a point.

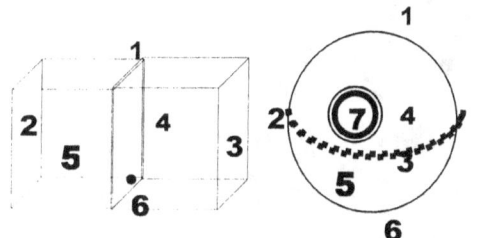

Kepler's formula also indicates that a sphere is within a cube that is holding a sphere

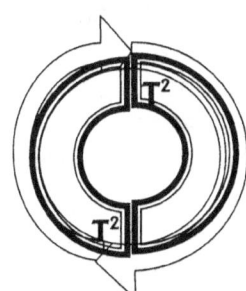

At this point Newton's second law come into affect. Motion by means of the Coanda effect introduced space as motion introduced time. For the first time ever time was interrupted when motion provided time the space to interrupt. From motion by the way of the Coanda principle gravity came about as a centre formed a point where motion surrounded space, By motion space-time was established in relation to singularity

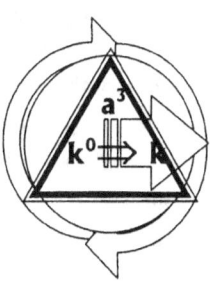

It then is the atom in the most centre part where space and time meets singularity, that Einstein found a Universe collapsing to a single dimension, and every atom at a point post of the proton where gravity initiates in according with the proton dimensional colas of $(\Pi^2+\Pi^2)(\Pi^2 \times \Pi \times 3) = 1836$

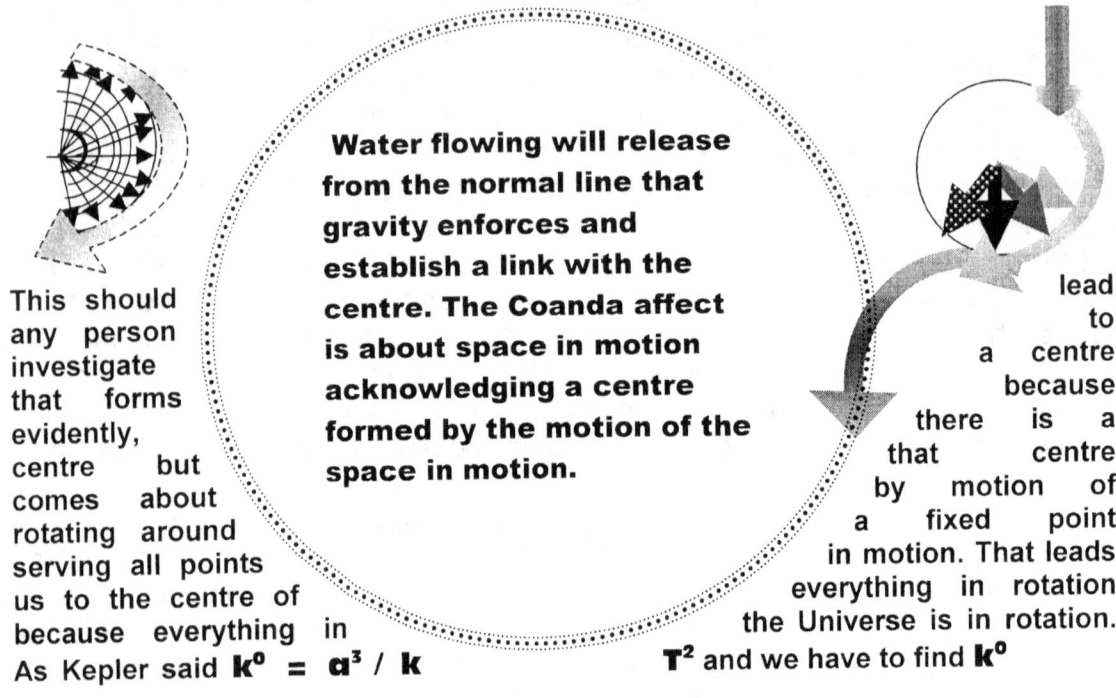

Water flowing will release from the normal line that gravity enforces and establish a link with the centre. The Coanda affect is about space in motion acknowledging a centre formed by the motion of the space in motion.

This should any person investigate that forms evidently, centre but comes about rotating around serving all points us to the centre of because everything in As Kepler said $k^o = a^3 / k$

lead to a centre because there is a that centre by motion of a fixed point in motion. That leads everything in rotation the Universe is in rotation. T^2 and we have to find k^o

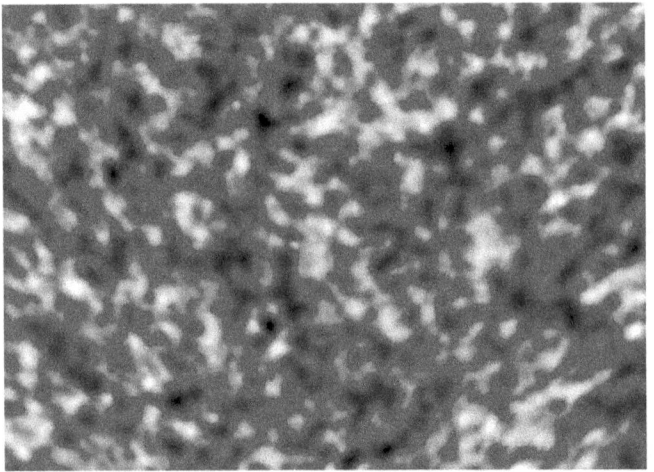

Every person with whom I have discussed the facts concerning creation recollects images in the trend depicted in a presentation as one may find to the above. That would be the most unlikely way Creation came in place. The recalling of pictures representing images about creation must have form, but to mathematics it had no form. From this thought the very opposite arises where Creation came from nothing but such an idea is mathematically simply not possible. The thought of nothing is just what it is, a thought of nothing and although it is in the nature of the human mind, to present nothing as a value in the recalling of something, nothing is a presentation of the figment in the human mind. There can be no number such as nothing and that was (possibly) Newton's biggest error. Nothing represents non-existing and that is just what nothing is, it is non-existing. In order to prove my point I wish to ask the reader to define the shortest line there can theoretically be. If he should answer anything but that the shortest line will be at a point where the beginning and is the very same spot he will be wrong. The shortest line that can ever be anywhere must have a start and finish holding the exact same spot. The line will be humanly impossible to create but we humans are capable of very little.

When the line has a beginning and an end at the very same spot and it wishes to extend the position as to further the possibility it has, which direction should it favour. Humans in the west would naturally think of extending from left to right while in the east humans may want to go from right to left.

Some persons will tend to go up or down, but all of the options are about human preference and not mathematical conclusions. Extending the line in any one direction will favour one direction without a conclusion about not extending in other directions. Such a conclusion has no sound mathematical foundation. The only option about extending will be in all directions equally in order to give a meaningful non-bias flow of mathematical equilibrium

The shortest line in the realm of possibilities must have a start and finish holding one spot and such a line will also be a dot or a circle. Not favouring one direction puts all directions at equilibrium meaning that any form what ever may be can develop from such a spot with the end and the start being the same. This reasoning prompted me to look for singularity in such a spot because if the prime spot from which all came was a spot, then the spot must hold the shortest line but more prominent it will hold the smallest form including the smallest circle. One possibility that the shortest spot can never have is having a starting point on the zero mark. If the mark of zero holds the start it must also hold the end because the end and the beginning has the same position. If the position of zero then is the beginning, the end will also be zero leaving the line without an end as well as without a beginning.The conclusion from this is that no line can start at zero because that will be a mathematical impossibility. A line or spot starting at zero would therefore be shorter than the shortest line possible. A line growing or extending from zero can never leave zero because of the influence of being zero disqualifies any possibility of growth. If the line then had to grow in all directions at the same pace the line must therefore be a circle. The value of the circle is Π, and that is where creation started.

Singularity by Motion

Singularity by Time

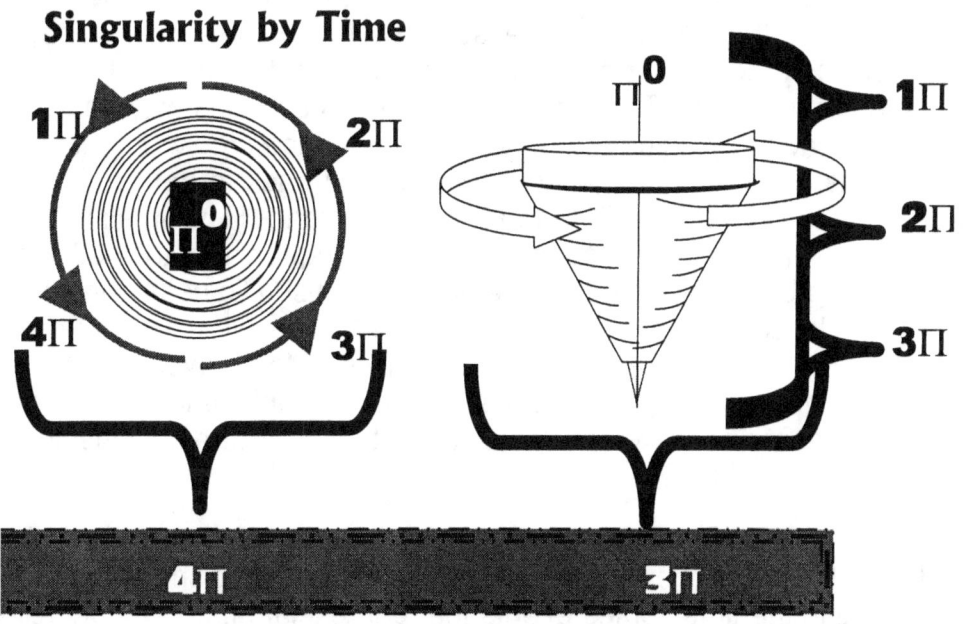

If the alignment is in ninety degrees to each other then Pythagoras has to apply strictly. Should my argument be sound and which it is sound we have to be able to use Pythagoras to determine the value of time in space. When the material rotates or moves the filling of the material is in perspective to the time. However, material can only be in one location in one split time. Since material has to cross over to the other side of the Universe in order to duplicate, which is how material moves, then the material, can be only on one side of the Universe,

Every time matter is generated and moves, it is singularity that is complying with it activating another point in singularity being charged with the motion. The Universe started from allocating singularity charged by heat into positions where such positions contributed to space- time. Every time the spot overheated the spot expanded into four dots and by expanding the spot cooled. In cooling the spot retained heat by which it spawned the dots allocated as time. In overheating objects expand and by expanding objects cool. That is gravity. Gravity is the expanding in relation with the cooling which means it is duplicating material in relation to a generated centre that is contracting the motion by cooling. Every inclination of motion is in fact motion and every movement be it contraction or expansion is moving to the other side of the Universe by bridging singularity because singularity is immovable. Therefore by being immovable, motion has to cross the division singularity applies and by crossing the division the factor that comes in place is $\Pi^2/4$, which results in the Roche limit. But such motion is three and the square of three in addition to the square of four brings about time in space.

As a school going youngster, I was fascinated by astronomy and in particular the cosmology aspect. In a long and strenuous process of self-education I was completely stunned by the behaviour pattern that the comet had in its relation as it orbits the sun. Please forgive my boyish way of presenting the following but it is important that I bring it across as I saw it as a boy and as a matter of fact still see it today as a middle -aged adult.

Science acknowledges growth as the Hubble constant and then refuses to put the growth in line with the solar system. The growth they reluctantly admit too, they refuse to connect that growth to the solar system in any way. They take a Universal year as a solar year being that of one cycle it takes the Earth to rotate the sun in the present day. Then they reflect on this as if this was going on since time began, because by doing that, there then is a nice crooked constant that fit mathematicians. Push this double standard applied back to before the Sun took its position and there was not Earth to indicate the year. How small was the year circle at that point in time and space. Take this right down to the:" Big Bang" where "the whole Universe were the size of a man's fist" (To use their words), how far did the circle goes to indicate a year then? The year was immeasurably smaller, shorter and faster than at present. This is logic even the Newtonians must accept. There is no space outside insanity to apply time to the past at the value it is at present and far worse, to use something so extremely insignificant as the Earth to measure it by.

Again I feel that the use of this type of constant just to fit mathematicians to corrupt the truth they in science are using such logic to rubbish the truth. There is just no rational in the time verses events that can explain facts without. Since the time of Newton, science has slowly nibbled at the truth to compensate for the game there is to play. It is as if the one crooked posture corrupts all in general. That is what Kepler's formula is all about? That is what Kepler indicated with his formula $a^3 = T^2 k$. The space of an object (a^3) is equal to the time (T^2), which it is in, in every given instant (**k**). If the space becomes smaller, the time duration becomes longer every instant of time's progress.

Singularity by Time

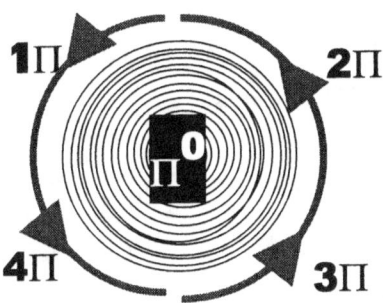

$$16 + 9 = 25$$
$$(25)^{1/2} = 5.$$
$$4^2 = 16$$
$$3^2 = 9$$

However because of dimensional duplication the square of time is ten and five will be on the one side of the Universe and five will be on the other side of the Universe. That then is why the Lagrangian system holds five positions in relation to singularity

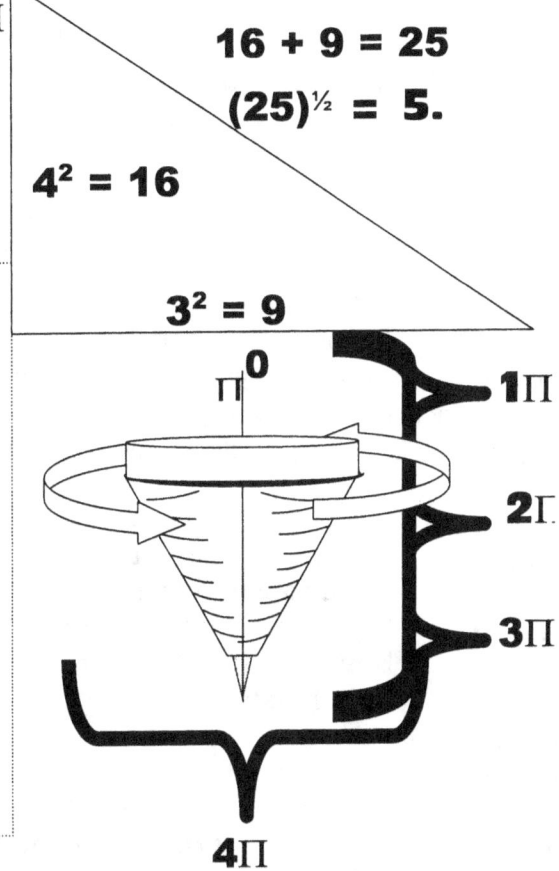

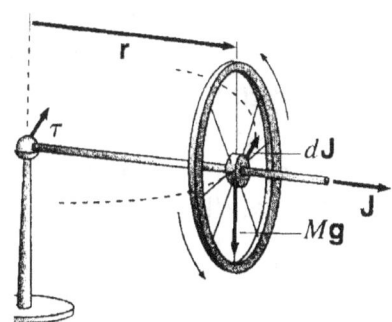

The pendulum not only stops at a precise point that the Earth holds as singularity presenting the singularity the Earth dictates at that given time. The relation there is in what Kepler discovered and that which Galileo discovered has gone by without many too my knowledge seeing such an extreme direct link. Kepler formulated space-time and Galileo implemented space-time. The space the pendulum swing through is representative of the space captured by the anchor singularity formulated by Kepler as a^3, the pendulum arm becomes the indicator **k** the swing distance of the arm becomes the time T^2. This is the recopy for time keeping since coming into the light from the dark ages. It is a half circle indicated by a straight line forming two sides where each side is holding a triangle in relevancy. This is in sharp contrast to Newtonian claims that the cyclic repeat the Earth has with the Sun and all the numeral Equalities derived from that by Using Kepler's formula still after one year comes to nothing.

Taking the argument back to Kepler's law,

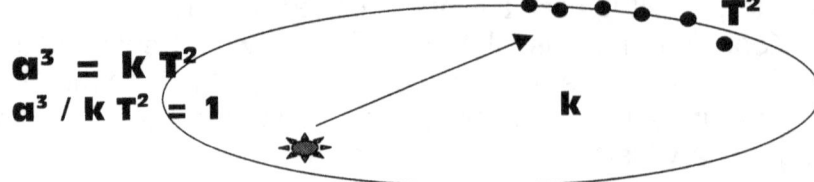

$$a^3 = k\,T^2$$
$$a^3 / k\,T^2 = 1$$

T^2

k

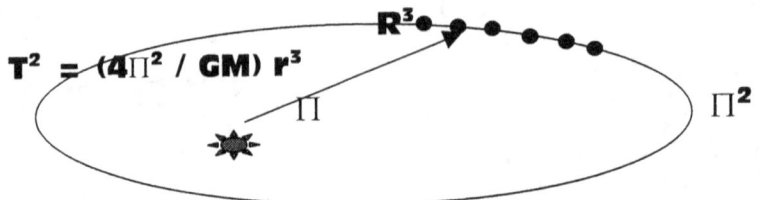

$$T^2 = (4\Pi^2 / GM)\, r^3$$

R^3

Π

Π^2

The spinning or not spinning is not part of the issue because at the point of absolute singularity the object never spins. Therefore spinning or not spinning does not apply to the point of singularity because singularity never spins in any event. In the whole structure with a pivotal centre as the control to the motion of the space the fact of Π is a natural outflow and any adding of Π is totally incorrect. According to Newton the result of spin is zero, however the top will tell a much different story.

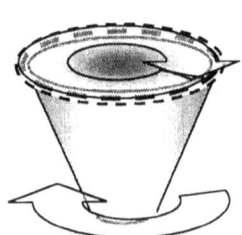

On the surface, at first glance the top is an ordinary piece of dead wood that is machined into a sloping shape. The top is normally fitted with a sharp needlepoint at the bottom and the sharper the point is the better will the spin balance be.

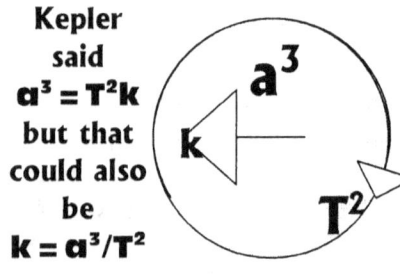

Kepler said
$$a^3 = T^2 k$$
but that could also be
$$k = a^3/T^2$$

a^3

k

T^2

When translating Kepler's mathematical expression into English we can see what Kepler said also read as $k = a^3 / T^2$ where k is one point from a centre point that is space a^3 relating to time T^2. From a centre comes space-time. The centre k brings space a^3 in ratio to time T^2, which is space / time a^3 / T^2. Reading this correctly cannot bring any dispute...yet it does...and it has been doing it for centuries on end!

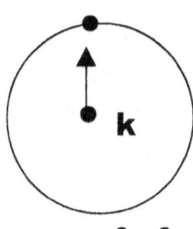

$$k = a^3/T^2$$

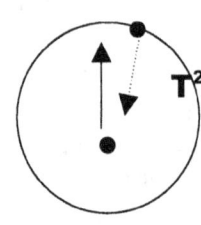

$$T^2 = a^3 / k$$

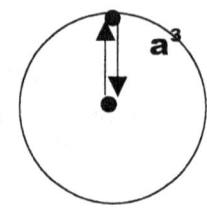

$$a^3 = T^2 k$$

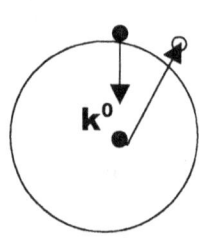

$$k^0 = a^3/T^2 k$$

If space were zero or nothing as Mainstream science so affectively teaches us, then Kepler's principle formula would need the changes Newton brought about. It is true and stands tested like no other research ever coming either before or after Brae and Kepler's work. By reducing the line to infinity and raising the line again back in the direction of space, the line would erupt as a natural sphere having Π as the natural basic value. That is the value Kepler interpreted. However not realising what he saw he chose to use different symbols.

$$k = k^{3-2} = k^1$$
$$a^3 = a^{2+1} = a^3$$
$$T^2 = T^{3-1} = 2$$

$k = a^3 / T^2$	$a^3 = T^2 k$	$T^2 = a^3 / k$
$k = a^{3-2} (T^2)$	$a^3 = T^2 k^1$	$T^2 = a^3 / k^1$
$k = a^{3-2} = k^1$	$a^3 = T^{2+1} (k^1)$	$T^2 = a^{3-1} = T^2$
$k = k^{3-2} = k^1$	$a^3 = a^{2+1} = a^3$	$T^2 = T^{3-1} = 2$

is the same as   is the same as   It is all the same

$k = k^{3-2} = k^1$ is in direct relation to $a^3 = a^{2+1}$ is in direct relation to $a^3 = T^2 = T^{3-1} = 2$. With this information staring mainstream science in the face and scream pleading at them to recognise the information they turn around and ask why can man not fly off to other galactica at the speed of light.

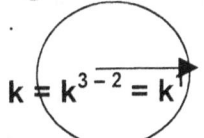
$$k = k^{3-2} = k^1$$

It takes time for space to fill **k** in the distance. In fact it takes the distance that **k**, developed since the Big Bang $k = k^{3-2} = k^1$ to fill the distance.

It also takes time $T^2 = T^{3-1} = 2$ to produce the distance forming k^2

It takes space $a^3 = a^{2+1} = a^3$ to form k^3 since coming from the Big Bang

When the astronaut is departing from space on Earth or filling Earth space it will take the departing astronaut k^2 time to reach k^1 and fill out k^3. At present and in this moment our most impressive astronautic engineers will devise an engine that would cut k^1 by say half. This achievement will come as they increase the power output say for argument sake to double what it is at present. There was no friction of particles destroying the frame of the craft because there are not enough particles in space to do it.

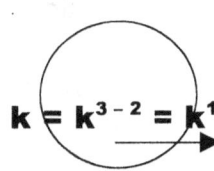
$$k = k^{3-2} = k^1$$

It takes time for space to fill **k** in the distance. In fact it takes the distance that **k** developed since the Big Bang $k = k^{3-2} = k^1$ to fill the distance.

It also takes time $T^2 = T^{3-1} = 2$ to produce the distance forming k^2

It takes space $a^3 = a^{2+1} = a^3$ to form k^3 since coming from the Big Bang

However, keeping Π as one ($\Pi^0 = 1$) we keep the Universe in the first dimension.

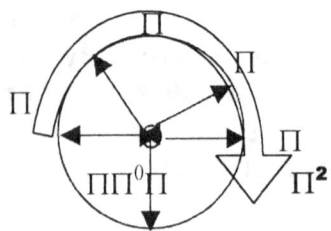

This point, which I now am referring to, is the point where Π is a fully appreciated value while the diameter D still remains a dimensional factor of one. This is the dawn of the second dimension where space was there but space was sparsely shared in some cases. It was when Π^0 shifted to become Π for the very fist time.

However, the equation looks far more sensibility when using the value of singularity

$k^0 = a^3 / T^2 k$ forms

$1/ k^0 = T^2 k / a^3$

$1/ (k^0 k) = T^2 k / (a^3 k)$

$1/ k = T^2 / a^3$

Expressing the equation by using the value singularity has instead of the symbols Kepler designated to the formula he introduced it makes far better sense expressed mathematically

$\Pi^0 = \Pi^3 / \Pi^2 \Pi$

$1/ \Pi^0 = \Pi^2 \Pi / \Pi^3$

$1/(\Pi^0 \Pi) = \Pi^2 \Pi /(\Pi^3 \Pi)$

$1/ \Pi = \Pi^2 / \Pi^3$

By taking **k** into a negative the space will reduce the time because the space cannot sustain the demand of space growth.

$k^0 = a^3 / T^2 k$

$1/ k^0 = T^2 k / a^3$

$a^3 / k = T^2$

$\Pi^0 = \Pi^3 / \Pi^2 \Pi$

$1/ \Pi^0 = \Pi^2 \Pi / \Pi^3$

$\Pi^3 / \Pi = \Pi^2$

In all my other work, I make exclusively use of the value of singularity Π since it makes a lot more sense, but when I use the value of singularity, which is Π then no one seems to have a remote idea to which I am referring.

$k^0 = a^3 / T^2 k$ forms

$k^0 / k = T^2 / a^3$ that becomes

$k^{0-1} / a^3 = a^3 / T^2 / a^3$

$k / a^3 = 1 / T^2$

The replacing of the symbols Kepler used with the value of singularity the mathematic equation comes into practise.

$\Pi^0 = \Pi^3 / \Pi^2 \Pi$

$\Pi^0 / \Pi = \Pi^2 / \Pi^3$

$\Pi / \Pi^3 = 1 / \Pi^2$

That proves that the establishing of distance **k** will produce space a^3 and set space a^3 in motion T^2 where such motion is in opposition to singularity, which means gravity or contraction is the deliberate opposite of expanding $a^3 / k = T^2$. In the beginning the expanding then also involved three more points all just outside the border of singularity but within the atom exclusivity. It extends k while it introduce a returning relevancy back to singularity k^0 by creating motion in spin and duplicating space by reducing space.

No matter how one looks at the Kepler formula, it signals the same principle. It shows how motion erects the Universe by mathematical equations. It puts singularity, as one in relation to six and that is the Universe decoded.

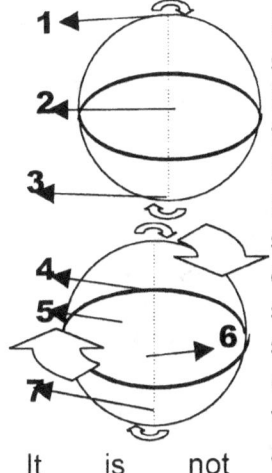

It is not possible for these points to have motion or to produce

By not having motion the lines also have no space as the space extends to form space forms space and the line includes serving the three points to the outside. Where there is no motion, there is no space and where there is little motion, there is little space. The only space the line may relate to can be a point that is on the border of the sphere that is crossing singularity and connecting the two edges on either side of the sphere that is forming the sphere. That means the line from one point holding singularity to another point holding singularity that line will cross the centre line which gives the line in singularity valid space-time to control. Singularity does not have the ability of motion therefore singularity does not hold space. Singularity is also eternally indifferent to motion and motion can excite singularity but singularity cannot be shifted by motion. Three points form the line.

spin since spin requires the refurnishing of singularity transfer or displacing space-time from one point holding singularity to another point holding singularity. Where space-time or space by motion is active, there the motion carrying space has to be transferred from singularity without motion and space to singularity being without motion or space. As the space transfers from one locality to the next locality there some energy has to be transferred with the space that the motion generates to a new location and such relocation is going about with the transfer of space-time. However, the motion excites singularity but does not shift singularity. Time shifts singularity because it is the exciting of singularity to get singularity to hold space for a period that forms time. Time is the flow of heat through space in space. **Space-time is conducted from one point to the next point just as if the flow of electricity is conducted. The conducting medium however is not similar. Should motion be required, there has to be a transfer of space-time from one location to the next location. This transfer is not that immediately instated but is a process that has to be developed. It takes time to break down space-time, relocate the position generating the space-time and then reinstate the energy that will produce space-time. The Newtonian view is that mass shifts but the truth is incomprehensibly more complicated than purely shifting mass.**

There is a breakdown of singularity attachment and then there is the motion transfer of space-time followed by the rebuilding and re-instituting of space-time that the motion generates, just like the motion in the Coanda effect is responsible for generating the pulsing of electricity. Generating electricity is in fact a process of creating and re-establishing space-time but only at the level of C by creating gravity at the level of C. Every point holding space is singularity being charged with receiving as much as distributing space-time at points that receive or let go of the next space-time it is charged with. The space-time, the point is charged with becomes the space-time it is displacing the charging. It is four points in receiving as much as the four points are in distributing

However Newton recognised just the opposite and even allowed a freezing of motion and therefore time.

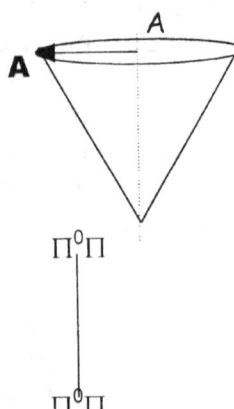

The moving of Π^0 to Π involved relegation and not motion as we consider motion. It was Π^0 getting a side and that is all. There was no true side but only a form that came into place. Singularity (A) received singularity (**A**) and no more of anything but the shift to comply with having a relevancy forming in relation to singularity. The dots had no sides, had no length or diameter. There was not measurable space or measurable time involved. The time could have been a micro, micro second as much a trillion millennium because time had no relevance. It was eternity interrupted by infinity, as it still is the case, however the line that eternity followed was no line because there was no space to hold the line. The line was momentarily interrupted by infinity, however with no one there, there was no one to notice. The lines were not lines but relations to sides being formed.

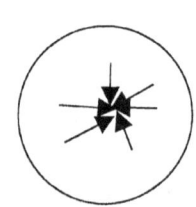

Inherent to the form the sphere offers, there is a specific location of singularity where the radius first goes single $r^0 = 1$ and then form goes into the realms of singularity $\Pi^0 r^0$. The cube also may have such a pint bur having such a point does not connect directly to six points located on the edges of the cube or any other form the is.

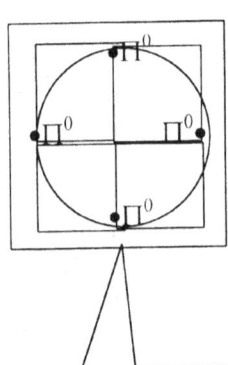

In relation to such a centre where $\Pi^0 r^0$ forms singularity there are always four cubes related to such a centre where the centre is part of seven points in total representing the sphere. Every cube gas lost one side to a point of the sphere where the sphere takes control of form and removes one side of the cube. In relation to the time factor that is inherently part of singularity by the extending of singularity there are five sides connecting to four points standing related to singularity by the Π^0 factor and that gives 5 X 4 = 20. That is always directly in relation to seven points singularity offers.

This only applies in relation to time because time is the square or then if you wish time is the flat to space being the cube. Time in the square draws space in the cube flat and that is the principle behind the effort gravity can apply to destroy space. Once gravity destroys space and time goes square then all factors preventing time to remain has gone single leaving time falling into singularity as well. This does not yet explain the three dimensions forming the six -sided Universe we find we have

How does one reconcile the behaviour of the top with the foundation of science?

All spinning matter has the point where the spin is still there but the radius is too small to measure by any means. That point in the very and precise centre of all rotating objects is standing still in relation to the rest of the body that is spinning around such a centre. In relation to that logic I do not except Newtonian science holding the radius of a spinning object unaccountable in the spin, whether the spin is applying or not.

Applying Newton's second law F = ma
One arrive at the formula
$GMm / r^2 = m (\omega^2 r)$

By replacing $(\omega^2 r)$ with $2\Pi / T$ we obtain Kepler's third law
This law predicts that $T^2 = a^3$

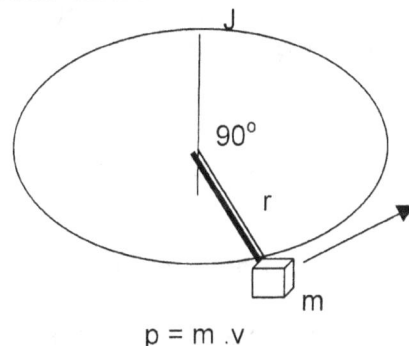

The mass (m) multiplying the speed (v) forms a new value J
AND THEREFORE j

CONTINUOUS TO IMPLY $J = I \omega$
  $= r \times p$ where $p = (V = r \times \omega)$

$J = r.m.v = m.r^2.\omega = I.\omega$ and becomes interpreted as $J = I \omega$
This establishes that $r = dJ / dt$

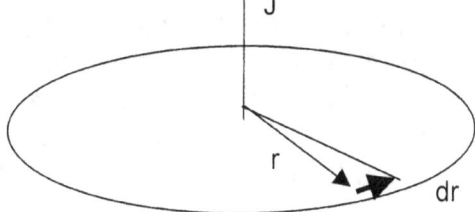

$r = dJ / dt$ In the case of planets in orbit around the sun r forms a value of zero because $dJ / dt = 0$.

What this statement implies is that r does not exist. When anything has a value of zero it is for all purposes non-existent. Only when an object is following s straight line can the radius be non-existent because the radius alters value through time development.

Newton had the revelation of all the above mentioned as an apple fell from a tree apparently very close to him. He was admired as an instant genius and the one the world was waiting for to be born. I do not, for one second, deny or dispute the revelation. What I do encourage is to place the event into its correct context. It was merely, and simply an apple that fell from its branch to its roots. The apple did not pretend to be a meteorite that fell from the heavens. If it were a meteorite, I am sure, with the man's genius, science would be somewhat different at this stage. However, as a young man, being very impressionable, as all young men are, and with the attention this brought about in the world of science, the matter overshadowed the fact.

I am not disputing Newton; I am disputing the relevance of Newton's scientific breakthrough. It was not two objects of cosmic proportions, colliding in a show of the spectacular. It was, after all, only an apple falling from a tree and not that big an event. With this miracle he revealed, Newton found he was competent to improve on the work of Kepler and what Newton saw about what Kepler found was to Newton's mind the proof of total mathematical incompetence. He (Newton) saw a circle and without Π there can be no circle. Further more, since he was the founder of the invert four square principal, the principle also had to be included the make the picture a smart Newtonian picture and with that remove Kepler as such.

$\frac{dJ}{dt} = 0$ Newton, and science, made one enormous blunder, from this stance.

They took the radius of a wheel not to have any influence on the wheel. In doing that, they removed the very fact that keeps the universal attachment together.

They put two objects in an attaching relevancy and then announced no relevancy. Doing that is breaking the most fundamental mathematical principle.

$\frac{dJ}{0} = dt$ or $\frac{0}{dt} = dJ$ This disputes mathematics.

DJ / dt can have any number from eternity to infinity, only excluding only one possibility; it cannot be 0. By placing the one in division of the other, you bring in relevance. You cannot then say there is no relevance. By doing such, you proclaim that one of the factors is non-existent. In both cases, one of the factors then does not exist. Such a claim is incoherent, because you proclaim that a circle has no radius, or a radius has no circle. When calculating a circle, you multiply either the square of the radius by Π, or the quarter of the diameter at a square by Π.

$\frac{dJ}{dt} = 0$ constitutes a circle and is also therefore $\Pi \times r^2 = $ CIRCLE

If you remove r it then is $\Pi \times r^2 / r^2 = $ CIRCLE.

You cannot then say $r^2/r^2 = 0$ and therefore $\Pi \times 0 = 0$. That is nonsense. $\Pi r^2/r^2$ will always be $\Pi \times 1$, and that is where Kepler placed singularity. By hiding this fact Newton went and threw the baby out with the bath water.

In the motion every wheel has to have a pivot around which the wheel turns. That is called the axis.

Taking the argument back to Kepler's law,

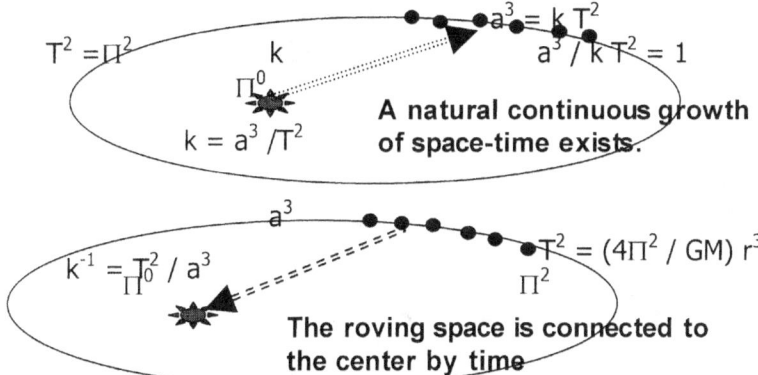

The spinning or not spinning is not part of the issue because at the point of absolute singularity the object never spins. Therefore spinning or not spinning does not apply to the point of singularity because singularity never spins in any event.

Since Newton became an institution forming the King bee of the academic cartel world wide The Brainy Bunch had Newton's vision written in the minds of the future generations almost at gunpoint...well definitely at an academic gunpoint.

$r = dJ /$

$r =$

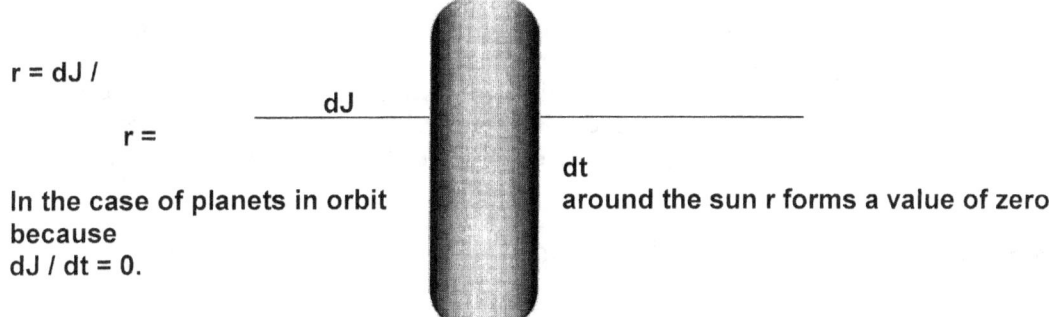

In the case of planets in orbit because $dJ / dt = 0$.

around the sun r forms a value of zero

I am not the brightest in the world that I admit, but one thing no one can do, not even if you are the one and only Isaac Newton, is that you cannot place any relevancy in a relevancy and then claim it not to be in a relevancy because such a relevancy does not suit your taste.

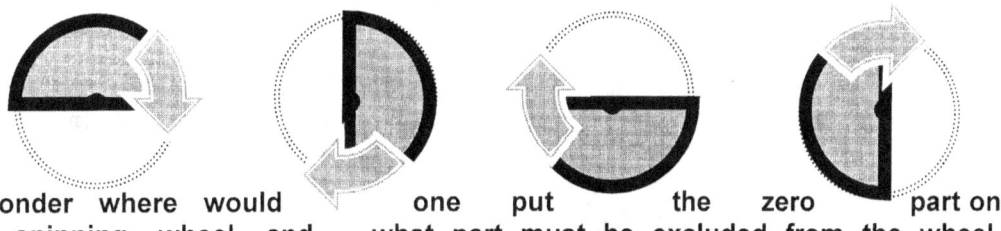

I wonder where would one put the zero part on the spinning wheel and what part must be excluded from the wheel. What Newton suggests, is a wheel has one side on top and no side at the bottom. While the wheel is spinning one may not remove the one side and then claim there is no attachment between the top and the bottom. That would mean in a graph the top is not connected to the bottom because a wheel spinning is a graph moving against time. It is the principle all driving is done and not the least electricity.

By denouncing Kepler and his formula, one must be prepared then to denounce all motion in that manner, and Newton more than most should have realised that.

You cannot put something in relation to another object and then decide there is no relevancy in the relevancy

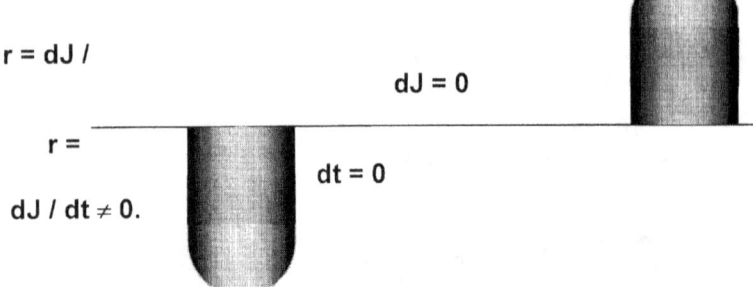

r = dJ /

dJ = 0

r =

dt = 0

dJ / dt ≠ 0.

If dJ = 0 then dt = 0 That is a mathematical principle, much larger than even Newton

Newton, and science, made one enormous blunder, from this stance. They took the radius of a wheel not to have any influence on the wheel. In doing that, they removed the very fact that keeps the universal attachment together.

$$\frac{dJ}{dt} = 0$$ This disputes mathematics. DJ / dt can have any number from eternity to infinity, only excluding one; it cannot be 0. By placing the one in division of the other, you bring in relevance. You cannot then say there is no relevance. By doing such, you proclaim that one of the factors is non-existent.

$$\frac{dJ}{0} = dt \text{ or } \frac{0}{dt} = dJ$$ In both cases, one of the factors then does not exist. Such a claim is incoherent, because you proclaim that a circle has no radius, or a radius has no circle. When calculating a circle, you multiply either the square of the radius by Π, or the quarter of the diameter at a square by Π.

Quite the very opposite is true and the rotating wheel in fact is the moving wave.

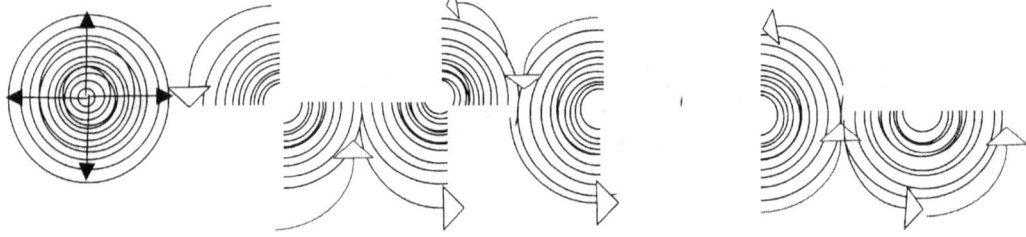

The graph which is a cornerstone of mathematics and which is used extensively in various calculating procedures would not function because at the end of a cycle there will be no cycle.

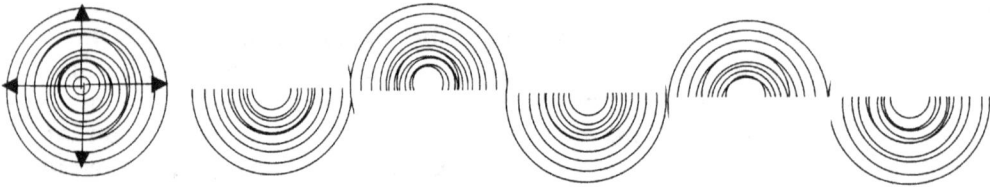

When looking at any rotating object, there has to be a point of no rotation and no rotation means "no rotation", not no existence.

No rotation means a factor of 1, not zero. That then is singularity. The eternal Π, the Π that may not have significance but still it is Π of value.

$$r^2/r^2 \neq 0 \qquad\qquad r^2/r^2 = 1$$

The relativity remains one, eternally one, but it cannot be zero. Therefore, dJ/dt cannot be zero.

dJ/dt can be eternal or infinitive or at the worst it can be dJ/dt =1 but dJ/dt ≠ 0

When explaining this to any child, they can immediately see that. Explain this to any Newtonian High Priest and he may have you removed forcefully from campus. I cannot find one Newtonian, large or small to accept that.

What is it the Newtonians fail to see? If an electron is orbiting around an atom, 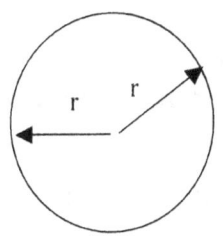 **the inside of the atom must be a circle. If the atom was not a circle, it then had to be a cube. The electron cannot rotate around a cube; therefore, the inside of the atom is a circle.**

In a circle, there is a radius that initiates the circle. The calculation of such a circle is Π X r^2.

The radius r runs from the circle outwards, from a circle centre point towards Π, the value of the circle. In the centre of the circle, there is a point where the radius starts. It runs outwards from that point in all directions towards the circle Π. Technically, there then has to be a point where r is infinite and not zero, an absolute infinite. However, the circle therefore remains Π. The circle does not disappear; it remains there for all to see. It is only the radius that almost disappears into the infinite, but it does never become zero!

$$\frac{\Pi r^2}{r^2} = \Pi$$

If one removes the radius from the circle, the circle remains, only holding the value of Π. By removing the value of r, Π becomes singularity with no place to be. Singularity is the place where there is no space to be in place. However, Π remains because once r receives the slightest of space Π will find space. Then the circle will grow to Πr^2 and r would determine the space. Without space, there is no r but there is a circle with the value of Π. Singularity is in every single rotating object, be it the proton or the combining effort of all particles in the Universe. That is what light and the photon is. It is concentrated heat that the Sun(or any other generator of electricity) connects heat to singularity where the heat receives either temporary connection to singularity or a small piece of individual singularity.

All spinning matter has the point where the spin is still there but the radius is to small to measure by any means. That point is standing still in relation to the rest of the spin. In relation to that logic I do not except Newtonian science holding the radius of s spinning object unaccountable in the spin, whether the spin is applying or not.

What this statement implies is that r does not exist. When anything has a value of zero it is for all purposes non-existent. Only when an object is following s straight line can the radius be non-existent because the radius alters value through time development.

The spinning or not spinning is not part of the issue because at the point of absolute singularity the object never spins. Therefore spinning or not spinning does not apply to the point of singularity because singularity never spins in any event.

$\Pi \times r^2 = $ CIRCLE

If you remove r it then is $\Pi \times r^2 / r^2 = $ CIRCLE.
You cannot then say $r^2/r^2 = 0$ and therefore $\Pi \times 0 = 0$. That is nonsense. $\Pi r^2/r^2$ will always be $\Pi \times 1$, and that is the eternal circle.

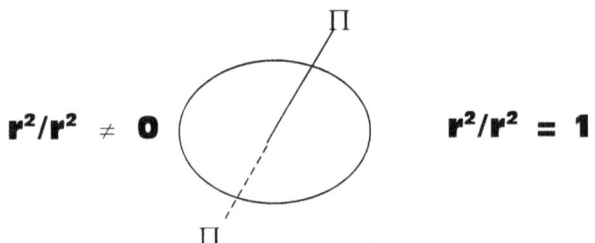

That then is singularity. The eternal Π, the Π that may not have significance but still it is a Π of value. The relativity remains one, eternally one, but it cannot be zero. Therefore, dJ/dt cannot be zero.

> dJ/dt can become eternal or infinitive or at the worst it can become one
> dJ/dt = 1

When explaining this to any child, they can immediately see that. Explain this to any Newtonian High Priest and he may have you removed forcefully from campus. I cannot find one Newtonian, of any significance being large or small to accept that. By not having a wheel rotate, the wheel becomes the factor of one, and the rotation becomes zero. The wheel does not disappear. In the cosmos, everything is rotating because nothing ever stands still. Therefore the mean equilibrium, the common factor there is to share, has to be one, eternity, the eternal Π, because all rotating objects has Π in singularity, and sharing singularity, gives every object in space a relation with all other objects in space. After trying for many years to bring them the candle, I concluded that Newtonians are incapable of realizing that mathematical principle as reality.

If Newton had said that dJ / dt = 1 then that is exactly what Kepler said when he said that in the centre of space–time singularity is allocated a position of control $k^0 = a^3 / T^2k$. Kepler also said the motion brings about the filling of singularity $k^0 = 1^0 = 1$ when he said that the space is filled by the matter in the motion through the time period. Motion establishes space in time and cannot be zero because THAT is what gravity is. It is the motion of space-time and that can't be zero. If gravity were equal to zero the entire Universe would stop existing.

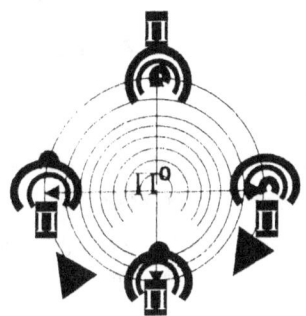

With everyone of the four rotating points duplicating the value of Π in relation to the centre Π^0 at a measure of Π / 2 and where Π^2 is the establishing of Π by the motion thereof therefore Π^2 / 4 became a limit in relation to the developing centre. One has to remember that the star of the present takes characteristics of the form from the era before space was a factor.

In the same manner the ring cannot remove, because the spokes will then still imply where the ring must be. The only way to cheat yourself out of the situation is to remove the wheel and spokes altogether, and you are left with what you say there is: NOTHING. But that does not apply in cosmology. The object rotates the centre structure and therefore there has to be a radius holding the circling orbit in relation to the centre structure.

Removing the radius from a circle does not be removed the circle, because the circle is there, securing the ring If the Universe started from a point of singularity, then there was initial spin at a pace where the spin did not apply and that spin included the entire Universe, still in non-existence. That is singularity. That is the only singularity there can be. The spin was going on for eternity because the spin does not apply, it has a value of infinity and infinity was running along the line of eternity.

By receiving the command, singularity received a value outside eternity as Π^0 received edges. Granted the fact that the edges were so small there still was no r to present a circle.

Having edges where Π^0 duplicates to present the edges, singularity lost the value of Π^0 to the value of Π^1 with the same value singularity had being Π^1 to the one side and Π^1 to the other side, the cosmos received the eternal value of the first dimension outside eternity. It was the square of Π^1 being Π^{1+1}. That was the first dimension outside singularity Π^0 where singularity has a value of Π^1 in the form of $\Pi^{1+1=2}$. The first claim to space had a value of Π^2. This applied to both sides of the claim to space outside singularity, and the double proton became the dominant factor on matter.

That, which formed the Universe was the growth of time expanding while overheating. It is the expanding by the measure of Π^2 that has much significance as where the expanding went Π and the expanding was limited to the value of Π^3.

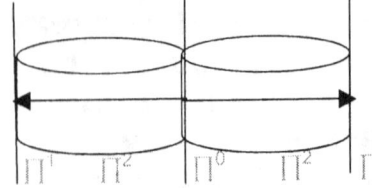

As singularity burst out into matter forming space as much as occupying space inside singularity, the protons started flying around, spinning around singularity, as each individual proton occupies matter in space

It truly makes me feel bitter thinking about the many times I tried to explain the facts to the Brainy Bunch with no luck. You know you are correct, but that person holding the establishment secured for Newton just push your argument aside, because he has the authority to investigate and lacks the interest to initiate change

From the spinning motion Π^2 does not stop at the end of the solid structure but the influence of Π extends and this then becomes the atmosphere. The influence of Π^2 or gravity does not stop at the end of the solid structure but the influence of Π is extending and eventually plays a most dominant role in the local cosmos, although not yet recognised and that factor is most crucial to a better understanding of the implications of laws governing the cosmos.

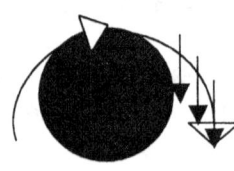

With the circle being Π^2 Π the Π^2 will reflect the circle in the square with Π forming the extending of Π^2. This is an extending of the six Π forming in alliance with the centre Π. This produces that any extension of 6 forming material one further extending goes into space and relates to a seventh dimension.

Newton, and science, made one enormous blunder, from this stance. They took the radius of a wheel not to have any influence on the wheel. In doing that, they removed the very fact that keeps the universal attachment together. They still insist that rotation results in nothing

$$r^2/r^2 \neq 0 \qquad \Pi \qquad r^2/r^2 = 1$$

$$\frac{dJ}{dt} = 0 \quad \frac{dJ}{0} = dt \ \text{or} \ \frac{0}{dt} = dJ$$

$\Pi \times r^2$ = CIRCLE If you remove r it then is $\Pi \times r^2 / r^2$ = CIRCLE.

You cannot then say $r^2/r^2 = 0$ and therefore $\Pi \times 0 = 0$. That is nonsense. $\Pi r^2/r^2$ will always be $\Pi \times 1$, and that is the eternal circle. When looking at any rotating object, there has to be a point in the infinite middle where the one side rotates in one way and the other rotates in the other direction opposing the opposing direction. That point in infinity is the point of no rotation and no rotation means "no rotation", not no existence. No rotation means a factor of 1, not zero.

That then is singularity. The eternal Π, the Π that may not have significance but still it is a Π of value. The relativity remains one, eternally one, but it cannot be zero. Therefore, dJ/dt cannot be zero.

dJ/dt can become eternal or infinitive or at the worst it can become one
dJ/dt = 1

When explaining this to any child, they can immediately see that. Explain this to any Newtonian High Priest and he may have you removed forcefully from campus. I cannot find one Newtonian, of any significance being large or small to accept that.

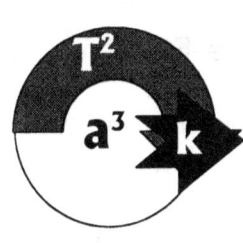

The spinning top is applying the Coanda principal to establish gravity in relation to the Earth centre or the controlling gravity. As soon as the top starts spinning at a specific velocity, the motion of the top secures a centre singularity that keeps the top upright. This is the motion the Coanda principle requires to produce a gravitational centre as to produce a specific space-less point in gravity. It is gravity that the top employs because only another gravity can interact with the Earth gravity to secure stability while the top is matching a specific spinning speed.

The comet rotates the sun, and the Sun by itself has a point of singularity where Π remains without r. The comet, holding the orbit, also has a point of singularity, but since there is space separating the two objects, they cannot share a mean point of singularity, the very point of existing. Since singularity means just that, being single, there cannot be two. The comet and the Sun have a mean point of singularity but the space they occupy divides their common singularity. That is why they orbit in an oval path, a path where the one structure holds on to more space from its point of singularity towards the space it claims. Since they do not claim equal space, BY THE DENSITY they hold, the space will not be in proportion. Singularity is a mathematical reality. Einstein may be the first to name it and Galileo (unwittingly) may have been the first to define it as Kepler was the first to formulate singularity, but in mathematical terms singularity is the most basic principle. It is singularity that attaches the top to the orbiting comet. At this point I wish to establish a fact that seems lost in all other grandeurs of cosmology. A straight line cannot begin at zero or nil it can only start at infinity/ Such a statement will hardly seem appropriate but the relevancy of this fact has no limits.

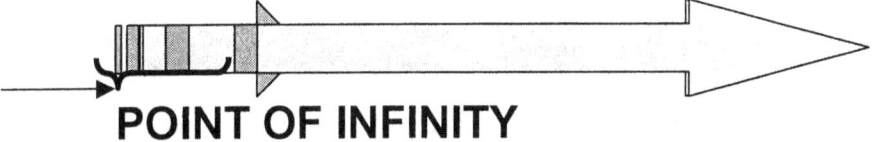

POINT OF INFINITY

If the line started at zero there was no line to start because zero multiplied by whatever results in zero as the answer. That must also be the cosmic starting point. Einstein introduced such a point and named that point singularity.

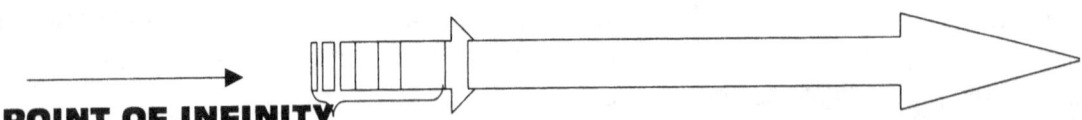

POINT OF INFINITY

The Universe does not change because there is not one single item that is in the Universe that can change. What the top evokes is what was established at Moment-Alfa when time was interrupted by space for the very first time. The spin or expanding that was introduces still present the very same principles as the spin or expanding did when it did for the very first time. The line of time is eternal and was interrupted by space with infinity bringing an end to eternity in time. However in infinity the line was interrupted extensively but briefly while the line continued intensely small but eternally long.

•⇒⇒⇒●••

Einstein introduced matter time and space and I can see where Einstein was heading with the three concepts forming one Universe but Einstein got his wires slightly crossed, because for one, what Einstein saw as space is time and what Einstein saw as matter is time in general as it is a connecting that singularity has with the flow of time throughout the universe. In that there are two strands of time that formed with one massive time delay that compacted the heat in the time delay and that time in delay compacted in nice units, which now forms the part that are the matter Einstein referred too. In that there is overall just time relating to time in time. However the only existing Universe is the Universe, which is not part of the Universe. The rest is creation by generating motion and it is not created as such in measure of time flow so very long ago and back in the most distant part. It is created by motion of time delay through time in time. By moving from one point to another point the flow of time goes square while the points duplicate. The duplication is a product of overheating in one specific spot where that spot exaggerate the space by expanding. However in duplicating there is cooling and cooling is reducing of heat. Heat is what there is becoming more of the same thing and therefore cooling is what there is taking away some of that.

The motion brought about the square of the value but in the square initially was only singularity.

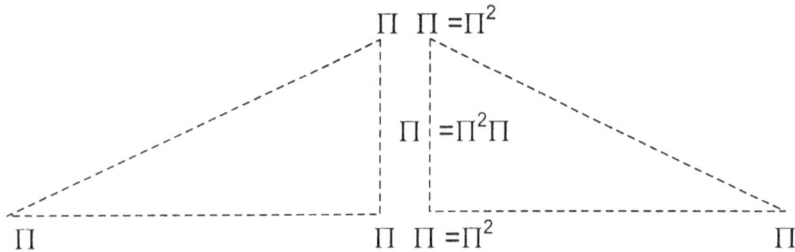

As we already determined on a previous occasion we now accept that expanding comes about from heat and only heat brings about expanding. By heating whatever one may find in the universe expand.

•⇒⇒⇒⇒⇒⇒⇒⇒⇒⇒⇒⇒⇒⇒⇒⇒⇒⇒⇒⇒⇒⇒●••⇒⇒⇒⇒⇒⇒⇒⇒⇒⇒⇒⇒⇒⇒⇒⇒⇒⇒⇒⇒●••

By expanding the space becomes more and the space in doubling cut the heat by half. It is an immaculate and genius way of controlling heat.

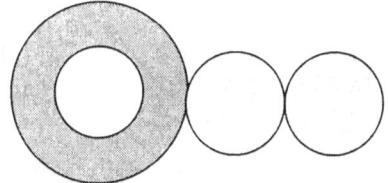

As it is still the case, the contraction results in duplication and by distributing the overheating space over a bigger area, the cooling contracts half (in the beginning but at present the proportionate) the heat back to secure the original singularity while the other position secures a new point that activated singularity. The distribution results in the contraction. When 1^1 expanded from 1^0 there was a linear motion established. The motion took what had no start away from what had no end. In that the expanding had a direction as lateral that connected to the cross of the lateral that connoted what remained of the eternal to the lateral.

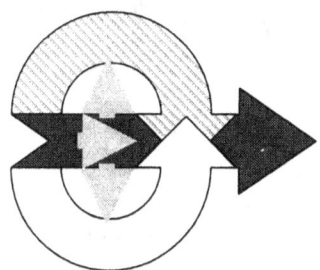

The lateral motion connected 1^0 to 1^1 but in reflection as a mathematical response there was too a connection with 1^1 the upper casing to 1^1 in the lower casing because there was an immediate contraction to the expansion.

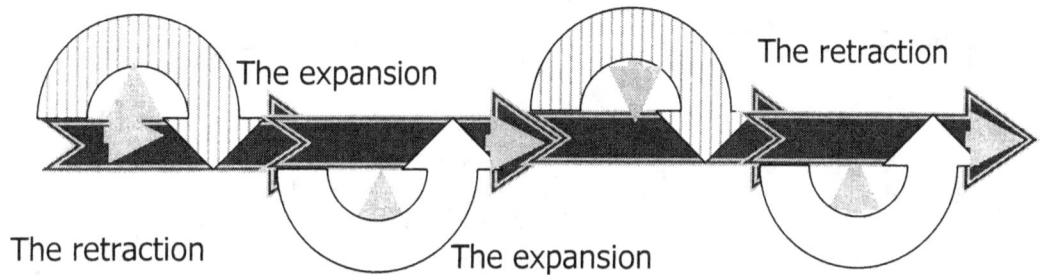

The flow of time from the past through the present onto the future had three dimensions resulting from the oblong time in the perfect eternity strayed from by going imperfect. This brought to the future where half confirmed the past, half conformed the future and believe it or not but half converted the future.

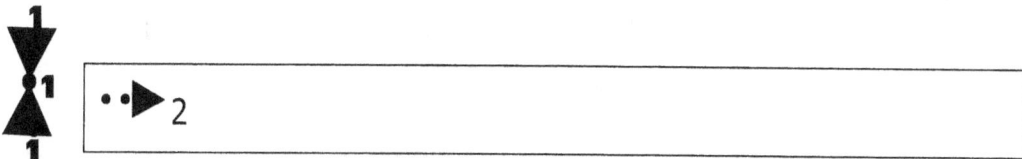

Therefore with one the Universe was in eternity and the value of the one was confining everything in singularity. By duplicating two relevant points forming three positions in singularity came into being a form in the universe. The number arriving in the Universe was two, but I am somewhat reluctant to say that what ever formed at this stage, was already part of the Universe. That is still miles off.

$T^2 = a^3 / k$ and $T^2 = k / a^3$. In this period of development the time associated with eternity much more prevalent than it did with a break on continuity in infinity.

Every time Π formed singularity brought about motion by gravity in Π^2, which still maintained a line. Gaps formed and that defined and became the atom in space.

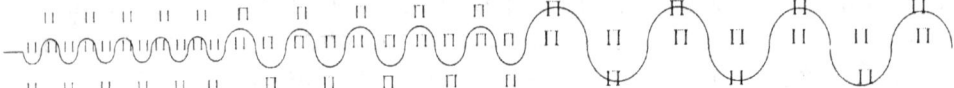

The line formed singularity by forming from the spot going onto become a dot. The heat surging moved as heat still does today. The heat expands as it moves into more space. That which formed was Π and the motion in the surge was Π^2.

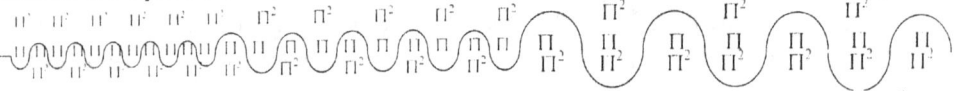

Every time Π formed singularity brought about motion by gravity in Π^2, which still maintained a line. Gaps formed and that defined and became the atom in space. The heat moving became the motion that became the gravity. The moving was a relation between $((7/10 + 7/10) \times 10) / 10 / 7$ and the moving formed Π^2.
$((7/10 + 7/10)$. Material forming as space becomes available provided by heat surging. It was material moving in time in space that was becoming time.
(10) The three in time-space or the space in which material has room to move.
$10 / 7$ heat cooling by surging back as the contraction retarded the responding growth.
In this was a line, which I now am unable to sketch since sketching firmly relies on a three-dimensional drawing, and at the time what was three-dimensional is very flat today.

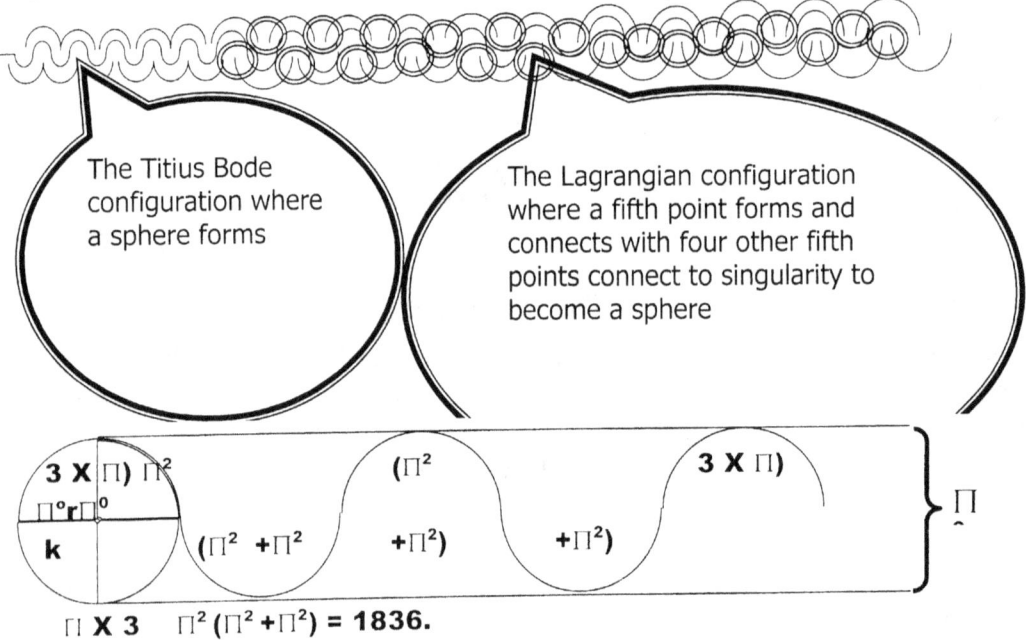

The Titius Bode configuration where a sphere forms

The Lagrangian configuration where a fifth point forms and connects with four other fifth points connect to singularity to become a sphere

$\Pi \times 3$ $\Pi^2 (\Pi^2 + \Pi^2) = 1836.$

Finally with all the heat retarding an interrupting of the line a gap formed in time.
▶ Π^0

All the while time is just a spotted and dotted line running along time as space duplicated with heat surging and cooling as cold contracted much similar to the actions of stars in the process of pulsating known by what ever name one wish to use. The star takes time back so slow we can see the pulsation of gravity cycles.

What happened back then is precisely what we are able to gauge from the behaviour we find the top shows because the Universe doesn't ever change.
This brings us back to the spinning top I presented at the beginning.

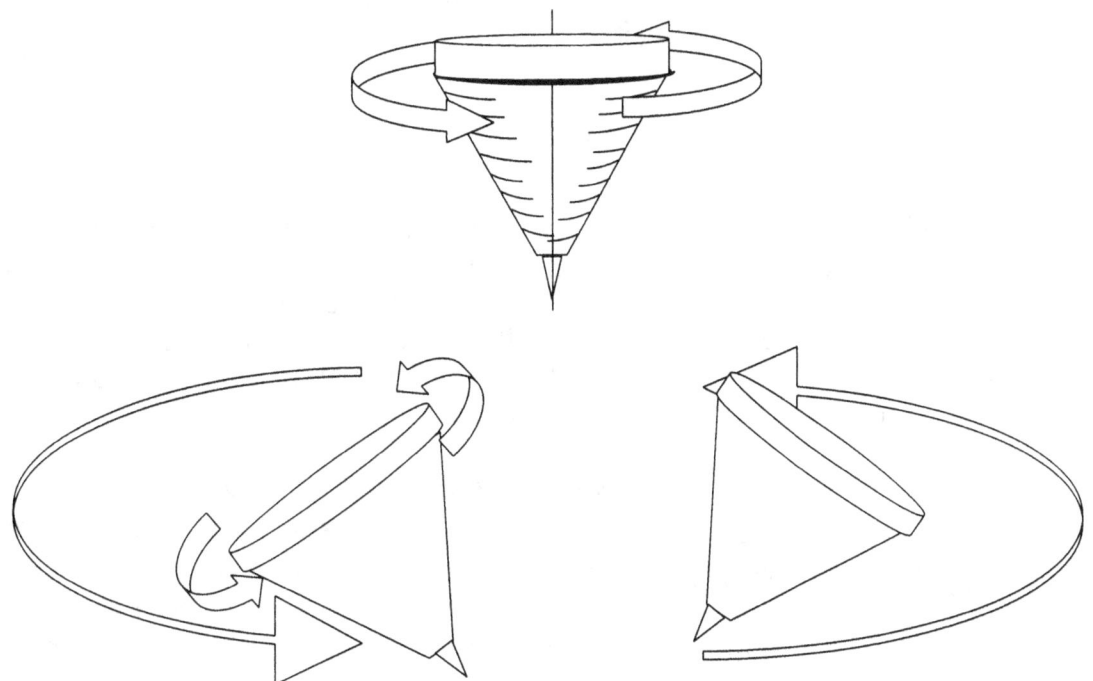

I have asked as many persons as I do not care to remember why the top sinning will remain spinning around one point while turning. The answer I receive from the most educated to the schoolboy is always about momentum. That is a very simple answer and to say the least a little too simplistic by further analysis. Why would the spinning top go off centre when spinning higher than a specific velocity and lowering the velocity it would stabilize and run square to the Earth only after that it will go oblong and then fall.

I could go on about different positions bringing across different momentum of thrust but I do not wish to insult your intelligence because I am aware that you are familiar with all the law. When the top is spinning it is spinning about its own axis and when it is not spinning it still remains spinning about the Earth's axis, therefore when it is spinning it is also spinning about the Earth's axis.

Therefore the limitations applying can only result as an influence coming from the Earth's axis. The second question now comes screaming across and that is in what manner could the Earths axis ever affect a spinning top since the spin and the spinning top is a gross mismatch to what ever standard the Earth may introduce. It is clear that spinning objects do influence each other in contrast to Newtonian opinion.

Every round object has a point establishing a very centre, a middle dividing one side from the other. That division determines the space from one side away from the other side. At one point there must be a point that does not fall on either side of the divide. Such a point will still be a circle, because from that side the circle divides into two sectors.

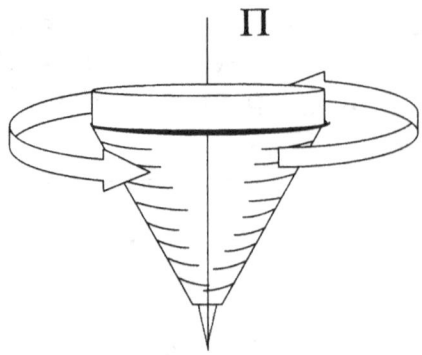

In every spinning object there is a point of infinity, a point that does not turn because it holds the dividing spin. From that point running in all directions the spin is opposing the other side. All spinning activity starts at that point diverting outwards and from that point the spin is either clockwise or anti clockwise in all directions. As I pointed out no line can start at zero because then there is no line and no rotating point can start at zero because then there is no rotation.

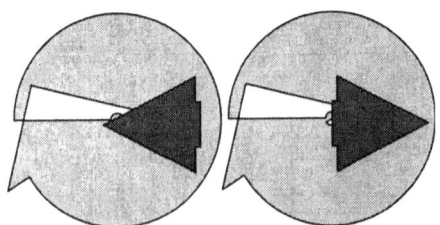

I have indicated that the motion produces the space and the space finds limits in the motion confirming the space while the space is conforming the motion.

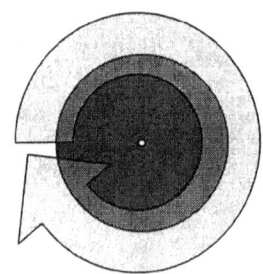

When 1^0 parted from 1^1 the motion that came about was 1^2. That in reality left little consolation because with $k = 1^1$ that left the space formed by the motion way outside the realms of the emerging Universe.

By reducing the one line the other line can never reach zero because then there was no such a line to begin with. That makes a straight line also inevitably always a potential square and that makes the straight line half the value of the square being 180^0. At a later point I shall continue with this argument, but for the mean while I wish to come back to the circle. This same principal applies to the cube and that means everything there is and ever will be is either a square being part of a cube or a circle. With the straight line forming half the value of a square $360^0 / 2 = 180^0$ in as much as being one line and reserving one line in infinity to eternity. The straight line is just half the value of a square. In that manner the triangle is also half a square and therefore holds the same dimensional value as the straight line being also 180^0

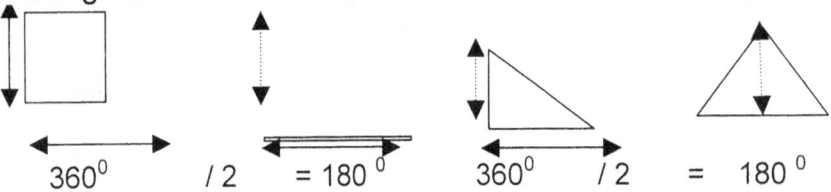

The circle is a square holding a round shape, as the straight line is a square holding one side to infinity. Calculating a circle involves two aspects where the one is either the radius or the diameter that is double the radius. The other is the factor Π

$\Pi \times D^2 / 4$ = circle and $\Pi \times r^2$ = circle

The point of singularity cannot be in space at large because space is not there and secondly what ever is there spin to slowly to have a connection with singularity directly.

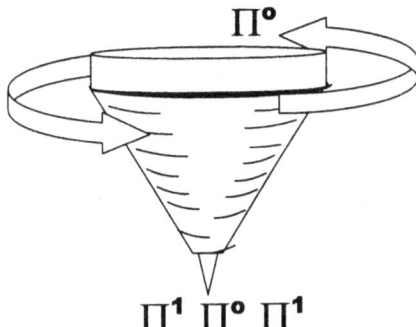

$$\Pi^0$$
$$\Pi^1 \; \Pi^0 \; \Pi^1$$

With everything in a cube or a circle or a potential of the two, brings about the implication of eternity in a form of singularity or the point of creation. Removing the radius of a circle does not remove the circle, because the circle is there, securing the ring. If the line (or imaginary line if you wish) holding the value of Π^0 = 1 there has to be a point where the circle is no longer in infinity but claims existing outside the imaginary. At that point the radius may be slightly more than infinity, but to all calculating purposes it still remains as infinity. The spin was going on for eternity because the spin does not apply, it has a value of zero and zero is another expression for eternity.

Having edges where Π^0 duplicate to present the edges singularity lost the value of Π^0 to the value of Π^1 with the same value singularity was being Π^1 to the one side and Π^1 to the other side, Π^0 must be the point splitting singularity into two parts of eternity, the eternal value of the first dimension outside eternity. It was the square of Π^1 being $\Pi^{1 + 1}$. That was the first dimension outside singularity Π^0 where singularity has a value of Π^1 in the form of $\Pi^{1 + 1 = 2}$. The first claim to space had a value of Π^2. This applied to both sides of the claim to space outside singularity, and the double proton became the dominant factor on matter.

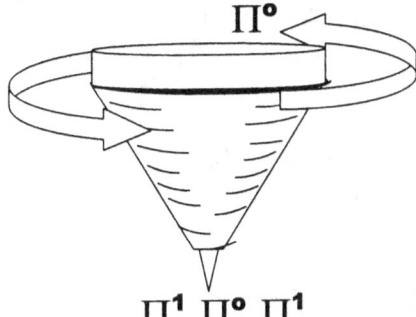

$$\Pi^0$$
$$\Pi^1 \; \Pi^0 \; \Pi^1$$

Right at the start before space and time became developed the motion produced space in the principle of the Coanda effect. By receiving space, singularity received a value outside eternity as Π^0 received edges. Granted the fact that the edges were so small there still was no r to present a circle. The manner that the top use to evoke singularity, which enables the top to maintain independent motion could be, traced right back to the very first line that came about as Singularity initiated spin

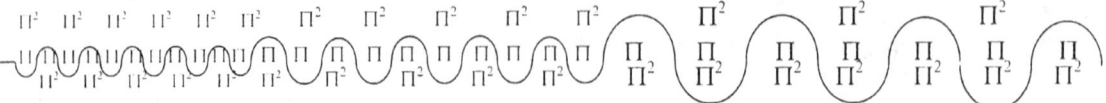

In the beginning there was no space in which to move so therefore the only way to move straight was to move in a circle. The movement k producing the line was the same as the motion T^2 that produced the circle and the space a^3 achieved was the compliment that two factors combined and that formed the space developed.

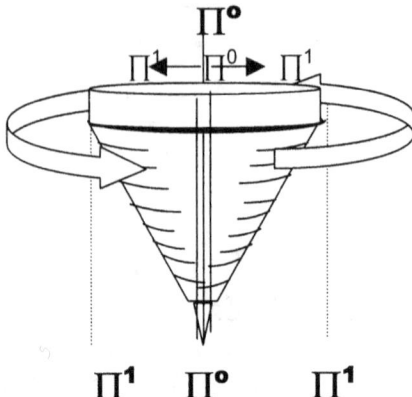

Taken from the point of rotation the two sides are in opposition to each other in every aspect that they may contain and with all that they hold.

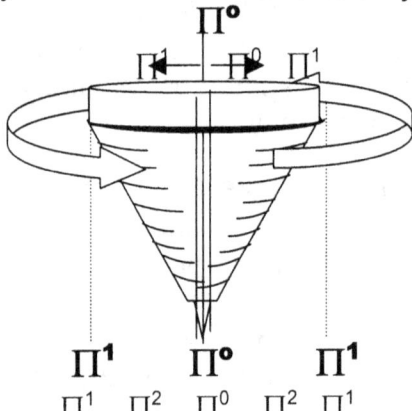

With Π^0 little more than a figment of the imagination there is actually to values of Π^1 facing each other in a relation combining Π^1 to hold the value of $\Pi^{1+1=2} = \Pi^2$ and with two sides being the very same but opposing each other there will therefore also be Π^2 to every side that holds Π^1.

At last I can come to the one part that I disagree with Newtonians, and what I regard as Newton's second biggest infamous or famous blunder. Science, made one enormous blunder, from this stance. They took the radius of a wheel not to have any influence on the wheel. In doing that, they removed the very fact that keeps the universal attachment together.

Singularity controls the Universe by establishing a Universe but that is done in a specific manner.

Also it is true that the entire form that is the sphere is controlled from a centre within the sphere. That centre holds the sphere in form and shape. Therefore the strong form is dictated from that space fewer centres where there is no space and no form left. The natural inclining is in the form of the sphere. It is part of the roundness that the overall shape of the sphere represents and this structural strength is carrying down to the very centre. Because the circle is forever reducing that reducing which is inherently part of the form of the sphere becomes a tool in distorting of space in the sphere and is eventually removing all forms of space from within the centre of the sphere. The very centre ends up as having no space because of the reducing that continuous down to become the space less inner centre. The all roundness is the ingredient that forms the backbone of the absolute strength that the sphere has and that is the component that the sphere is so famous for. The form the sphere has allows the sphere to have a control that is coming from the centre deep inside the sphere where the space vanishes and being without space seems to keep the entire structure rigged. From the centre the sphere shape shows strength that the shape as tough as it is. How does it work in its most basic analyses?

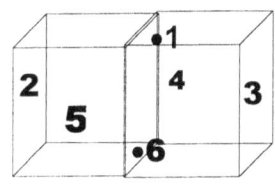

There is one more point in the sphere in the centre forming an addition in the sphere. That point holds gravity secure.

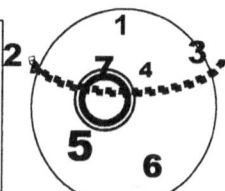

The cube has sides and the sides form a rather weak and flat surface that connects four corners. The flat surface produces a rather indifferent contact point with no special features on the surface. The corners connect to other sets of corners and those corners form a weak structure without any direct support coming from the other five sides. Without material to fill the body of the cube the cube has no direct connecting between any of the sides other than corners connecting at the edges of the sides.

Taking the vantage from the point the sphere is holding from the centre out into space there are ten points connecting to the centre. In that are the dimensions of singularity connecting to space where five connects to space in the second dimension of singularity, and five connects in the third dimension of singularity. On the other hand, the cube does show a very different characteristic, which involves only six sides (at least) connected.

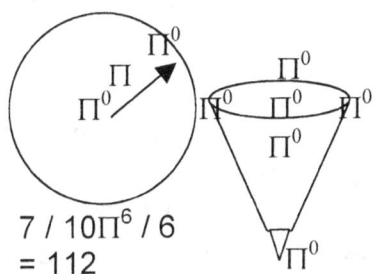

$$7 / 10\Pi^6 / 6 = 112$$

The spinning of Π^0 around the centre Π^0 establishes Π and Π is what produces the form gravity has. Still it is the relation or relevancy there is between the centre Π^0 and the spinning Π^0 that gives status to the form that Π represents. In out Universe we are accustomed to and are familiar to the rules we want to place seven points holding singularity to the centre holding singularity in a relation of $7/10 \, \Pi^6 / 6 = 112$. In that Universe everything less that a duplication ability to the value of 112 protons fit but only atoms to a maximum of 112 protons fit.

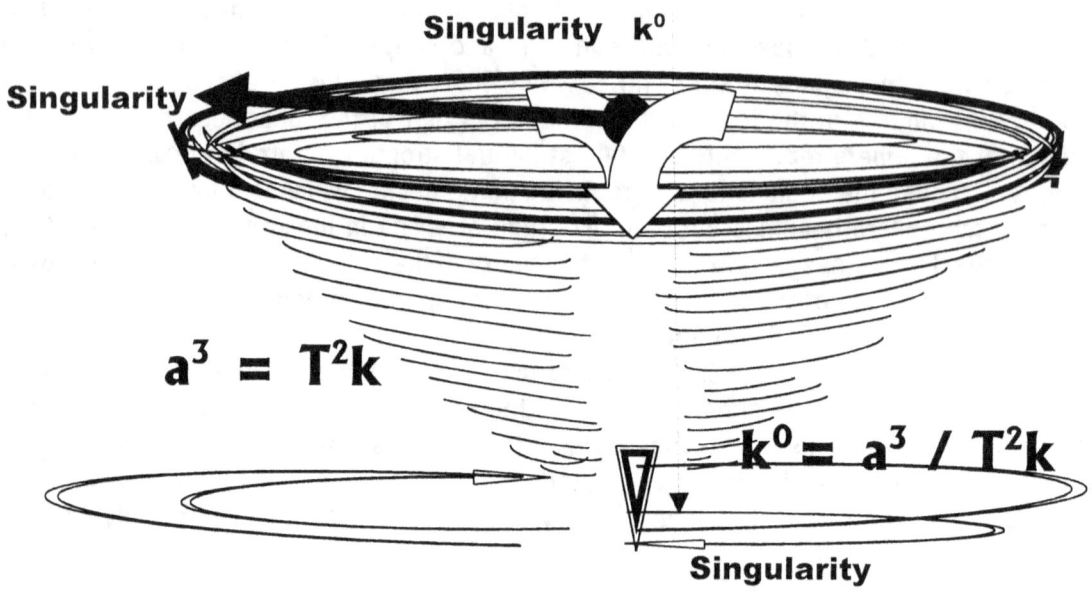

Singularity k⁰

Singularity

$$a^3 = T^2k$$

$$k^0 = a^3 / T^2k$$

Singularity

A liquid in relation to a marked solid that is without relative motion always does the motion

Material produces the dismissing or the concentration of space by applying the motion. Surrounding all elements is a layer we call the atmosphere and even Pluto and the moon must have the atmosphere because they have gravity. In this, the relevancy of ten to seven forms this layer and it results in forming a circle, because the combining of the motion duplicates the singularity factor Π, forming from that gravity as Π^2

THE PROCESS PARTED USING THE ROCHE PRINCIPLE

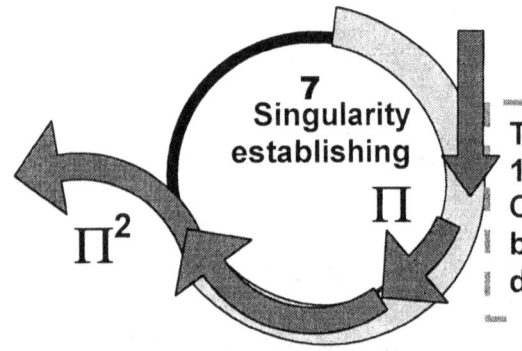

7 Singularity establishing

Π

Π^2

The liquid applying motion forms the 10 disciplines. No motion leaves no Coanda as well as no gravity because gravity is motion that duplicates singularity.

Because of the Coanda principle the motion that light provides will bring a bigger gravity affect than what normally apply to space-time. The higher the motion is the more does the centre gravity contribute to the repelling of space –time by creating space in the confining of a^3. Gravity cannot bend light because light at best is a fluid. The Coanda affect places through the extreme motion that light provides an even bigger restraining of the flow of space-time on the light passing by. The relation forming the duplication of singularity is a duplication but applies as a dimensional forming of Π and placing 7 in relation to 10 forming Π^2

Singularity is a mathematical reality. Einstein may be the first to name it and Galileo (unwittingly) may have been the first to define it as Kepler was the first to formulate singularity, but in mathematical terms singularity is the most basic principle.

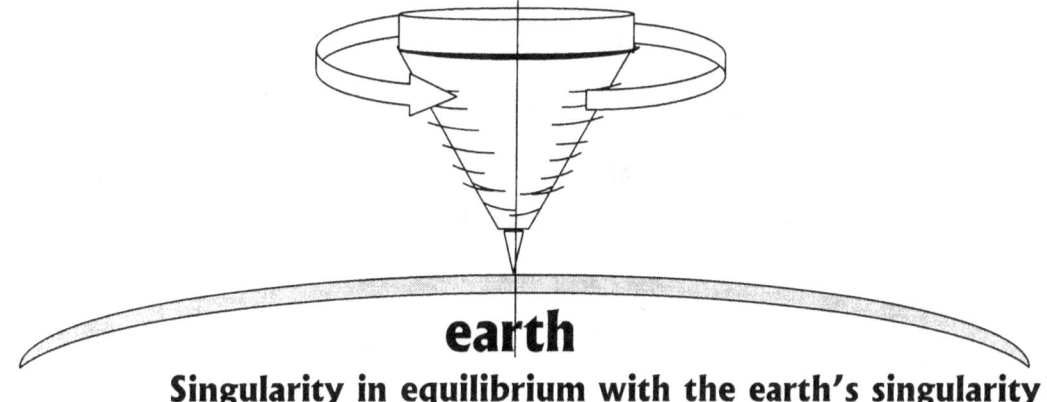

earth

Singularity in equilibrium with the earth's singularity

Singularity applies gravity by charging motion and the motion (not the pulling) of space-time is gravity. It is motion that moves the gravity that moves the Earth and it is also motion the moves the top that forms gravity. Gravity is not the pulling of but the motion or tendency to move to the centre of the Universe, which is the next domineering, point holding singularity in a control or a governing mode. The motion or tendency to move is that which forms gravity and mass is the occupying of space and therefore restricting the motion of gravity.

A child's toy holds the mystery to what the greatest minds in the world misses. It shows how singularity is charged into "life" from "nothing" by the dynamics of motion.

By rotational motion the top creates a line confirming singularity running down the line and by generating the line the line charges gravity. The gravity is what drives the top as the top and as long as the top spins. There is an influence generated by the spin of the top that keeps the top upright while the top is spinning.

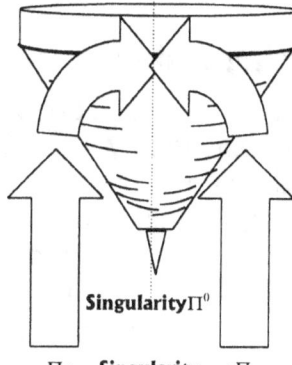

SingularityΠ^0

$\Pi \Leftarrow$ **Singularity** $\Rightarrow \Pi$

The line is generated but the line is far from magic. The line is where the centre of the Universe is which the Universe is then that what the top fill by particles from the line to the edge of the sphere. The particles in motion generate motion by electing a centre from the centre of every particle in the spinning top. Such an elected centre becomes the centre of the Universe as far as the top relates to a Universe because all the atoms in motion elect the centre of the Universe.

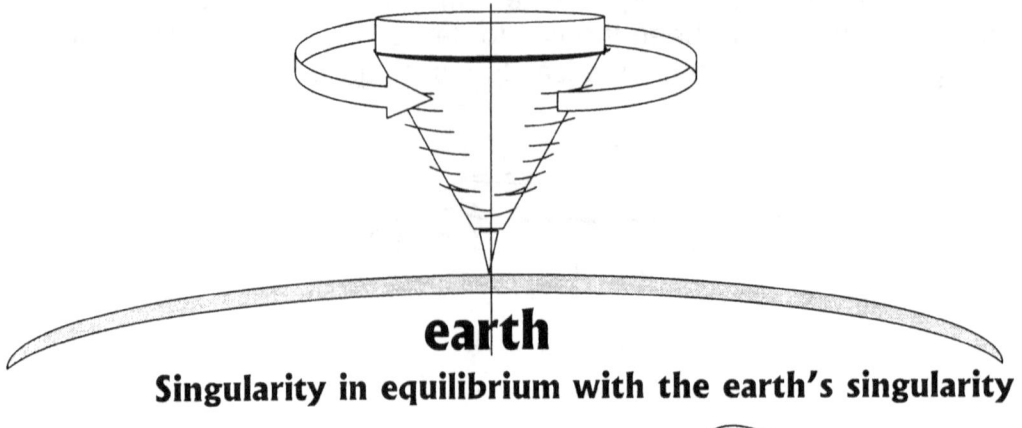

earth

Singularity in equilibrium with the earth's singularity

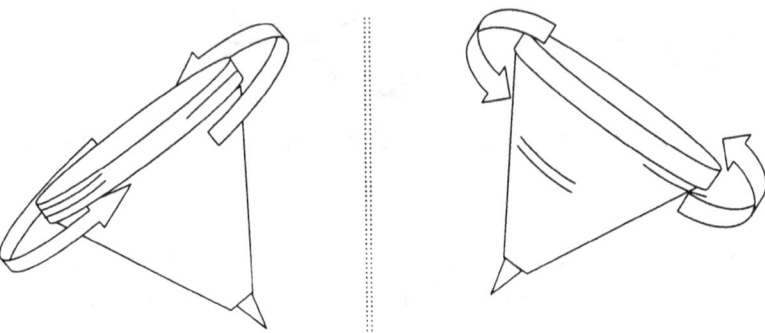

Singularity of the top exceeding the earth's singularity

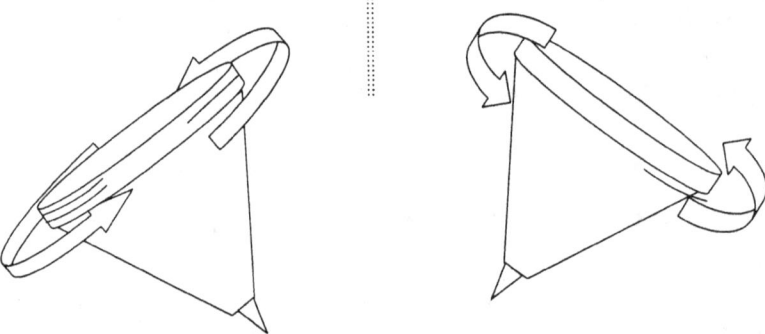

The earth's singularity dominating and exceeding the singularity top

The centre may or may not spin and the fact that it does or does not spin is all the same because that centre part never spins in any case. Therefore the boundaries set by the spinning motion does not depend on the spinning motion of the object but has to stand related to another bogy bringing about a larger spin influence. Granted the fact that the influence the Earth has on the top may be that of gravity but if that is the case then surely the Sun has also influence on the Earth and other rotating objects through gravity. It needs more investigation because it may bring about evidence we are not aware of.

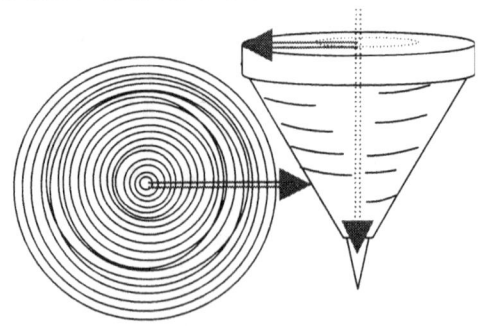

This observation places a much bigger question mark on the statement of Newton where he proclaims no influence on two rotating cosmic structures.

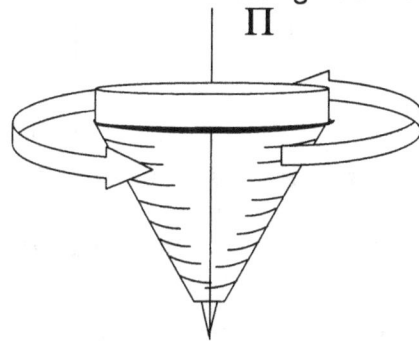

By rotation such rotation is a duplication of what singularity retracts. The rotation involves the three factors in time coming from the past through the present and onto the future, which holds three positions excluding the one position allocated to singularity. The four I named the eternal motion because singularity being without motion was contracting as well as expanding at the same time it was not moving. That placed singularity in eternal motion without ever moving being singularity.

Every quarter provide a distinct value that indicates the progress of the flow of time from the one point Π to the next point Π.

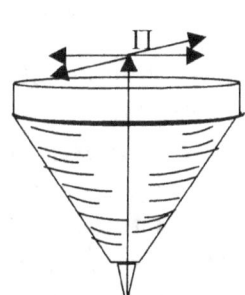

Any changers occurring in Π will lead to a an unequal triangle providing two different values to r and will alternate the link between r and Π² bringing about different form (Π) and time (Π²). When singularity forming the lines of the triangle is not in equilibrium the triangle will destroy the matching of half circle.

The sectors provide individual singularity a means in sustaining governing singularity by which provision comes through maintaining governing singularity the required spin in maintaining cooling. If this process did not apply, there would be no connecting individual singularity to major singularity

In every sector the directional flow will provide a distinct meeting of Π linking r to Π² and this allow the time component in the rotation.

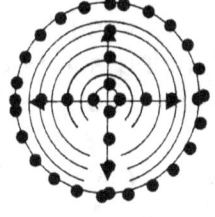

The issue is that the rotation activated the time component where four points set about by motion. This formed the basis of the Coanda principle as the rotating top then became liquid although the wood is a solid. The point on which the rotation turns becomes the solid. This is evident in the fact that the center is activated and not activating the spin. As the spin reduces its dynamics the turning still fight for balance

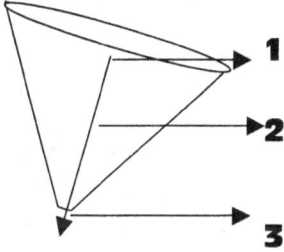

The time activates a line from which space begins. The line forms a space that turns the top, but just as important is the fact that the line extending singularity activates a dimension in time in which the top can spin.

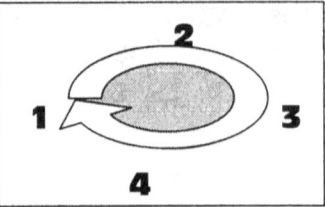

It is the time that relates to the spin of the space filled with material that allows the position the top has when spinning. When the co ordinance between the space spinning and the timeline keeping the balance falters that the top looses independence and fall.

The sectors provide individual singularity a means in sustaining governing singularity by which provision comes through maintaining governing singularity the required spin in maintaining cooling. If this process did not apply, there would be no connecting individual singularity to major singularity

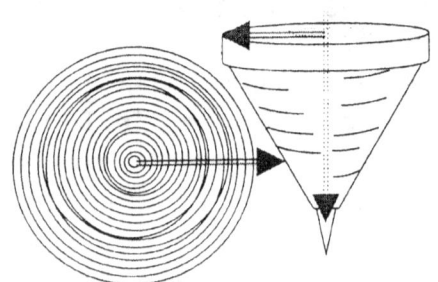

In the **precise middle** of all **objects in rotation** is a precise centre dividing the object in sectors that will **start the spinning initiation** from that centre point. As the rotating direction moves inwards, the rings will become smaller and smaller. Thus, the spinning object **will have a middle point**, a very specific **centre point that does not spin** and only holds Π as a specific value because no radius can apply. But also the one value such a line **cannot have is zero** because the line **is there and holds contact** to the rest of the material bringing about that **zero does not start any** line and therefore the **value of the line must be infinite,** just as described in **accordance** and by **the definition of singularity**. There must come a point where the ring is infinitely small, where it can reduce no more, where it reached its ultra limit, but at that point it cannot be zero, because the point is there

That point albeit hypothetical, is also as much a reality none the less and is placed where that point must be standing still because every line running from that point in opposing directions are also in opposing directional spin the other or opposing side. Move the rotating line progressively to the middle by reducing the length the line have from the edge to the middle. At one point all further reducing ends. In considering the spinning motion in the fraction of time in the detailed instant every aspect of rotation will turn in every instant of change in time. Although the points had the same characteristics only one instant before, they oppose the characteristics it had just before and just after the very instant in which they are and to which they relate by similar points also in rotation. The fact of the graph proves my point in quarterly opposing dimensions and values.

This only applies in relation to time because time is the square or then if you wish time is the flat to space being the cube. Time in the square draws space in the cube flat and that is the why the Universe holds the sphere in place.

Understanding all the following is connected intimately and all conditionally to the fact of accepting that all individual particles in the Universe use motion and therefore spin.

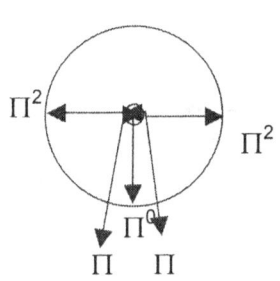

In dimensional terms, which explain later on the value of 2Π relates to Π². Then that relation extends to the next value where Π² relates to Π , which relates to Π⁰. The first space in the circle will then be Π²Π. From the centre being in infinity one can realise by applying mental power the single dimension factor not seen but present all the same. Extending that into the 3D comes six Π and any one of the six will further extend to form a seventh point as Π²

From this line of reasoning I dismissed the theory of the presence of a force being gravity but rather consider it as a dimensional changing contributed by the spin of the Earth and the spin comes from singularity located in the centre of the Earth. It is all about dimensional changing that influences space as a factor of ten to reduce to $Π^2$ on a continual basis from point forming new dimensions through billions of such points.

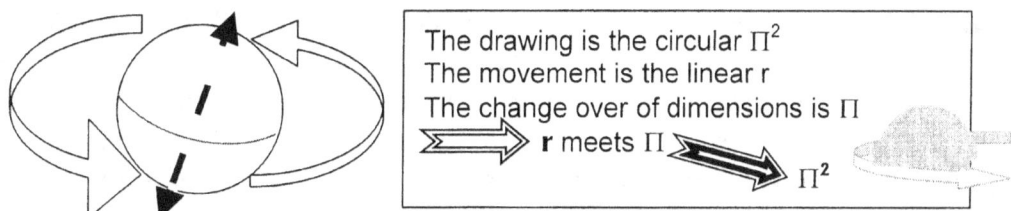

The drawing is the circular $Π^2$
The movement is the linear r
The change over of dimensions is Π
r meets Π
$Π^2$

The result of the five to one relation comes the Titius Bode principle.

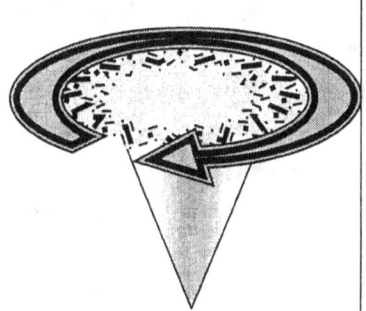

The spin under normal conditions can only come about as a result of more heat. With that aside the spin normally caused by heat will bring on a linear gravity running towards the centre of the top. This is then a product of $k = a^3/T^2$. But to counter this (Newton's law on action and reaction), another balance comes about $k^{-1} = T^2/a^3$ that centres the material in line with the progressive spin and the extending of the motion that should be because of a liquid heat adding to the material.

More spin increase both lines that force gravity by the increase of T^2 extending k, k^{-1} as well as a^3. The space wants to exceed its boundary because suddenly the motion allows the space to become extended. The gravity line running to the centre wants to extend for the same reasons and so does the gravity line running towards the liquid that should be there and that should be enforcing this sudden living up to better standards.

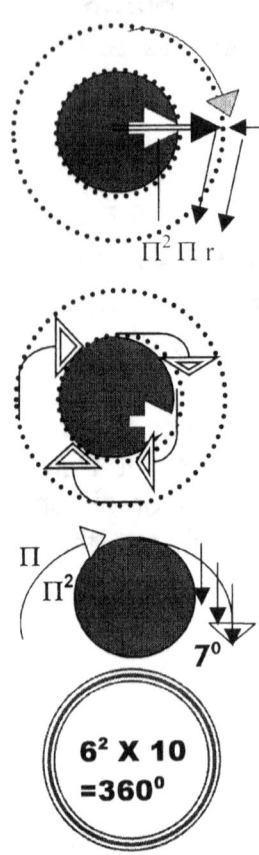

In the circle $\Pi^2\Pi$ the space surrounding the rotating object will also extend by Π as the concentration of the spinning motion draw or drag on past Π^2 extending the influence of Π^2 by the value of Π. This extending of Π^2 to accommodate Π we refer to as the atmosphere, but physics apply to this extending in the normal fashion. From the spinning motion Π does not stop at the end of the solid structure but the influence of Π extends and this then becomes the atmosphere. The influence of Π^2 stops at the end of the solid structure but the influence of Π extending plays a most dominant role in the cosmos, although not yet recognised and that factor is most crucial to a better understanding of the implications of laws governing the cosmos.

With the circle being $\Pi^2\Pi$ the Π^2 will reflect the circle in the square with Π forming the extending of Π^2. This is an extending of the six Π forming in alliance with the centre Π. This produces that any extension of 6 forming material one further extending goes into space and relates to a seventh dimension. The extending of Π will not end immediately but will carry to the surrounding space the circle influence through rotation. The influence immediately above the circle will have the biggest influence and reduce gradually as the value of Π reduces in the leverage that the space has on Π and a gradual but definite change from Π to r will affect the extending of Π progressively more. The decline of Π will follow the same contour of the circle at 7^0. Every one of the dimensions indicates an individual significance as I shall show later and the increase into space runs by 7^0.

In the circle using $r^2\Pi$ the r has to have distinctive qualities placing it as a factor apart from Π. Where the growth shows no separate distinction but a continuous flow from the precise centre to the precise edge the flow would become in relation with Π depicting the circle and Π replacing r as reference to any point on the circle. By using r as a distinction in the circle division is possible but by using Π there is no distinction possible making it a solid flow. Any object being in outer space floats and such floating is seemingly random with no specific detectable interfering favouring a movement in a particular direction. Such a devise is depending on influences not in our scope of detection. But then the object comes closer to the Earth and reaches one specific point where the six dimensions that influences the object suddenly changes. At one point, one of the six dimensions falls away as it disappears and the object quite latterly falls to the Earth. The support of one side disappeared and the centre point of the sphere took over the control. At that point the object is under the influence of one centre point in the sphere and we all also know that in such a centre point one will always find the strangest or the controlling gravity.

Space-time is a four dimensional position of the Universe where the position of an object is specified by three coordinates in space and one position in time. This evidence we find as matter grows into the dimension we now share with billions of stars in the cosmos.

With the dimensional change from space in the cube to space in the sphere a relation of 5 to 7 comes about depicting gravity on one side of the divided Universe. The principle of 5 sides in space relating to 7 in the sphere holding matter forms the basis of the Titius Bode and the Lagrangian principles.

The TITIUS BODE Principle Outside the sphere

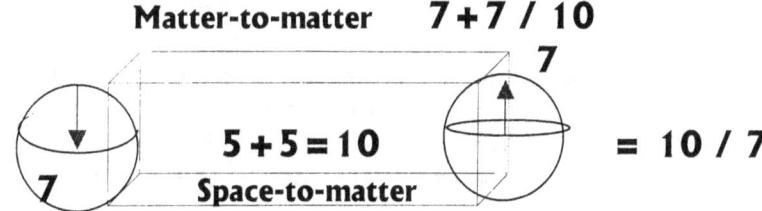

$$7 / 10 \quad + \quad 7 / 10$$

$$5 \quad \cdot \quad 5 \quad = 10$$

Matter-to-matter $7 + 7 / 10$

$$5 + 5 = 10 \qquad = 10 / 7$$

Space-to-matter

The Titius Bode law is an extending dynamic deriving from the law of the gravity dimensional factor where the space factor in a square of ten relates to a matter factor in the square by half (half since nothing can be in two places in the Universe simultaneously) of the matter factor of π^{7+7} or the square of space (10) relates to the matter factor of 7. From such a point every other point will be opposing any other point not pointing in the direction to which the first point is pointing, whereby it extends the direction it holds. No matter what the point is or where the point leads, such a point holding a specific direction will be unique in the direction it is rotating because at that or any other specific point wherever, it will be directing not in the direction it spins but in the direction flowing from the centre point outwards.

When the foundations were laid in place with singularity expanding even before it was growing The Roche limit became one condition. But while that was taking place another principle came about which is as secured in the foundation of the third dimension as the sides supporting the third dimension. Sides came about through the dimensions that are framing the dimensions, as we know them.

There was the dot. The dot had no borders therefore there was no separation and still we know there were more than one in a group of one. The evidence of this is very present in the cosmos at present and one can find such evidence all around us. The dimensions personify the Titius Bode principle and understanding the relevancy between the dimensions will also mean the understanding of the interlinking values of the Titius Bode law.

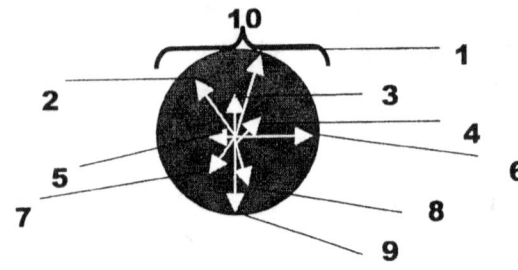

With the cosmos still in a single dimension there were no limits as we know limits to form in the Universe we use and no borders indicating limits because after all it is the single dimension where there is only one dimension holding so much diversity. The borders were part of development because we can witness the legacy of such borders in the present day holding the 3D in place.

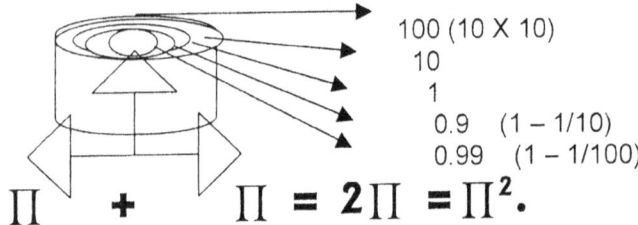

$$\Pi \ + \ \Pi = 2\Pi = \Pi^2.$$

In the manner that the Roche limit comes about from singularity one can trace the same development by studying the dimensions, as they had to develop. In this singularity there can be no sides and without sides there can be no drawing showing the explanation by means of illustrations. Where I do just that, I ask for your forgiveness because being human means I have no capable means of performing an explanation. Yet I am forced to do just that and I have to allow the means of sketches, well knowing the implication that such act is not allowed.

Bode's Law:
A numerical sequence announced by J.E. Bode in 1772, which matches the distances from the Sun of the six planets then known. It is also known as the Titus-Bode law, as it was first pointed out by the German mathematician Johann Daniel Titius (1729-96) in 1766. It is formed from the sequence 0,3,6,12,24,48,96, and 192 by adding 4 to each number. The planets were seen to fit this sequence quite well – as did Uranus, discovered in 1781. However, Neptune and Pluto do not conform to the 'law'. Bode's Law stimulated the search for a planet orbiting between Mars and Jupiter that led to the discovery of the first asteroids. It is often said that the law has no theoretical basis, but it does show how orbital resonance can lead to commensurability. The importance that becomes known is the sequence the Titius Bode law saw in the number arrangement of 3; 6; 12; 24; 48; 96 etc. The incorrect application of the Titus Bode law lies in subtracting the figure of 3 from 10 leaving 7. The other way of reasoning is to add four each time to the first value of three starting with 3 and so on. The true significance of the Titus-Bode law is that it points directly to a circular growth of 7 stages. The 7 relating to 10 is a precise derogative of the Roche limit or the Roche limit is a precise derogative of the Titius Bode principle because the two systems interlink.

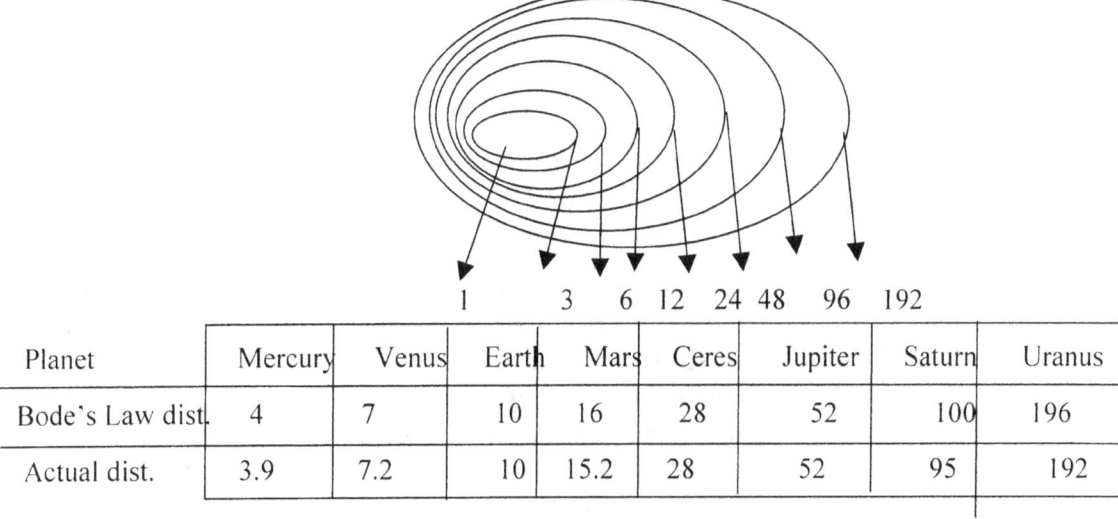

Planet	Mercury	Venus	Earth	Mars	Ceres	Jupiter	Saturn	Uranus
Bode's Law dist.	4	7	10	16	28	52	100	196
Actual dist.	3.9	7.2	10	15.2	28	52	95	192

Gravity produces mass but mass is only the result of gravity. Mass do not produce gravity and the manner in which science uses mass can only apply when using the calculations in terms of the Earth. However applying it to stars as science indicates by their formulae used in their calculating of gravity on structure beyond the solar system is very inaccurate. Heat stored in motion produces gravity. Any one not in agreement convinces you by comparing the neutron star with the massive red giant. To calculate a Black hole they go and throw C^2 next to the dividing radius and throw the square onto the C that presents the speed of light. Then they sit back and feel smart in the way they manage to cheat once more to prove their incorrect views correct because after all who will ever fly down a Black hole and return to support or deny their calculations. The Gravity of the Black hole is a speed because the entirety of gravity is speed or better said it is motion. Then the speed that light has is gravity. The gravity of the light can be gravity as much as it at that very same time can be antigravity. What the hell has C^2 got to do with a Black hole because you can pop what ever nuclear device far away from a Black hole and it would be at the most and at the worst very much insignificant. The light will not even escape form the gravity of the Black hole. When this became apparent that the radius of stars reduces as the stars develop through progress. It is some time ago that someone was supposed to say: hey there is a dead rat I smell. For my saying so I am the clown in the courtyard, the one with the two dead brains cells and have no more to use as spare.

The overall picture resulted in a ring and all rings hold Π to secure the form. The only form that existed then was Π and therefore even today the borders use Π to indicate positions. But in the single dimension such definitions were far from clear and the only distinctions came from securing singularity in preserving the position of singularity to apply gravity and thereby absorb all anti -gravity. But anti gravity could not control expansion by counter acting contraction through gravity so the overheating continued forming non-existing borders in some thing infinitely solid just as Einstein predicted because this took place before light came about and therefore before the speed of light became part of the cosmos. The cosmos formed a partnership with one side overheating forming antigravity by expanding into space through the applying of the overheating and the other side formed gravity or contracting of space

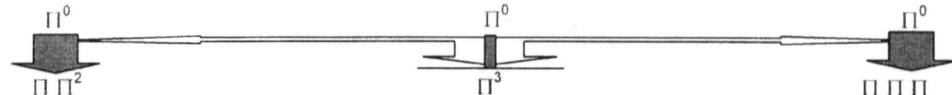

Singularity split the universe into two parts that under no circumstances can ever meet. The one side of the universe perform a balancing act to the other side of the universe that duplicate but never double. The dot started overheating while the dot remained cool by activating gravity

ΠΠΠ **Space**

Π   ΠΠΠ **Time**

ΠΠΠ **Matter**

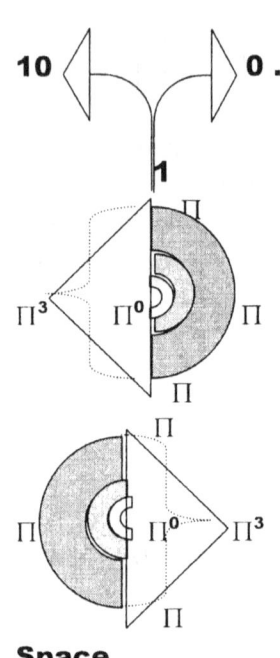

Space

ΠΠΠ   **Matter**

Π   ΠΠΠ

ΠΠΠ

Time

With the first dimension came matter, but also came space and came time splitting the universe in segments of matter relating to space filled with matter and time influencing the spinning matter.

Taking the queue from the numbers line that runs in opposing directions Π^0 going larger as well as smaller. But the centre takes a value of one. It is a private choice preferring $k^0 = 1$ or $\Pi^0 = 1$ but that splits the universe into two part, being smaller and being larger.

It is apparent that one cannot substitute the correct formula used to measure the area of a circle by using $a^3 = \Pi r^2$ because if k is the diameter then the formula must be $k^2 \Pi$. But k cannot be Π because in Kepler's formula k takes the value of the radius. In that case what will the value be of T^2? That places the formula outside the normal use of mathematics practised in the normal sense of $a^3 = \Pi r^2$.

By using the Kepler Formula $a^3 = T^2 k$ it is good to change the values to Π and see what pans out. If $k = 1$, k at the same time would be k^0. By replacing $a^3 = \Pi^3$ then on the other side of the universe $k = \Pi$ and $T^2 = \Pi^2$. But to secure this k in the centre must be 1 leaving $a^3 = 1(1 X 1 X 1) = 1$ and $T^2 = 1(1 X 1 = 1)$ That complies with Einstein's definition of space-time being: Space-time is a four dimensional position of the universe where the position of an object is specified by three coordinates in space and one position in time.

If **k** is the middle being $k^0 = 1$ then $a^3 = k^0 = 1 = T^2$ When time is in a shift freezing then $a^3 / T^2 = k^0 = 1$ In order not to overstep my limits by changing valid formulas I changed Kepler's formula to $R^3 = T^2 = 1$. But the book being written in Afrikaans the R stands for Ruimte meaning space and T is time. From that I deducted that the space used in a specific location will equal the time meaning the density of the heat in space. That brings the proof that space equals heat and space is the same as heat. Heat deforming or exploding is the equal to the space created. Also it confirms the substitute between Kepler and Π is correct

In the way space and the sphere connects the sphere will have 7Π points holding a relation to 3Π points not within the sphere forming the 10Π that creation started with. This will mean there is a division forever, and such a division may run smaller everlasting. With fluids connecting it is simple to recognise the sphere as

Π for the form will indicate Π as the form of the sphere. By gas forming the connection there are the three points of space being apart and not forming Π, but still holds a relevancy to Π^2 through the value of Π.

In single dimension seen from one aspect, with single dimension contacting aspect the sphere will still keep the seven positions because the sphere remains a solid structure though apart because of singularity. In the core of the sphere connects the proton alliances Π^2 $+\Pi^2$ with the solidity of the neutron holding Π^2 as a second forming value. From the centre to the outside is a connecting of Π, in relation

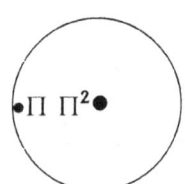

to the Π^2 that brings about the liquid or neutron form. In the centre of all the atoms space is relinquishing a position through the dissolving of heat by means of maintaining singularity. But through it all, another singularity forms in the very centre of the structure, claiming the position where space is the least available that bind the singularity of all the atoms sharing space as a unit. With that evidence I realised there are a connecting of singularity and that connection is electricity. In the cosmos all objects form a sphere. Some solids do not seem to be a sphere and space is no sphere, but the truth is hiding in the way of connecting. At the centre connects Π^2 forming the base of the solid. At any one specific given point forming the surface of the sphere is another marker holding the connecting relevancy of Π. When there is no sufficient heat to form space that will part Π from the other holding of Π, the two will combine in a solid joining connecting as 2Π that translates to Π^2.

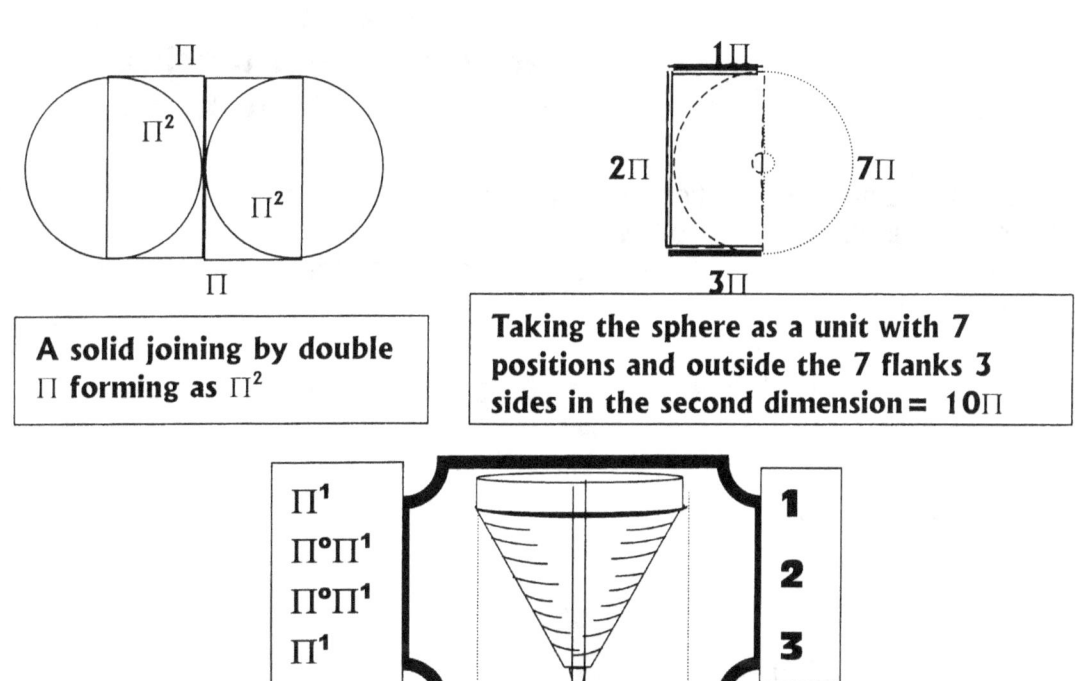

A solid joining by double Π forming as Π^2

Taking the sphere as a unit with 7 positions and outside the 7 flanks 3 sides in the second dimension = 10Π

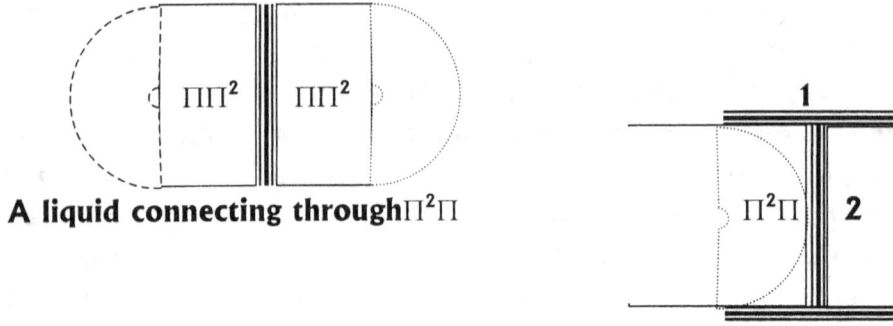

A liquid connecting through $\Pi^2\Pi$

Total connecting relevancy of the sphere forming matter connecting to space = $\Pi^2\Pi$ 3

How many dots there was is a question no person can answer because everything was un-dividable solid and yet it did group together to form every atom located in the 3D. Individual singularity and governing singularity and group singularity enhancing the gravity every time singularity find an accumulation.

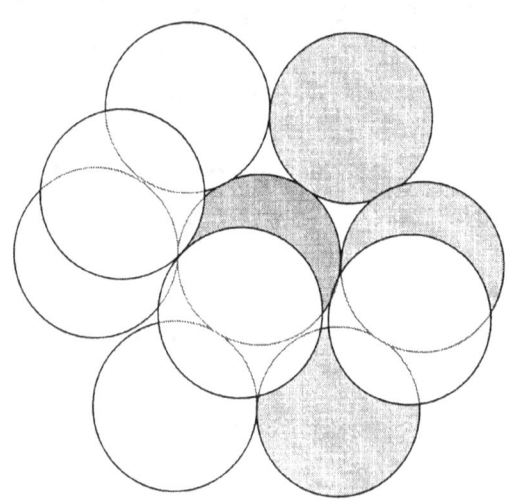

On the one side of the Universe in relevance to all the dots that came before, three dots landed forming one side while three dots formed the second side and three dots formed the third side, all relating to a centre dot which in turn related to the original centre dot from which all the dots came and developed.

The Universe came into position by deploying dots supporting other dots and some dots remained dots while other dots went on to become dots of hybrids as it was supporting dots through claiming dots of lesser density and pass that on to dots with larger density.

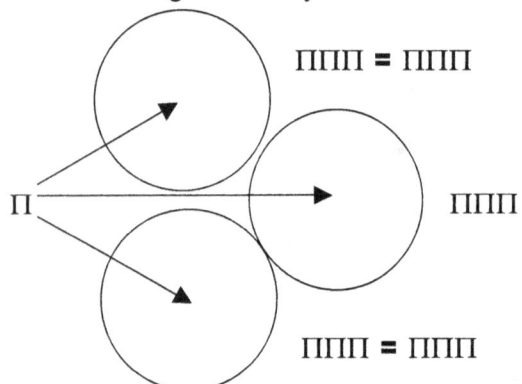

$\Pi\Pi\Pi = \Pi\Pi\Pi$

$\Pi\Pi\Pi$

$\Pi\Pi\Pi\Pi = \Pi\Pi\Pi\Pi$

Π

Matter formed where matter had to have $\Pi\Pi\Pi = \Pi\Pi\Pi\Pi$
space to occupy since it was to be in some space $\Pi\Pi\Pi\Pi = \Pi\Pi\Pi\Pi$

therefore ΠΠΠ met with ΠΠΠ to form the proton in $\Pi^2 + \Pi^2$ because the matter is within the space it holds and another Π^2 employs Π as a representative of singularity. This then placed the seven positions of singularity as the ending of matter and the three squares ($\Pi^2 + \Pi^2$ and Π^2) of singularity as the limit of material. The last ΠΠΠ became $\Pi^0 \; \Pi^0 \; \Pi^0$ and that became the space producing heat without occupying matter in order to allow heat to be restrained inside the dome singularity provide.

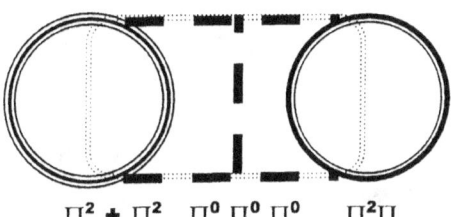

$\Pi^2 + \Pi^2 \quad \Pi^0 \; \Pi^0 \; \Pi^0 \quad \Pi^2\Pi$

From that the effect of gravity as a restraining on the exploding of space came into effect.
It is all about relevancies applying the relations gained and lost through relations. If one place $\Pi^2 + \Pi^2$ on one side then $\Pi^2\Pi$ is related form where $\Pi^2 + \Pi^2$ is in the other side of the Universe being on the other side of the relevancy. Then $\Pi^0 \; \Pi^0 \; \Pi^0$ will again relate to the other two factors forming the "outside" of the other two being the "inside".

The Universe divides into two separate issues because of singularity. Nothing can be in two places at the same time the rest has to confine to the law applied by singularity. Objects can only be in one side of the Universe holding three parts or in the other side of the Universe holding three parts. From the totality three will be a double with six sides too shows, but that forms 3D. From singularity it is flat with three sides forming on either side of singularity as the formula used to measure the sphere indicates..

Newton said a sphere is $a^3 = 4/3 \; \Pi \; r^3$

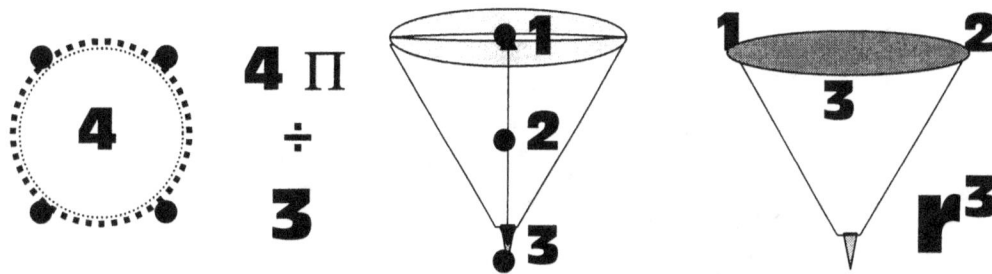

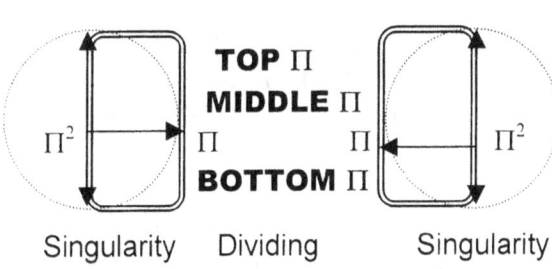

TOP Π
MIDDLE Π
BOTTOM Π

Π^2 Π   Π Π^2

Singularity   Dividing   Singularity

The names I use in TOP, MIDDLE and BOTTOM must not be viewed as sides but merely as terminology using names to implicate divisions. Direction depends on positions and positions form a value only when the observer forms part of the cosmos and not part of the observing.

At first when material presented one side of the Universe matter had three sides to show. Matter had to have space to keep matter somewhere in some part of some Universe and that made up three positions. Between the two Universes k and T^2 placed a value but since only singularity applied any values the value therefore was $\Pi^2\Pi$ where $T^2 = \Pi^2$ indicated time coming from 7/10 in relation to

10/7 and $\Pi^2/2$ (proof of that is somewhere in the book) and k = Π valued by singularity. When space-time developed 3D the dimensions falling outside the sphere becoming space-heat formed as $\Pi^0 = 1$. The electron holds a relevancy of 3 relating to the Neutron being $\Pi^2\Pi$ and the three keeps the electrons in different Universes relating to separate or individual singularity.

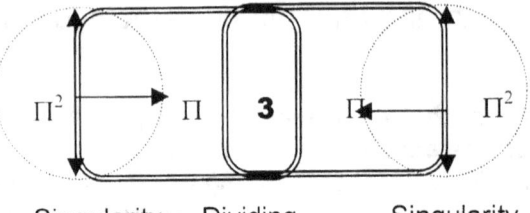

Singularity   Dividing   Singularity

The relevancy between the two particles secures individual positioning between the opposing particles, which positions the material that sufficient space secures cooling and preventing overheating.

Π^2 Π **3** Π Π^2 Becoming Π^2 Π **3** Π Π^2 Becoming Π^2 Π **3** Π Π^2

As the relevancy between the particles promote overheating or applying antigravity (overheating) to the responding cooling or applying of gravity, the one repels material into space-time while the other is collecting material into space-time. The one loses material and sustain a model of preventing overheating while the other gains material and sustain a model of overheating. The one we named the Hubble constant where overheating produces space and the other we called gravity where gravity is demolishing space, but both phenomenon is at present dominating the flow of time in the Universe and will do so until equilibrium again comes about.

Keeping these factors in mind it is clear that Π^2 are the choice of gravity and not r^2.

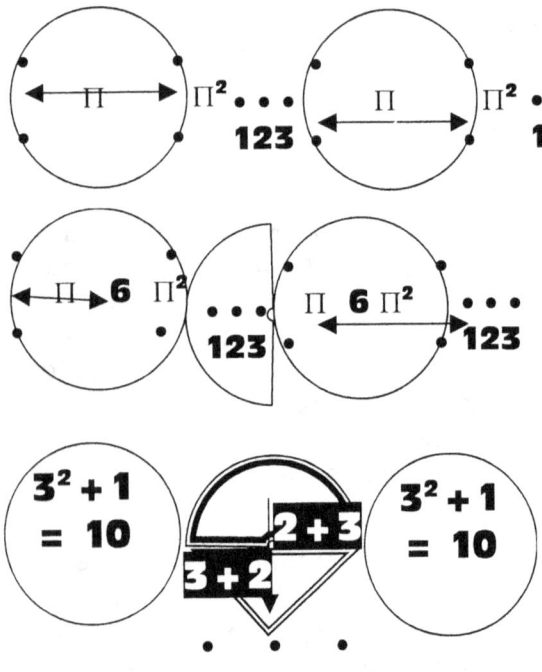

Material formed at a position of six points from singularity. That is three on the one side of the divide and three on the other side if the divide. It is one centre one on either side

In relevancy from one another material held five inclusive positions of two in time including the three positions as material. That made being in one quarter of time five in all. That makes the Lagrangian system dominant.

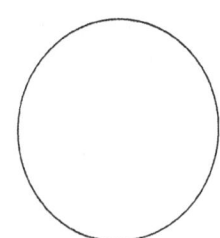

Why would a water drop floating in a space capsule in space, in micro gravity always form a sphere when left capture free form? We all accept that the true cosmic form would be and most probably will be the sphere...but why would the sphere form as the original form when matter is not pre-cast to have any specific form and therefore take on by cosmic pre-cast the sphere as form? We know gravity is there, but qualifying gravity as a force lets the process of investigating science a bit off the hook. Thing become rather simplistic in the modern age when all else is so highly investigated but gravity is merely defined as a force influencing matter. Why would we find in space, where there is supposedly nothing, something we named micro gravity and would bring about micro gravity when gravity is not present? What would cause gravity up to one point and from such a point in the area there supposedly is nothing there is micro gravity.

$$F=G \ (M_1 \ x \qquad m_2)/r^2$$

Is this truly the   **answer**...?

In the investigation of light and gravity and objects and gravity, the mathematical rule of the invert square law must apply without question. But according to the observation of Roche that is not the case. From what one gathers through the Roche limit implicating two orbiting structure the opposite is applying. One must accept that although k proves as an indicator it is also much more when complying the thin influences brought about by singularity in the values carried on by singularity.

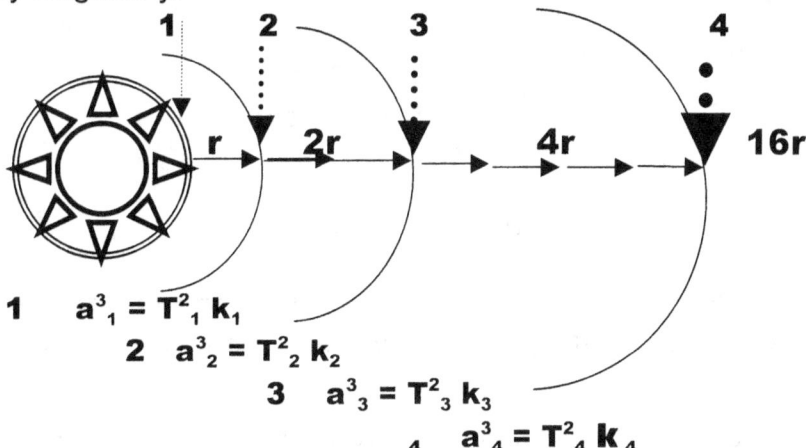

1    $a^3_1 = T^2_1 \, k_1$

     2   $a^3_2 = T^2_2 \, k_2$

         3    $a^3_3 = T^2_3 \, k_3$

            4   $a^3_4 = T^2_4 \, k_4$

In drawing a most basic picture of light passing the gravity lines extending from any structure, I felt it was most insightful that the brains in cosmology was not able to see why light does not bend in the presence of increasing gravity. More surprising was that I found the mathematicians had to call on Einstein for advice regarding an ordinary problem. Light does not bend when passing large objects. It is Kepler's formula applying, and the evidence is clearly in front of the searcher for truth. But one has to go back to Kepler to re-apply what Kepler formulised and change the significant from Newton's significance.

As a^3 increases, so does T^2 as well as **k** increase and with that the influence of gravity per space unit increases with the concentration demise of a^3. But why would that be and what are we missing? Light shows there is an influence out there in outer space, that redirects light's route through space when passing large gravity fields. It is about the relevancy of k influencing the a^3 to allow the T^2 of light to divert in route because of influences established by k on a^3 and slowing down or increasing the line diverting. In this measure one may also find the Roche limit applying, but to truly understand how the Roche limit comes in place and how the Roche limit works one has to replace Kepler's factors with singularity and singularity extending being Π^3 $\Pi^2\Pi$ and **3.**

In the Roche limit the space factor provides space to a solid structure and therefore the value of r is replaced by the value of Π bringing about a square in half of Π. The cube holding 5 to either side removes allowing the extending of Π to indicate position to space.

5/2

Five sides divided by two spheres.

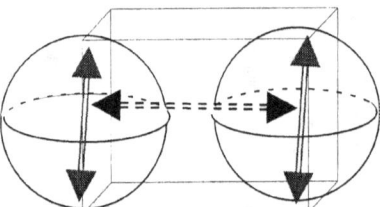

Where Π extends to lock onto the next sphere's extending indicator, Π has to connect to Π forming the square of space and translating that to the half of Π being $(\Pi/2)^2$.

The Roche limit 5/2 becoming = $(\Pi/2 X \Pi/2)$ = 2.4674 as singularity interferes

The space between the spheres divide in half, but because of the extending of Π and not applying r as ordinary mathematics will suggest where Π replaces r the singularity extending from Π^0 will be half of Π in the square of Π = $(\Pi/ 2)^2$ = **2.4674.** In this lies the dynamics why planets have a positional (be it rather a dimensional) relation of 7/10

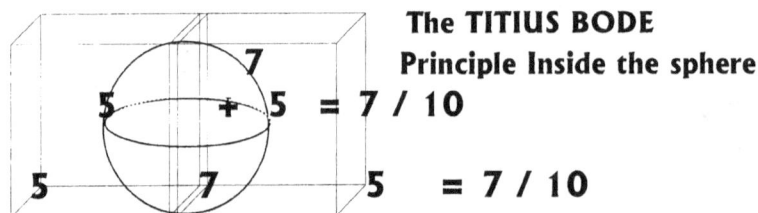

The TITIUS BODE Principle Inside the sphere

$5 + 5 = 7 / 10$

$5 \quad 7 \quad 5 \quad = 7 / 10$

Space-time is a four dimensional position of the universe where the position of an object is specified by three coordinates in space and one position in time

The Titius Bode law must not be seen as some obscure event that took place just before and / or after the Big Bang or when the solar system formed it fell into place. The Titius Bode law applies when the top is spinning, when an atom is spinning, when a motorcar wheel runs on the tar, when a jet engine fires up. It takes Place whenever the Coanda principle comes into effect and the Coanda principle is wherever there is motion in relation to singularity in a centre forming a centre.

This is why we can use degrees measuring the circle by (6^2) (forming the square relating to matter through singularity) X 10 (square if space) = 360^0 however it is always in motion. That proves no point can be static or constant, though it may seem that way to outsiders. Although matter is matter, matter can also be anti-matter and moreover form its own anti-matter at the same time. This degeneration of structure is very likely to occur with overheating. Revaluing Π to Π^2 will bring about a new contact point where Π meets r forming another relation in Π^2. Every time material swaps sides it also qualifies as anti matter to matter because if it goes out of orbiting rotation frequency. It has the ability to collide with the same matter it forms union with but is located on the other part of the spin. It then becomes in a situation where Π revalue to r. Time is the changes in relation where Π contacts a different r not withstanding the many r points there may form because every r constitutes a different value to the Universe through other ratios and relevancies brought about by heat and light. Time is the duration it takes Π to rotate between any two given points of r and therefore must always amount to a square (T^2) moving from point to point through the cube of space (a^3) in that duration of time (k). With that it proves Kepler's a^3 (space) = T^2 k (time in the instant of motion) but motion must continue through a specific value in space where the space-time is maintaining relevant equilibriums throughout singularity connecting.

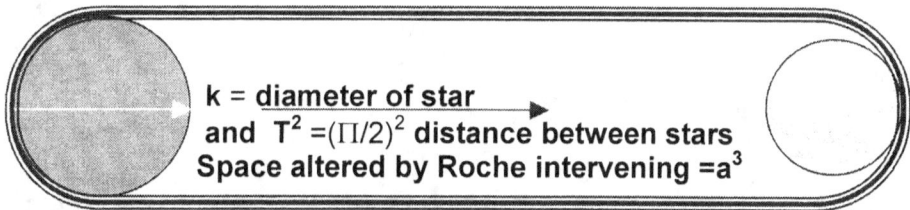

k = <u>diameter of star</u>
and $T^2 = (\Pi/2)^2$ **distance between stars**
Space altered by Roche intervening = a^3

In the Roche singularity apply all three components

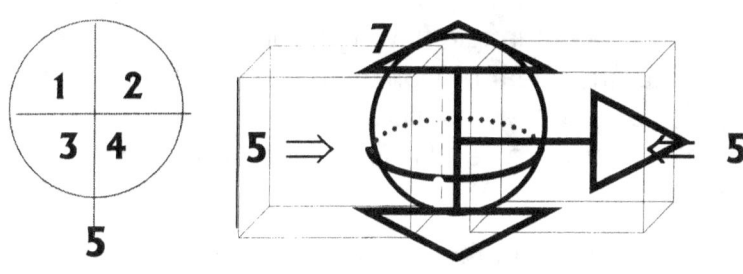

In the Roche limit the **straight line** forms part (1) and the **half circle** is part (2) and the **triangle** forms part (3) to **singularity** (4) Holding 5 points outside **singularity**

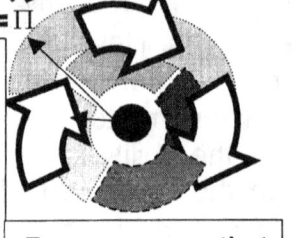

Coming from $= \Pi$

Singularity $= \Pi$

Singularity: a mathematical point at which certain physical quantities reach infinite values for example, according to the general relativity the curvature of space-time becomes infinite in a black hole.

With no line starting from zero because there is no zero as a mathematical fact, then all particles hold the point of infinity and not merely the Black Hole,

From that argument one may conclude that all stars will become Black holes depending on the gravity increase they may generate.

Where singularity holds position in the centre of any and all rotating objects as a value of Π merely applying movement (in the form of atoms) qualifies all matter to be space-time. It does not only fit the description of space within Black Holes but it fits all stars where singularity becomes part of all the stars from the minute to the largest cluster of matter.

Through rotation encircling the point of singularity and matter is (1) coming from, (2) being at, (3) as it is going too in one movement in relation to the specifics of the centre point being singularity, all matter then qualifies to form space-time.

The influence of singularity as the extending of Π into space links Π^2 to r and forms 2(5)+2(5) =10+10=20

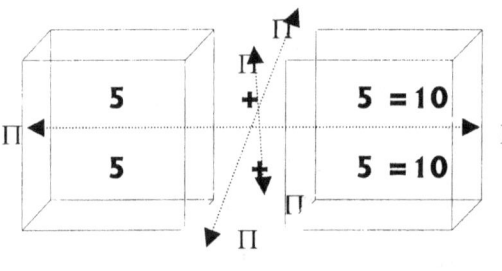

From the position of singularity there are different values in Π where each indicate a position. The value it represents being $\Pi\Pi\Pi$, Π^3, Π^2, Π and Π^0

From there it influences singularity in the triangle flowing through to the half circle. It is an interaction between circular and linear motion as the value of Π continuous past Π^2 (at the end of the solid) and every cosmic structure holds an individual and specific singularity. The field where Π extends we call the atmosphere having a value of 21.991 / 7, which is Π.

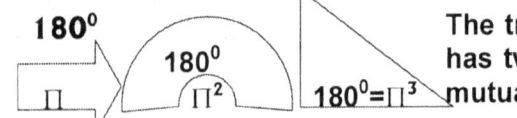

180^0

180^0

Π Π^2

$180^0 = \Pi^3$

The triangle, the half circle and the straight –line has two things in common, they share 180^0 as a mutual value and they are part of singularity.

Using the concept that gravity applies Π as the circle factor Π as well as Π^2 replacing r^2 the replacing by Π brings two values as Π and Π^2. That I found is the case with gravity and will be apparent when explaining the sound barrier as well as the Four Cosmic Pillars. In order to create a distinction I remained using r as the indicator of the cube or non-circle that has vacant space and by vacant space I refer to non-solid structures. In the solid structure I use Π as a value for reasons that will become apparent in due time.

Gravity does not apply mathematical equations to the letter as we would like, but rather use Kepler's thinking by enlisting an average gravity applying through out because it never favours and is equal every where. In gravity one find the extending of Π implementing Π^2 on average as a unit and not the radius r as a specific.

Looking at the affect of gravity it shows the precise quality of no distinctive point as gravity never seems to end at a point but flows all over affecting all that holds a position in its sphere of influence. The gravity coming from China meets the gravity coming from America at no particular spot but intermingles without distinction. This takes mathematics back to another fact beyond normal explaining.

We take a line running between two points as being 180^0 and the rest of the explaining is saved in the accepting part of mathematics. Any one of the two points the line starts or ends at is a point in infinity, The start and the end depends on the viewer putting the relevance to favour the side of choice. That puts the point of end or beginning in the spectrum of choice and not fact. Any direction is as equal as all other directions.

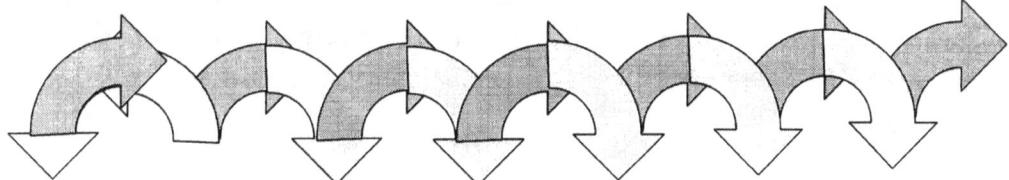

Following the flow of any line such a line is an extension of the previous dot in infinity to the next dot in infinity without any ability to skip or bypass any of the other dots in the connecting line. Any direction change including the remaining of travelling in the same direction is in relation to a line travelling all being the very same. Change does not affect the line.

A straight line, triangle and half a circle will always have equality in dimensional capacity providing equilibrium being 180^0 because each one shares a common denominator in singularity to the value of Π. As the straight line averts a zero it holds another straight line in place to set about such an averting where the two lines will always carry a relevancy in relation to progress (the triangle) and a common denominator in the start from singularity. This concept we apply as the graph or the vector. By going back to a line, any lines and all lines, the line is a connection of dots in infinity, running from one specific to another specific and avoiding zero or dots. At every point in infinity it dips into infinity coming out on the other side by choice of direction and the direction is unforced and change presents any angle including the straight line, which incidentally is just another angle.

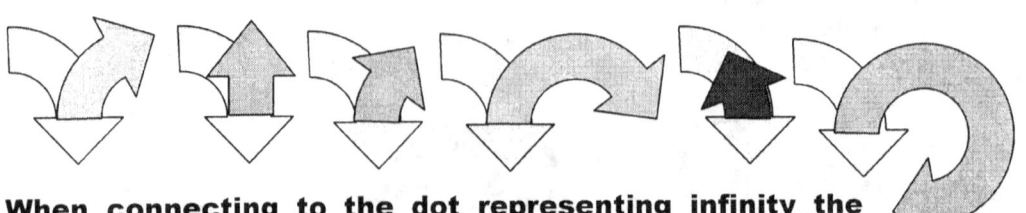

When connecting to the dot representing infinity the flow **can be in any and all possible directions, including in the** same **direction. We all live in a graph as the universe with all in it is nothing less that a three-dimensional graph flowing according to time. That means in the case of Pythagoras the mere fact that the line shows changes in direction does not implicate or affect the line as a tool of mathematics. Whether the line changes into a half circle meeting at the other end again or meeting in a triangle in forming a half square by joining the point where it began, the result still indicate a line flowing between points.**

The line **dips** into **infinity every time it** passes **infinity** when **it** cuts through infinity. **The line going into infinity comes natural as the line progress because all lines are infinite dots linking from one point to another point. That brings about that coming from infinity might change in angle bit that directs the route and not the form. The form is all the same**

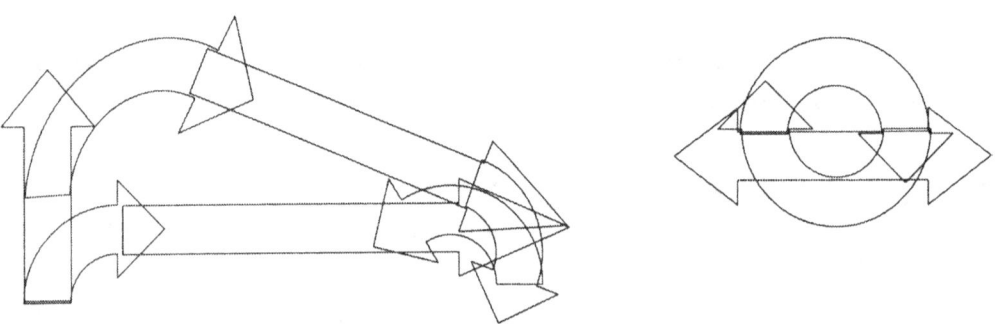

In that way a circle is a straight line following a loop as it comes out of singularity at a different angle and a triangle is a straight line that dipped into singularity but at three stages changed the angle with which the line then left to follow different directions at specific points. From the point singularity observes it still remains a straight line because there is no direction alternation in the first dimension and in that dimension it still remains a straight line in which we on the outside may experience as three forms but is in fact one single line. Only when the direction changes completely in reverse the line doubles in value but comes from multiplication for instance 2Π become Π^2.

But the Lagrangian system proves much more than dimensional interlinking, it proves Pythagoras in principle.

LAGRANGE (-TOURNIER), JOSEPH LOUIS DE (1736-1813)
French mathematician, born in Italy. In celestial mechanics, he studied perturbations and stability in the Solar System. He examined the three-body problem for the Earth, Moon and Sun(1764) and the motion of Jupiter's satellites (1766). In 1772, he found the particular solutions to the problem that give rise to the equilibrium positions called Lagrangian points. Lagrange also studied the Moon's liberation. LAGRANGIAN POINT One of five points at which small bodies can remain the orbital plane of two massive bodies; also known as liberation points. Three of the points lie on the line joining the two massive bodies: L_1 lies between them, while L_2 and L_3 have the two bodies between them. These three points are unstable, slight displacements of a body from then resulting in its rapid departure. the fourth and fifth points (L_4 and L_5) each form an equilateral triangle with the two massive bodies, 60° ahead of and behind the smaller body in its orbit around the larger one. A well-known example of bodies flying at the L_4 and L_5 Lagrangian points are the Trojan asteroids in Jupiter's orbit. Among Saturn's satellites, Telesto and Calypso lie at the L_4 and L_5 Lagrangian points in the orbit of the much larger Tethys. In similar fashion, tiny Helene precedes Saturn's satellite Dione, keeping 60° ahead of Dione. The Lagrangian points are named after the French mathematician J.L. de Lagrange, who first calculated their existence.

LAGRANGIAN POINT:
The Lagrangian points
are five equilibrium points
in the orbit of one body
around another, such
as a planet around the Sun

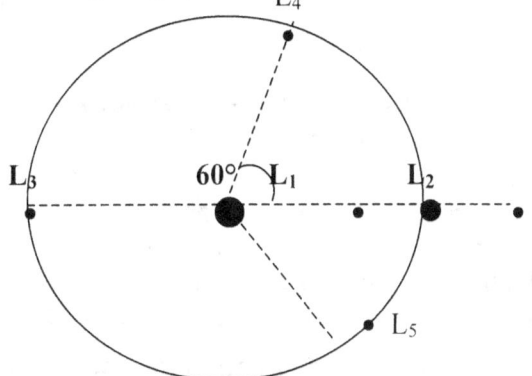

The Lagrangian System implicating the five positions extending from singularity

Singularity dividing the cosmos
Each triangle claiming a side of the universe

1 Half circle	$=180^0$	L_3 L_4 L_5
2 Triangle 1	$=180^0$	L_3 L_4 L_5
3 Triangle 2	$=180^0$	L_3 L_4 L_5
4 Straight Line	$=180^0$	

The half Circle $=180^0$ combining as a sphere when comprising
Singularity in the matching of the value of the straight line forming
the half circle and combining as the triangle and all are equal 180^0

The second one also fits in the singularity influence on the Universe.

The second one also fits in the singularity influence on the Universe.

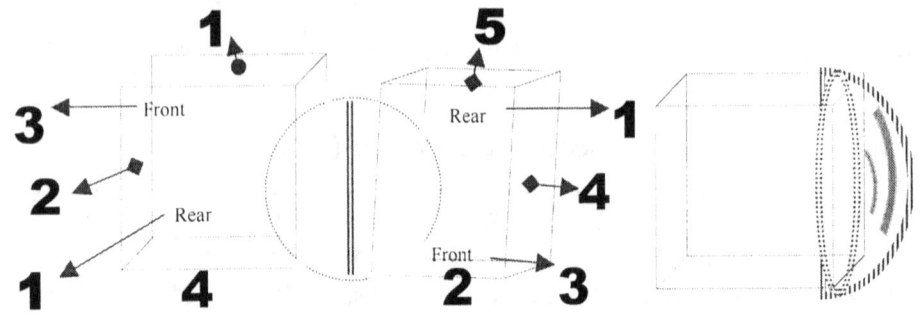

1 Relating to 5

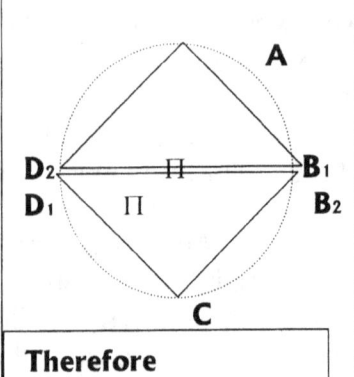

$(D_2 A)^2 + (B_1 A)^2 = (D_2 B_1)^2$ (PYTHAGORUS)
$(D_2 A)^2 = (B_1 A)^2$ (EVEN SIDED TRIANGLE)

$2(D_2 A)^2 = (B_1 A)^2$
$(D_2 B_1)^2$ (DIA. OF CIRCLE) AND ABCD EVEN SIDED SAUERE WHERE AB = BC = CD = AD

$(D_2 B_1)^2 / 4 = (AB)^2 + (BC)^2 + (CD)^2 + (AD)^2$
$2(D_2 A)^2 = (D_2 B_1)^2$ BUT $(D_2 B_1)^2 = \Pi^2$ (Replacing r^2)

$(D_2 A)^2 + (D_2 A)^2 = (D_2 B_1)^2$ $[(D_2 A)^2 = (B_1 A)^2]$
THEREFORE $4(D_2 A)^2 = (D_2 B_1)^2 / 4 = (\Pi / 2)^2$

Therefore
The Roche lobe is
$= (\Pi / 2)^2$

The value of singularity stems directly from the law of Pythagoras or Pythagoras is the result of the average of singularity. With the shortest line being a dot, all lines must start from a position implicating Π. A circle is a square without corners implementing Π and a half circle is therefore a triangle without corners. The corners are the factor that confused every one in the past. When replacing the value we normally attach to circle being r with Π, the law of Pythagoras becomes quite meaningful and mathematical.

By placing a connecting circle on the sides of the triangle half a circle forms. By implicating Π as a relevancy and not the straight-line r, two values of Π applies to each circle, and the straight line is no longer r, but is Π^2. This will bring about that each circle holds the square value implicated to the allocated conditions applying to Π in that specific instance. By adding the two half squares forming the two half circles and then calculating the square root of the total that then forms the average diameter, an average of Π in the connecting line will come about. As both lines are the straight line forming singularity coming from one line being Π, the connecting line then must be the average of the two lines as Π^2. That is what **the law of Pythagoras says.**

A STRAIGHT LINE, TRIANGLE AND HALF A CIRCLE WILL ALWAYS HAVE EQUALITY IN DIMENSIONAL CAPACITY PROVIDING EQUILBRIUM BEING 180^0 BECAUSE EACH ONE SHARES A COMMON DINOMINATOR IN SINGULARITY TO THE VALUE OF Π. As the straight line averts a zero going down infinity it holds another straight line in place to set about such an averting where the two lines will always carry a relevancy in relation to progress (the triangle) and a common denominator in the start from singularity. This concept we apply as the graph or the vector.

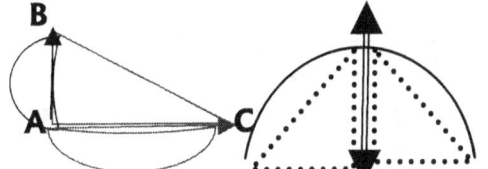

BC EITHER RELATE TO **AB** OR **AC** AT ANY GIVEN TIME OCCUPYING SPACE AS MOVEMENT DICTATES DRECTIONAL CHANGE THROUGH DIRECTIONAL FLOW

With the normal extending of singularity it will always form the triangle in a half circle whereby Π relates to the cube by 5 points to either side of the line singularity forms. Thus there is 10 standing related to seven and visa versa. By calculating the 4 squares in the circle with the dimensional changing of space (5) becomes the twenty

The normal flow will allow singularity extending to 10Π but when singularity blocks another sphere in singularity the two will form a joint value and by this joining the larger will dominate the space as well as the time of the lesser taking control of the surface and the atmosphere. Through this the Roche lobe comes about with all its other dynamics I describe further on in the theses. The principle is the same, which we know as the conducting of lightning and Jupiter uses it extensively to implement this action.

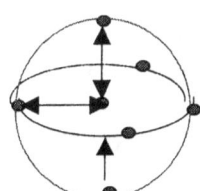

In the sphere there are never only one direction implicated in movement. Movement are always in relation to the centre position because as a line goes up it also goes in or out. When a line goes north or south, it also comes towards the centre or going away from the centre.

There is always relevancy present in movement. As this moving indicates direction it also apply Π^2 for indicating value forming the time factor.

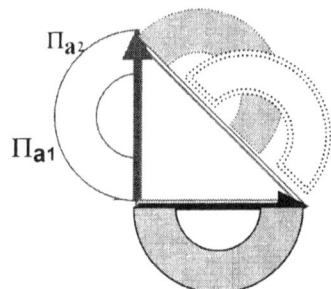

$(\Pi_{a2} \times \Pi_{a1}) + (\Pi_{b1} \times \Pi_{b2}) = (\Pi^2_a + \Pi^2_b) / 2 = \Pi^2 = \text{gravity}$
and that is proven by Pythagoras.
Gravity is the average movement of matter through space in time determent from the position where matter in the sphere meets space in the cube from a point of Π to a point of Π^2 In this the figures of 2(5) = 10 (space) stands related to 7 from singularity as (matter)

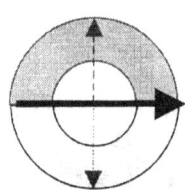 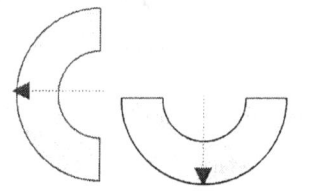 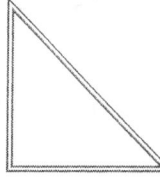

From the star holding a dominant point or most valued point in singularity it affirm all three other structures, each holding singularity individually and in a compliment of 5.

The network of individual singularity not only provide spinning through governing singularity in the sphere but also provide spinning in the geodesic through out the cosmos linking all matter to matter in a network no one will ever come to understand in full. In the sphere the four squares forming the triangles linking the lines to the half circles holds space in time maintaining singularity of different assortments. In view of the matter-to-matter Roche factor where the factor consists forming relation between particles occupying densified space-time of where (Π / 2 X Π / 2) relating to the foursquare triangle the value of gravity Π^2 comes in position as Π^2 / 4 X 4 = Π^2.

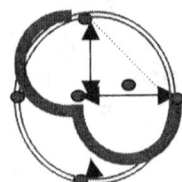

Because every moving line represents one quarter of the sphere in relation to the rest of the sphere and the line also indicate the relevant position between the point indicated and the point in the centre it is a relevancy of singularity in progress. By connecting the line, as Pythagoras will suggest the singularity within the sphere become a specific value indicated representing one half circle.

No object can be in two spherical quarters in the same time, but has to alternate in aliens to the space in accordance to time rotation.

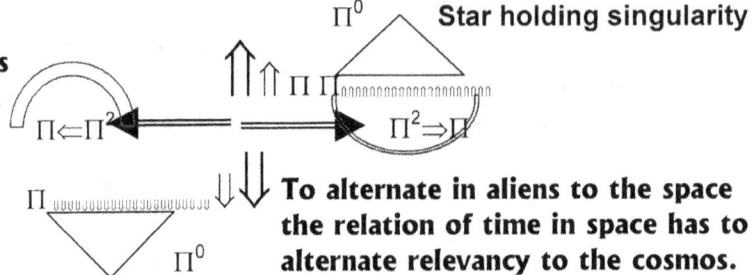

Star holding singularity

To alternate in aliens to the space the relation of time in space has to alternate relevancy to the cosmos.

Singularity holds five dimensions inside and five outside singularity as matter and space forming space-time. The ten dimensions I named the atomic relevancy is also showing the double value of singularity as singularity extends into as well as beyond space. The atomic relevancy is $(\Pi^2 + \Pi^2)(\Pi^2$ X Π X 3$)$ = 1836 that is the mass relation between the electron (3) and the proton. Proton = $(\Pi^2 + \Pi^2)$ Neutron = $\Pi^2 \Pi$. The atomic relevancy holds the dynamics of singularity control.

The TITIUS BODE Principle Outside the sphere

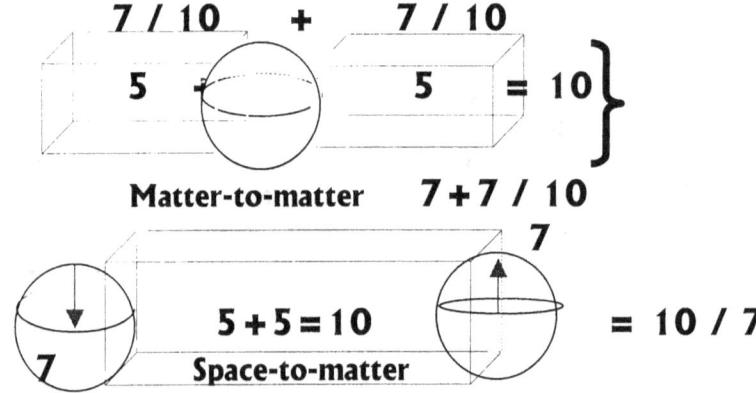

From the dimensional implication comes about, not only the Doppler's effect, but many more of phenomenon not yet understood. The dimensional relevancies formed between matter as six, matters end at seven and space at ten, comes the value of Π.

The process is all intermingled and stands in relevancy to one another. The relevancy compliment holds such attachment that none of the factors can even stand-alone. It is the way that science places every aspect in the cosmos as individual and not related to each other that launches the problems of miss understanding. The Value of singularity appreciates or demises by ten fold. For instance, the value of Π will increase by ten every time singularity applies another layer.

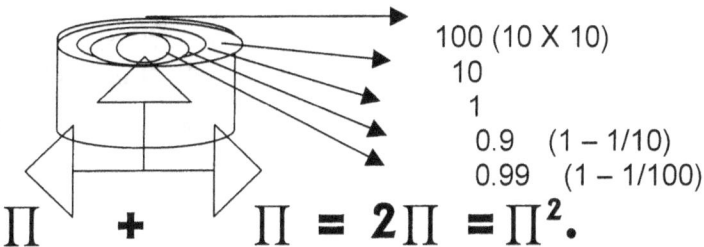

100 (10 X 10)
10
1
0.9   (1 – 1/10)
0.99   (1 – 1/100)

$$\Pi \quad + \quad \Pi = 2\Pi = \Pi^2.$$

The normal flow will allow singularity extending to 10Π but when singularity blocks another sphere in singularity the two will form a joint value and by this joining the larger will dominate the space as well as the time of the lesser taking control of the surface and the atmosphere. Through this the Roche lobe comes about with all its other dynamics I describe farther on in the theses. The principle is the same, which we know as the conducting of lightning and Jupiter uses it extensively to implement this action. In the Roche limit the straight line forms part (1) and the half circle is part (2) and the triangle forms part (3) to singularity (4) Holding 5 points outside singularity. Every aspect connecting to the Universe changes everything it holds totally and becomes the anti-matter to which it was matter 180° previously

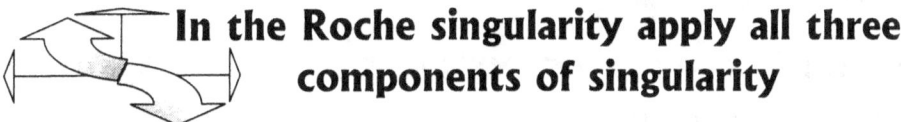

In the Roche singularity apply all three components of singularity

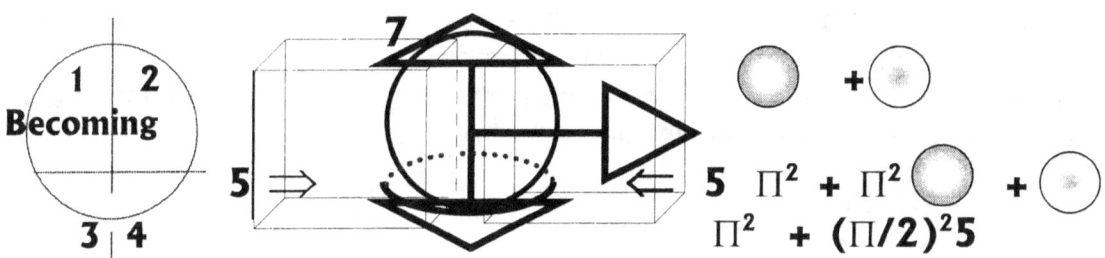

1	2
Becoming	
3	4

$$5 \quad \Pi^2 + \Pi^2 \bigcirc + \bigcirc$$
$$\Pi^2 + (\Pi/2)^2 5$$

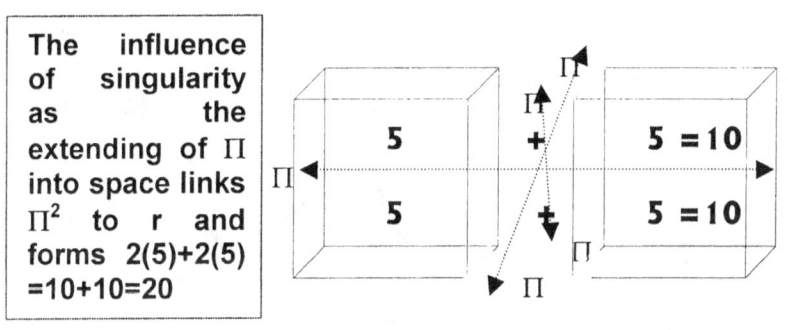

| The influence of singularity as the extending of Π into space links Π^2 to r and forms 2(5)+2(5) =10+10=20 | 5   +   5 = 10
 5   5 = 10 | From the position of singularity there are different values in Π where each indicate a position. The value it represents being $\Pi\Pi\Pi$, Π^3, Π^2, Π and Π^0 |

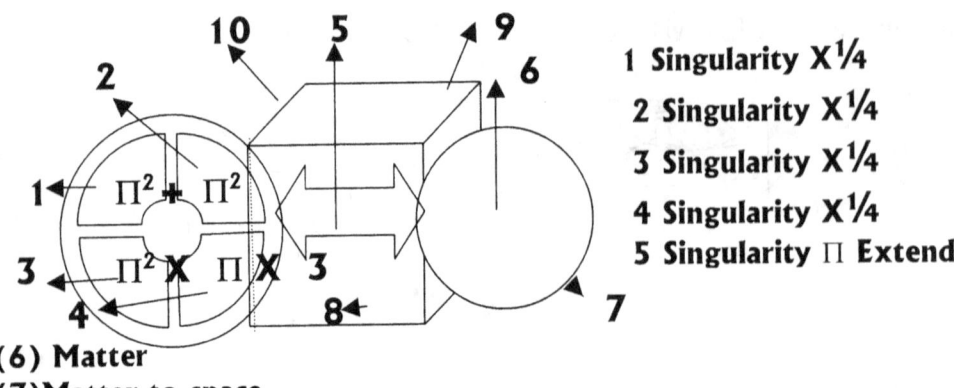

1 Singularity X¼
2 Singularity X¼
3 Singularity X¼
4 Singularity X¼
5 Singularity Π Extend

(6) Matter
(7)Matter to space
(8,9,10) Dimension1,2,3) in the cube's six sides

Gravity is about a relation established when time begun between particles we know as material and particles we know as free or unoccupied space. Gravity reduces space to apply to fit the form of the sphere and later accept the form of the sphere.

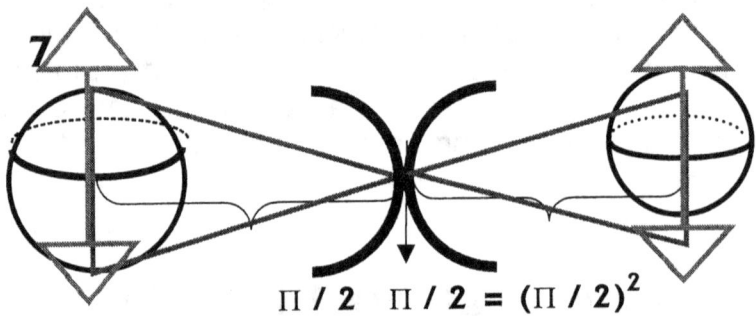

$$\Pi / 2 \quad \Pi / 2 = (\Pi / 2)^2$$

SINGULARITY MEETS AND COMPLIMENTS EACH OTHER.

The diameter of the cosmic structure holds the value of r and singularity holds the dimensional value of Π meaning that the radius or diameter (r) extends to become the diameter multiplying the value of singularity. But since r already consists of the square of space holding a definite positional relation with the value of singularity being Π the diameter comes into effect. Π extends each to an individual value to a point where the singularity on each side meets, bringing about a mutual Π^2 to the value dominance of the larger singularity control.

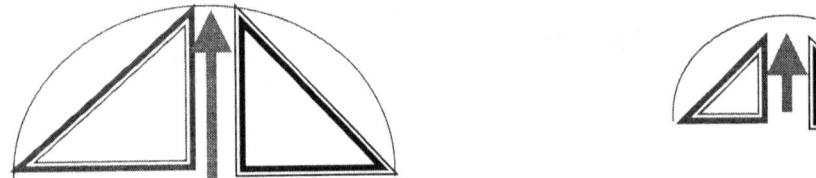

At this point the equality of the straight-line dimension to the triangle and the half circle holds prominence as a straight line, a half circle and a triangle is dimensionally equal. The common denominator will bolster all factors to an equivalent ratio.

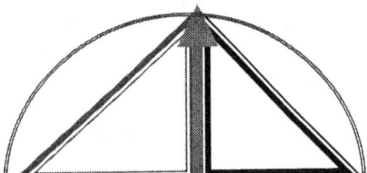

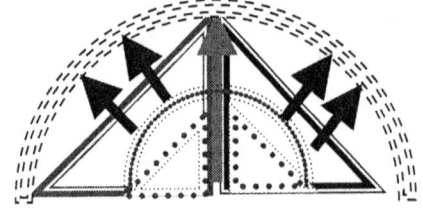

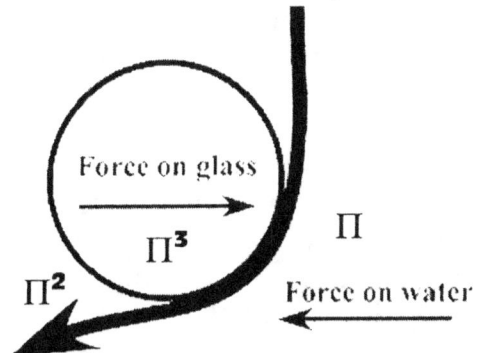

In the sketch one may take delight in the Newtonian application of forces forming. There is a spook or a force for the water and another spook or force for the glass and then Newtonians say they do not believe in spooks! More seriously though the Coanda principal indicate how motion forms the limits to space. The motion of the water attach to the surface of the sphere or circle which then forms what Newtonians describe as the curvature of space-time. It is only the form singularity applies than manifest as the roundness we see. In that is the proof about singularity taking on Π as form. The water attach to the sphere as much as giving the sphere the gravity with which it does attach. In that manner we know that that was the way particles formed combinations just after the arriving of moment-Alfa (a name I gave to the very first moment when heat parted form cold). Singularity brought the Universe but also singularity brought the divisions between the many Universes that followed the immeasurable many Universes that came after the flooding of Universes to follow the leaders. At this point mathematics renders it useless. Every slightest point in space became an opportunity of establishing a Universe with most different functions and ingredients there might form. This is apparent from the fact that it still takes place at the present moment by motion attaching new singularity through duplication and through duplication release previously attached singularity from serving the purpose of duplicating by motion.

When singularity by the straight line increases the singularity by the triangle it will also bolster giving equal potency in singularity by the half circle. As the singularity of the major component revives the lesser singularity to equality, the **triangle in singularity** will match the performance and so would the half circle respond in precise ratio setting equilibrium in order. The major partner's singularity in the straight line excites the minor partner's singularity in the straight line affecting all other aspects holding singularity in both objects to match equilibriums in all aspects of singularity. That is the Roche lobe.

From this the lesser partner will fill by the extent of the larger partner and as soon as equilibrium sets in the growth will duplex to matching in both accounts, normally to the fatality of the lesser partner, as the lesser partner will be capitulating under the strain of the dual. In that way the inner planets came in place as I explain in part 7 *of Matter's Space In Time The Theses*.

The Titius Bode configuration in accordance to orbiting formation holds a slightly different explanation to the explanation that applies to cosmic structure surrounded by space. It is moreover the individual singularity in maintaining the major singularity, which sustains the governing singularity providing equilibrium in space-time. Not only does atomic individual singularity maintain self preservation, but in doing that it also sustains a governing singularity holding structural composition and form within a cluster of matter for example a star. As there is between stars so there is in the same manner a mutual or bonding singularity between atoms in stars, which we see as fusion. From this one may freely deduct that gravity is not forcing material closer but is destroying space whereby it converts the space to a density the senior partner has in the atmosphere of the senior partner. Where does all the information given thus far take us you might ask? For one it can help to explain something Newton science can never understand. To start with we have to realise that the Coanda principle is the manifesting of Kepler's gravity and we have to accept Newton's version of gravity is a load of rubbish.

All objects are classed by heat either being in motion through duplicating (overheating and expanding) or being in motion through heat contracting (heat being reduced through motion removing space), but most of all is that all material is about motion forming the space-time and classifying the space-time. This is most pivotal in understanding cosmology notwithstanding Newtonian views.

In the past no one even thought of placing the Coanda principle in line with gravity. By following the Kepler's formula we find that the Coanda affect, in fact is gravity. Gravity is moreover the Coanda principle than gravity is any other variety of forces or concept of contractions. The Coanda principle is gravity. It is the how all gravity is charged and distributed. That forms the basic principle of Cosmology

Kepler said

$$a^3 = T^2 k$$

The dot indicates a centre point from where the gravity is centralised and is controlling the Coanda principle.

k

T^2

a^3

The Coanda affect is proof of the functioning of gravity inside the atom. It proves that motion (T^2) of the neutron establishes a centre in line where the compliment of material forming the atom will secure a controlling singularity that is governing the entire atom. That forms the centre of the Universe. Singularity then finds a position at the distance of (k) and such motion claims the space (a^3), which is the atom by construction from a centre within that motion (T^2). The motion (T^2) creates a centre at the line of (k) and a centre of the space (a^3) the motion (T^2) establishes a gravity field all along the lines and at the distance of (k) in the space (a^3) that the motion (T^2) created.

The motion of the liquid which the neutron is proves to be the time (T^2) aspect because as it increases it claims the space (a^3) in the at a distance k of time (T^2) that the running has increased. The faster the motion is the stronger is the gravity that the motion generates in the space it claims by the gravity it generates. In the Coanda affect we can read how the atom became the Universe. The proton is the substance performing as the solid on both sides of the Universe, The neutron being a liquid is what establish the gravity, which helps the singularity secure space-time by heat compensating for overheating. With the neutron forming a liquid and the proton providing the spin, the neutron established the space that forms the atom to the inside of the electron.

All objects are either cold and reduce space-time in relation to others being hot and expanding space-time. Being cold puts the object in the role of conserving space-time in contraction and that puts the object in a position of being a solid.

Then in relation to the first conserving factor there is the overheating factor, which brings into the relation the expanding, or moving away from the singularity.

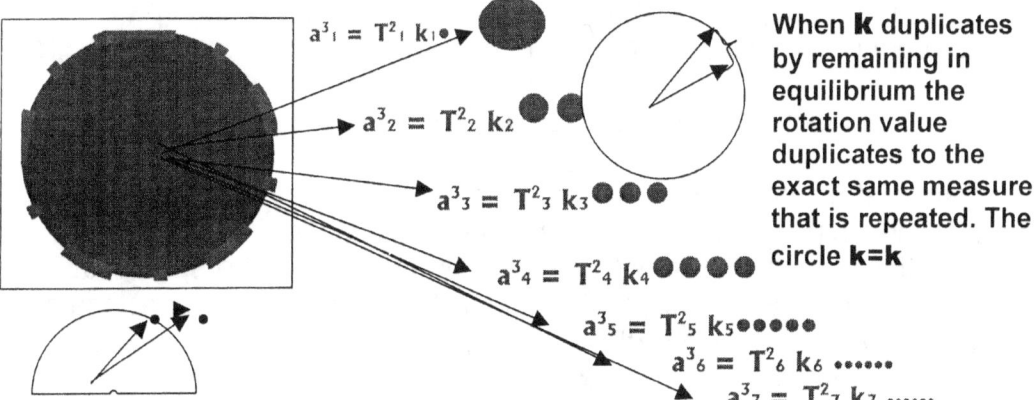

$a^3_1 = T^2_1 k_1$

$a^3_2 = T^2_2 k_2$

$a^3_3 = T^2_3 k_3$

$a^3_4 = T^2_4 k_4$

$a^3_5 = T^2_5 k_5$

$a^3_6 = T^2_6 k_6$

$a^3_7 = T^2_7 k_7$

When k duplicates by remaining in equilibrium the rotation value duplicates to the exact same measure that is repeated. The circle k=k

The duplicating requires a repositioning of the aligning of k from a certain position to a more forward position in relation to and in that k will also have to extend a value when moving from k_1 to k_2

The Roche limit is:

The region surrounding each star in a binary system, within which any material is gravitationally bound to that particular star. The boundary of the Roche lobes is an equipotential surface, and the lobes touch at the inner Lagrangian point, L_1, through which mass transfer may occur if one of the components expands to fill its lobe. It names after the French mathematician Edouard Albert Roche (1820-83).

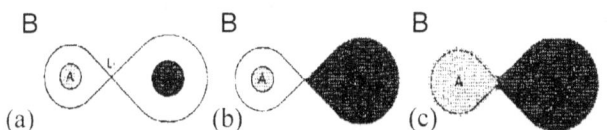

THE ROCHE LOBE: In a binary system, the Roche lobes of components A and B meet at the L_1 Lagrangian point.

(a) In a detached system, neither star fills its Roche lobe. (b) In a semidetached system, one massive component, B, fills its Roche lobe. (c) In a contact binary, both components overfill their Roche lobes and share a common envelope. Lets explain the importance of this Roche limit and how the Universe used the Roche factor to produce the Big Bang. That is where it all started...

This letter you are reading is my effort by which I hope to interest you in reading my Introducing letter an open letter To Selected Academics ISBN 0-9584410-9-X The book on offer has the title of an open letter To Selected Academics ISBN 0-9584410-9-X and is the actual letter I sent to various establishments.

What brings about the expanding?

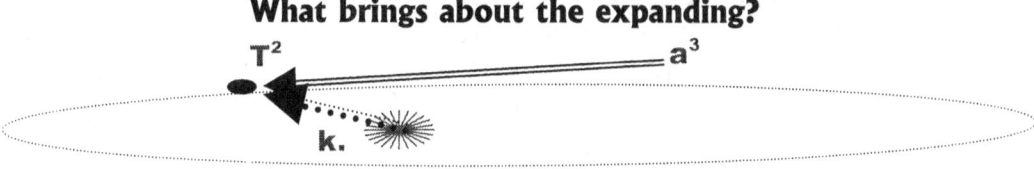

T^2 a^3

k.

Kepler was the very first person to mathematically introduce space a^3 centre k and time T^2. Not only did he introduce space-time a^3 / T^2 but he also placed space a^3 and time T^2 in a relevancy long before Einstein did and placed gravity in space-time a^3 / T^2 even before Newton named gravity. Kepler was the person who placed gravity as the ingredient in the Universe that determines space a^3 and time T^2 and much more. Kepler was the first one that saw that gravity comprises of two factors being **k** or linear gravity and circular gravity or T^2 as gravity keeps space in form while all is staying together.

Since gravity also influences the space outside the sphere, the space we call outer space has seven plus three points bringing about ten positions of gravity influencing space.

The influence inside the sphere also captures the space outside the sphere.

This means that in the cube at the point of contact between the cube and the sphere the cube experiences such a contact point as if the "bottom falls out" of the cube and without a "bottom" to support objects they fall to the sphere as objects does fall to the Earth. Remember that a body "floats" in space, but at one specific point it starts to "fall" to the Earth. That is gravity and it is a dimension change much more than any force. I shall explain this last remark later on. That too is the Lagrangian system with five cosmic structures holding relevancy to the centre structure where the centre structure stands in for seven positions diverting from centre and the orbiting structures standing in for five positions in space.

Gravity has all to do with dimensional changing and reforming of forms to re-affirm alliances supporting the centre. It is the reforming of space converting space to more concentrated heat.

The Universe is in the three dimensions using twelve dimensions that is visible to us and indefinite number of stages in size differences ranging from the immeasurable small to the immeasurable large where mathematics becomes a short fall to the next and the previous dimension.

Up to now every one in science is normally acting as if gravity is a commonly explained factor, which every one knows every aspect about all principles that are involved in gravity down to the smallest detail. In truth, no one in science anywhere remotely knows what brings gravity about and I used Kepler to unravel this mystery called gravity. But no one in science will admit this fact about Kepler being the one who formulised gravity decades before Newton came and gave gravity the name.

Newton did not underwrite or define gravity and even today the most informed in Science at best can only assert their suspicion on a rumour presumed about what causes gravity to perform as the part interlinking the cosmos but no one can go any further by explaining the concept. Newton started this realising of gravity but it had and still has no more substantial proof than a rumour has and Newton admitted to it being a concept he could not explain. In Newton's ignoring to test Kepler's findings, Newton missed the opportunity to find what gravity is. Since Newton every person in science also ignored Kepler and every one is guilty of missing the opportunity Kepler maid available.

By my efforts of studying the implications that results from Kepler's finding I can now un- emphatically declare I know what gravity is. Gravity is the entire following locked into one compiling unit:

Gravity is not being some magic force found between particles grabbing onto everything. I mathematically explained the following phenomena:

Gravity is singularity as a factor forming space-time

Gravity is finding space-time

Gravity is proving space-time and aligning space-time with gravity

Gravity is the working principals behind all cosmic occurrences that pre dates the Big Bang period.

Gravity is the Roche limit.

Gravity is the Lagrangian system

Gravity is the Titius Bode law

Gravity is the Coanda effect

Gravity is the sound barrier

By being able to pin point prove what Gravity is that enabled me to unravel the other entire phenomenon that forms gravity. Each of the phenominon I mention above has one part or role in what is forming the totality that which we know as gravity.

Should you think this is rather a wild presumption I challenge you to spend a little more time and please think about what I say when you read about what I say in the next few pages. The first thing you should admit in private is what study did you personally so far made about the work of Kepler?

Still, to this day nobody in science at present will denounce the principle of gravity as vaguely researched. Gravity has never been explained as a principle. Even when one is considering what the importance of gravity is, gravity never yet has been understood. It is by now very clear that little if nothing of all objects are pulling closer in outer space. Comets are missing the Sun on a regular basis and no planet has come much closer toward the centre of the Sun. Still everyone in science acts in a manner as if Newton's gravity ideas are the best detailed proven fact and only occasionally does someone quietly admit that even Newton admitted not knowing what gravity is. No one ever comes to the front and boldly state that gravity is just a rumour spread by scientists pretending to know all there is to know and knows little to nothing about what there is to know. Newton admitted that much when he introduced the name (not the concept). Mistakenly Newton corrupted the concept he named as gravity.

Going according to what Newton introduced Newton's concept will by now have the moon much closer to the Earth than it was in the time of Kepler's studies, yet we know the moon is moving away instead of coming closer. By the same measure Kepler suggested that the space a^3 is content with the motion kT^2 as long as the motion T^2k is equal to what the space a^3 will allow. Kepler suggested motion of space remains in equilibrium as long as motion of space a^3 duplicated space a^3 by motion thereof T^2k. That is much more true than objects rushing towards one another by the pulling power of mass. Newton agreed that he could only declare gravity as a vague concept. This fact was at that time drowned by the man's stature and was relieved from the manner of requiring the proof that later

Academic science became an absolute necessity. The proof one would demand now a day was never given to put Newton's rumour beyond doubt. When Newton announced a force he also admitted the force could be anything. No one ever came after Newton and proved the fact better. That still underlines the fact that the force to this day can be anything. Not once could one person in the past or present provide substantiating proof on gravity as reality by defining the very principles thereof.

That includes every one since Newton as well as including Einstein and even Hawking. Scientists can declare gravity was a factor at 10^{-43} seconds after the Big Bang but what brought gravity about or why gravity became or still remained, as a presence is still tightly concealed information which all are speculating on. Even in the best and most informed circles and amongst the most educated there is no one that knows what gravity is because they all ignored Kepler and for them to ignore Kepler the price they pay is not finding the principles bringing about gravity. Using Kepler even makes the method to follow and understand Einstein's discoveries shockingly simple. Gravity is the motion of space relating to time in movement.

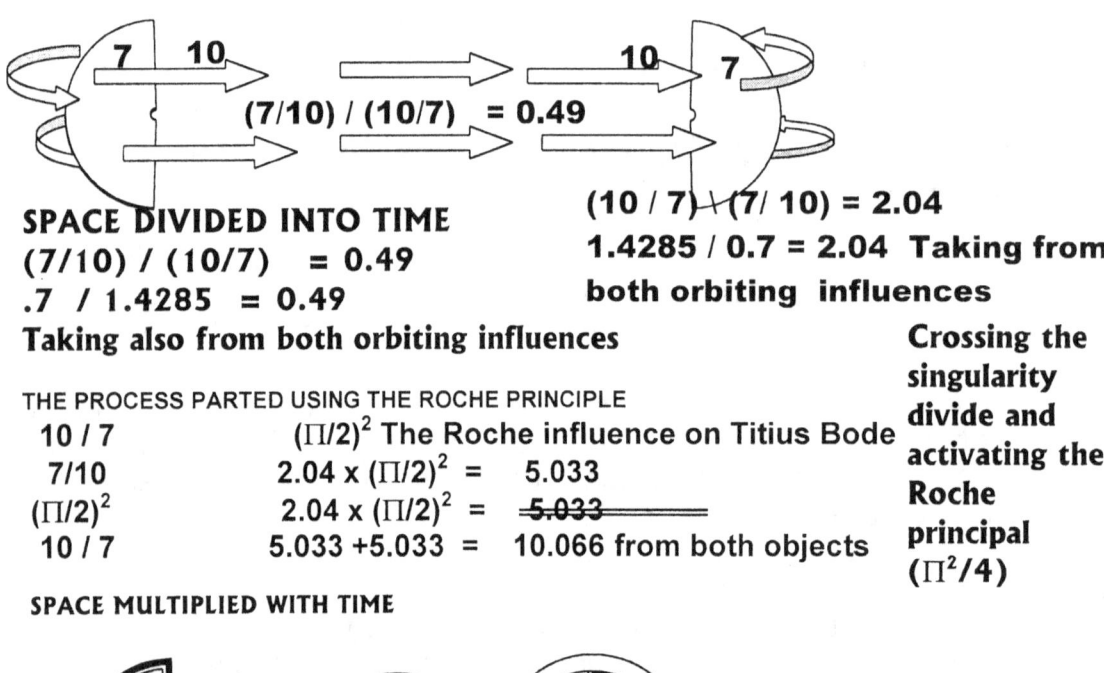

(7/10) / (10/7) = 0.49

SPACE DIVIDED INTO TIME
(7/10) / (10/7)   = 0.49
.7 / 1.4285 = 0.49
Taking also from both orbiting influences

(10 / 7) ⅹ (7/ 10) = 2.04
1.4285 / 0.7 = 2.04 Taking from both orbiting influences

THE PROCESS PARTED USING THE ROCHE PRINCIPLE

10 / 7	$(\Pi/2)^2$ The Roche influence on Titius Bode
7/10	$2.04 \times (\Pi/2)^2$ = 5.033
$(\Pi/2)^2$	$2.04 \times (\Pi/2)^2$ = ~~5.033~~
10 / 7	5.033 +5.033 = 10.066 from both objects

Crossing the singularity divide and activating the Roche principal $(\Pi^2/4)$

SPACE MULTIPLIED WITH TIME

(Π/2)

10 / 7

7/10

(Π/2)

7/10 / 7/10 = 1 and   10 / 7 X 7/10 =1

From dissecting the formula I prove that:

I prove gravity is strongest where space is least (not where the Universe goes flat as Einstein promoted).

I prove there are relevancies that are all applying equally and without such relevancies in balance there are no gravity

In the relevancies there are opposing motion where each participant provide a motion contradicting the other relevant motion. I follow on what Kepler introduced when he introduced the fact that space is half of the motion and the motion is the other half of space. Therefore space cannot be if not in motion and motion is only about when it applies to individual space moving in relation to other space. That relation is the time factor and time depends on space moving from one point to another point. To shortly condense my view I would explain my view as follows: In gravity there are two opposing quantities each representing singularity within one unit. The motion of the one is getting away and the motion of the other is reeling the running away in. From the view the onlooker has it may seem as if some rabbit is pulling some dog all over the area that is remaining in one place where the dog holds the running rabbit in one area circling around the dog. This has another angle where as from the dogs point of view the dog will love to dismiss all space and capture the rabbit.

The rabbit on the other hand would love to leave the dog at a distance where the rabbit will never again see the likes of the dog. While the space between the two is a merely a common fact, it unifies their differences. Both in relevancies have to appreciate their differences by the space in the unit that is keeping them apart. The space is the factor that has to resolve the issue being the differences in motion but cannot because different relevancies sustain equilibriums. I prove that as much as there is Newton's pulling there is Kepler's running around and the running around is equilibrium of the other factor providing the running away part.

The angle science is looking at the issue science either dismisses or cannot explain the characteristics or principals, which is there none the less. The explaining of the phenomenon is quit impossible when using the pulling rope magical attachment idea in the manner science tries to explain gravity. Therefore instead of dismissing the rope they dismiss all other factors present by gravity unleashing free motion but they would not release their idea about the rope. Gravity is motion between two particles that brings about mass. In the book I explain this in much detail but frankly there is not enough room to explain this in this web page.

Mainstream science knows about that gravity has never been defined, the Bode principal that is there in all the planets and even the fragmented planet, the Roche limit, the Coanda affect, the Lagrangian system and the sound barrier, but cannot explain any of the phenomena all though the presence of these phenomena is without dispute. It is the explanations about what causes the phenomena that is part of the dispute but in science the manner in which they defend Newton, science would rather discount the obvious phenomena than question the legality of Newton's cosmic views. I only dispute Newton as far as his cosmic principles are inclined. The phenomena being there or not becomes disputed. Science fails to give acceptable explaining of such occurrences we see in the phenomena and therefore disputes the validity of the phenomena and this failing to explain the presence becomes disputing the presence thereof. I on the other hand found a way where these explaining of the phenomena took me past the Big Bang era and introduced me to the start of all starts. Science cannot get past one specific date because they do not accept or understand the phenomena, which I prove started the Universe.

In such a light Scientists must somehow realise they are barking up the wrong tree with the information they have to use to do some explaining. They cannot refuse the phenomena and not realise they must have the cat by the tail as far as cosmology goes. Please remember that with this I am referring to cosmology and not general physics. There is an Earth versus a Universe with huge difference between the two concepts but Newtonians fail to see that because Newtonians cannot appreciate the differences thus they're not able to understand cosmic gravity; they go about blurring the understanding of gravity. If there are that many phenomena (it represents all there is in cosmology) to explain and such little ability to explain (science fails to explain even one) by using the information Mainstream science is using to explain the cosmos, then someone somewhere has to realise there is something drastically wrong in the way they present the knowledge they claim to have. One cannot be serious about science but defend your view by dismissing the validity of all unknown indicating factors presented as such. There then is some gross incorrectness in the way one reasons. The Roche limit is there and no denouncing thereof can remove it from the cosmos. They may refer to evidence received from the Hubble telescope as "the star is blowing bubbles" for the lack of explaining what is occurring but occur it does. One cannot say it is some unknown gesture presented on occasions because by not explaining the pictures present certain foolishness. It leads to tragedies in aviation and the tragedies they are incapable to understand or explain. For fifty years they lost many pilots but still has no idea what brings the sound barrier about, or find the link gravity holds in the process we call the sound barrier. Instead they try to interpret some effect established almost two centuries ago with steam trains that is travelling at the same speed that horses run. No further investigation with the science in hand brought them closer to new facts! That they should rather see as a sign telling them they are going about incorrectly because by ignoring the cosmos one produce a fantasy and not science. Nevertheless, it does not because tell them anything because Newton did not say so. When I first came upon the amount and the totality of the unknown quantities in cosmology as well as the complacency those involved has about such unknown factors being discarded, it stirred a sense of disbelief and I decided to respond.

All principles I use in the theory I introduce with the publishing of this book. All principles I apply are part of nature. I base my theory on heat stabilizing through space using motion to produce cooling. That is gravity.

I believe some of Creation remained as some particles formed by applying gravity in motion and the lack of motion in others became the lack of gravity, which inspired overheating which then formed plasma. Plasma is the result of heat where light is the epitome of heat. How light became plasma is rather obvious, which again I believe (within reason) I do prove. I believe heat is the destructed form of material and this information the atomic thermo explosions give us.

Analysing Kepler's formula without Newton interrupting Kepler's work helped me realise science has been running on an error for the past three hundred and fifty years. Please let me explain: Tycho Brahe and later Kepler made a study of outer space as never repeated afterwards. From this Kepler concluded that $\mathbf{a}^3 = \mathbf{T}^2 \, \mathbf{k}.$

We all know that a^3 is space and with the space indicated as being in the third dimension and the third dimension is unmistakably a cube that forms volume, which by definition is presenting space. We also know from the way calculations come about by using the formula of Kepler that T^2 is the duration of a specific period of time relating to a specific centre. On the one hand we have space a^3 and on the other hand in direct relation to the space Kepler introduced motion coming from a centre that forms time T^2 k. Kepler gave us space-time a^3 / T^2 centuries before Einstein gave the concept a name but no one ever took any notice. In the formula is space a^3. In the formula the space a^3 has direct relation to time T^2 If k is a^3 / T^2 it means that from the centre holding the gravity is space-time. Space is a^3 and the motion of space a^3 we accept as time T^2 k and such accepting is part of our understanding for the past three hundred and fifty years. Kepler gave us gravity before Newton named it as a force. Kepler gave us space-time long before Einstein named the notion. With Newton's meddling he missed Kepler introducing gravity as $k=a^3/T^2$ space / time.

I believe that I achieved an all time breakthrough success because I can now explain what gravity is. Remember that not even Newton could explain what gravity is or where it comes from, but Kepler did that without any person ever noticing. Scientists over the years paid the price of ignorance about gravity by their unwillingness to investigate the father of gravity, which coincidently is not Newton but Kepler. From such explaining what Kepler said without Newton changing formula on Kepler's behalf, I prove the Titius Bode principal also known just as the Bode principle. I prove that the Bode principle forms gravity when using the Roche limit. These phenomena were never explained or understood by Mainstream Science although they appear more than regularly in the cosmos. In the same breath I might add that Kepler also was never investigated. My achievements came from my effort where I separated Kepler's work from the opinion that Newton formed and that he (Newton) gave his compromised views about Kepler's work to the world. For instance from Kepler's work I can explain the operation of the Black Hole, which not even Prof. Stephen Hawking understands. That is because Hawking ignores Kepler. In my opinion my explaining of gravity makes much more sense than the accepted force of Dark Age proportions…and the best part is you do not have to be a genius to realise or understand it. Even a simple person such as myself can see it clearly! From my view a force is just motion applying and that is what Kepler said gravity is. Kepler said $a^3 = T^2$ k. I dissected k as a factor in the Coanda effect and found that the Coanda effect is proof of my view about gravity and the ability of establishing gravity by centralising space, which forms singularity that produces the Coanda effect. The Coanda effect is the establishing of individual space a^3 by applying motion T^2k. Where the Coanda effect is producing gravity and such producing is stronger in a small space than the gravity produced by the Earth in that spot I use that principle to show that there was some manner in which the reducing of k brought about a stronger T^2 just as Kepler said. This was a crucial part during the Big Bang and therefore had to play a part during the period of the Big Bang. Einstein came to this conclusion but failed to refer his view back to Kepler.

As presumptuous, as it may be on my part of trying to disprove Mainstream Physics, such a presumptions does not change the truth about Mainstream science being incorrect about gravity. After all they admit they do not know what gravity is. I am not disproving anything because they agree they do not know, which paves the way for my showing what gravity is. By admitting not knowing what gravity is they then also admit there is a chance that they can be incorrect about gravity but unfortunately mainstream physics do not see it that way (yet). The question in hand is finding what role gravity played when the Creation came about for the first time. I had to find a method that would allow me to explain why gravity played a role.

My ambition is proving the Universe not coming from nothing and therefore outer space cannot hold nothing. By taking Kepler's $k = a^3 / T^2$ and using k as a line I show through using the line as an example that the cosmic Universe holds everything and all concepts. However the only thing it does not hold is also the only aspect not present in the Universe at all. That is the value of nothing or zero in as much as carrying the definition of the absolute absence of any value. This means the Universe is filled to the point it is overflowing which we call the Hubble constant and not there is not room to be empty. With the line that light uses to flow the lines eliminates any such a possibility of nothing being present. Mathematics is a means of communication about matters concerning the cosmos. As an intercultural language spanning across race and ethnicity or as a principle as such mathematics cannot have zero because mathematics indicatings lines, which is about not applying the numerical number or value of nothing. Everything came about from singularity and Einstein proved that. From singularity nothing ever had the chance to enter space. I challenge any person that disagrees with this statement to show mathematically where nothing as a factor ever entered the mathematics of the Universe. If there is any one there believing there is nothing in outer space I challenge that person to prove mathematically where nothing is a factor in the cosmos. Your attempt may either be before or after reading my work but my challenge will stand since mathematically nothing cannot be part of mathematics. Multiply whatever with zero or nothing and such multiplying results in nothing where nothing is then, can establish no multiplication. Kepler gave us the relation between cosmic objects as $k = a^3 / T^2$. From the formula k forms a connecting straight line filling the first dimension and not the single dimension because k in the single dimension is not zero. It is unproven how k can backtrack to become $k = 0$. I deliberately press this point and make it an issue because that removes all the theory of mainstream science from any logical base they have in support of their views that space is, holds and comprises of nothing.

In the book an open letter To Selected Academics ISBN 0-9584410-9-X the book only and exclusively deals just with the fundamental basics of my theory. I do not elaborate or explain the broader aspects or form an overall view. I found if I do that before a solid understanding of the basic concept is established no concept becomes established. The most basic to explain is that the line cannot start at zero because then there can be no line to follow zero. The cosmos has lines forming cubes and lines forming circles, which in applying 3D manifests as spheres. Between the circles and the cubes run lines, so the key to understanding the Universe is the following of a line. The Big Bang was a time when the Universe was incredibly small making the running lines small. Understanding the Universe is taking the line connecting particles through space back to its limits where such limits were during the Big Bang. But the reducing cannot go to zero because zero removes the line all together. By reducing the line to where the line will not reduce any further we will find at that point that all points land on the same spot. The spots all share one position because that is the only position there is to hold in the form singularity presents. That is singularity being one to all but it is not zero. Finding form in that point shared by all will give a value of singularity. Extend that value received to a Universal centre and bring that value to align with Kepler's $a^3 = kT^2$ and understanding the Universe by finding the centre of the Universe makes the Universe simple as can be The Universe becomes sensible making the entire different yet unexplained phenomenon as easy as children schoolwork. There are suddenly no more mysteries in the Universe. It is only possible when we see gravity not as a grabbing force instead of seeing gravity reducing the space between particles. Gravity is not being some magic force found between particles grabbing onto everything.

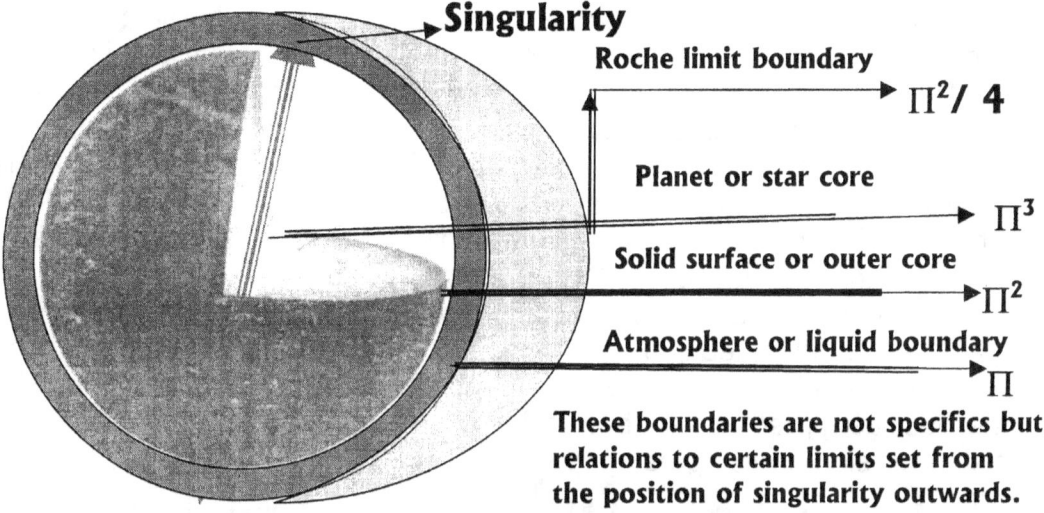

Singularity

Roche limit boundary

$$\Pi^2 / 4$$

Planet or star core

$$\Pi^3$$

Solid surface or outer core

$$\Pi^2$$

Atmosphere or liquid boundary

$$\Pi$$

These boundaries are not specifics but relations to certain limits set from the position of singularity outwards.

It proves that there are dimensional implications all around and that the dimensions are valid. The same implications are validating other principles in the cosmos such as in the case of the Titius Bode phenomenon.

We stand on the outside 150×10^6 km from the spectacle and from such distance we judge the sun. We don't even judge the sun from what we can see but we judge the sun from what we feel. We feel heat coming from the sun and from that we argue that the sun is hot. We see the sun has heat rising from the surface as a liquid soup. That puts the hydrogen layer as the outer layer in a liquid. Hydrogen freezes on Earth at a temperature, which is the coldest amongst all other elements. Yes, the sun is 6500 ^{0} and that is on the outside. To a human that is hot but a human has no mind judging the sun. If the sun squirts pure heat turned to liquid from the surface and the heat falls back into the surface the sun is a lot colder than the Earth is. The earth requires an enormous effort to cool hydrogen down to a liquid state. We must mind the way we think of the hydrogen in liquid. The hydrogen remains a solid. The element is untouched by temperature differences. It is the heat environment surrounding the hydrogen that changes from a gas to a liquid to a solid. One removes or one amplifies the heat in which the hydrogen is and that turns to liquid or solid or gas. The hydrogen is untouched in the elements worth.

Yet we see the heat flow amongst the hydrogen as a liquid. Nevertheless we remain adamant that the liquid   is a gas and the hydrogen is in a gas and the sun is a gas bowl filled with hydrogen because to our mind hydrogen must be a gas.

After all, our element table classifies hydrogen as a gas and that is the way we think of hydrogen. We do not consider hydrogen to be in a liquid state when we see the heat is flowing just like a liquid and shows all indications that it is a liquid. No the sun is hot because the sun feels hot.

In the Universe there are no hot or cold but a state of differentiation produced by time. The Universe parted by parting heat from cold when eternity parted from infinity, when Π^0 singularity parted from Π singularity, when 1^0 parted from 1^1. There is no hot or cold but there is a relevancy where one factor cools and another factor overheats. By retaining the sun is the coldest space in the solar system and outer pace is the hottest there can be.

From since the time that man discovered intelligence (if he ever did) man has been with the presumption that the sun is the hottest centre in the solar system. Later on in the present time, it came to someone's attention that the sun also holds the solar system in gravity. The Earth by its standard and dominating its sphere of which it can control with influence is the hottest centre in the space of its domain and it holds the moon centred to the Earth. The gas planets are the hottest centres in relation with the most heat and they all hold their satellites captured by a hot centre. All space structures hold in every centre there is that is confirming their independence at that point of securing independence the centralizing of the most heat it is able to concentrate and from that centre holds all material captured or controlled in the domain of what that forms the independence of the structure. I can go on and on but heat in the centre couples gravity to space-time, just as if Kepler said before he was spoken for on his behalf and without his permission or his agreeing to it.

$a^3 = (T^2 k) = a^{3 + 2 + 1 = 6}$ with the sphere presuming the position of singularity as part the of $k^0 = 1 = $ **singularity**. Einstein proved that at the point where space reduces and such reducing reaches a point where space as a factor in the third dimension disappears into the single dimension (space going flat) gravity is overwhelming. Einstein interpreted this, as the complete Universe going flat but while it may be true that the Universe is going flat, that can only be within singularity since singularity represents the Universe as flat as it can get.

The centre of any sphere has to be at the very point where space completely falls away. It is at the point where all the points of line centres meet by the crossing the centre of their individual connection coming in to contact as a group. In that way one may assume that the lines connecting the controlling points on the other end are crossing on a centre point that all that is participating in the constructing of the sphere is democratically electing such a centre. Please note this conclusion very well because this forms the heart of the Coanda principle. That will put that position where the lines cross, which in itself is centralising all space in the sphere at that point, such crossing point will become very distinct and controlling where that point forms in the single dimension and singularity is the single dimension. Kepler also solves another riddle that truly got Newtonians unstuck. This, to which I now refer, is what is referred to when they refer to the Hubble constant.

The growth we see in the Universe is an adding of space in every cycle completed by every cycle, which all the protons complete. The adding is the smallest addition that can come about in the shortest period of repeating by cycle rotation there can ever be. This growth of space-time next to singularity confirms the growth of singularity as singularity recalls the space it uses to grow in the time it grows. The margin of growth will be by the extension of **k** in the formula $k = a^3 / T^2$. Every cycle completed in the relation to space by the initial value of **k.** $k = a^3 / T^2$ leaves ultimately a^1 extending as space or as Kepler chose to indicate it as k^1. That too has to be compensated by the duration of time reducing the time aspect by the margin that the space expands. This confirms what is evident in the Hubble Constant. The further one looks at time the more time seems to race because time has the invert properties we give to space.

There is a position that is in motion that is forming the very edge of the outside. To be in motion the position must be in relation to a point from a centre. From the centre, there must be a specific allocated space ending at the object in motion and starting from a centre that has no dimensions. The object in motion determines the one limit and the centre with no sides and no space, which is standing still in singularity, determines the other limit. By that we can see there are only one way of looking at what we can observe and that is from the outside in.

The atom must be the utmost coldest and the proton is even much colder because when that cold escapes it turns to heat forming space that no one can understand. When the spin of the atom allows the cold of the atom to release the heat it had it had frozen to space the atom holds but when this heat releases from the containing form of the atom it brings about much more heat than the Human mind can cope with. One may not look at the material and judge the surroundings. The fact that hydrogen remains a gas and so does helium in outer space must serve as enough proof that outer space is hot, regardless of our interpretation of the temperature gauge telling us what we wish to hear. One must look at outer space and judge outer space from the findings only considering outer space. If helium remains a gas, it is hot. The removing of heat makes the centre of the Earth cold although we see it as being terribly hot. The only reason why it can seem to be hot is because it is cold and in such a cold environment, the heat can gather and space can collect heat because the particles find the surroundings extremely cold.

The cold in the earth centre causes the concentration of heat by space reducing, as all cold surfaces tend to do. If it was hot, the space within the Earth would expand and the space within the Earth where we think so much heat is concentrated does not expand therefore it must be cold. To gather and accumulate the space in a liquid means it became much colder being a liquid. Finding the surroundings terribly cold will allow the heat to gather and not expand but when the surroundings are hot, it will not tolerate more concentration of heat and thus will expand to rid the balance of excess heat within space. Look at the sun and see how the sun turned the hydrogen to a freezing cold liquid at 6500 K. Hydrogen is in a fluid state within the sun and is colder than the hydrogen that is in a gas form in outer space. The sun is the coldest place in the solar system. That is when the protons oversupply the removing of space to produce the cold that is so apparent. By the reducing of space, it can concentrate heat to a fluid state by producing the opposing cold that finally freezes the heat to a solid state.

The expanding of space is a way of duplicating space without reducing space and by duplicating in the form of expanding it becomes just the opposite to duplicating by motion therefore reducing space by halving space in time. That is what gravity does. By motion, space duplicates and by space, halving it removes heat in space as well as by dismissing space. In all the applying of gravity, space dies. The density of the protons brings about space dense enough to harbour the heat in such quantities and visa versa applies in outer space.

The application of gravity that condenses space and bringing about heat by the compressing of space we apply in the way we go about tapping into the energy that nature provides. Internal and external combustion engines all rely on this application for harvesting motion by driving power. Compress space even today with a piston in a cylinder and then pump the compressed air into a container and such confining of space will increase the heat by the piston effort to reduce the space brought about in the container. The heat coming about inside the cylinder has no relevance to particles colliding because all compressor cylinders cool down. The walls become colder because when that cold escapes it turns to heat as the heat releases from space forming a secondary form of material forming space that no one can understand when the spin of the atom allows the cold of the atom to release into uncontrolled space. This release and unification with space that heat does is the heat it had frozen because the motion of spin to space that the atom holds, remains in a frozen state under the guard of the spinning electron. When this heat releases from the containing form of the atom frozen by the spin of the electron, it brings about much more heat than the Human mind can cope with. One may not look at the material and judge the surroundings.

The fact that hydrogen remains a gas and so does helium in outer space must serve as enough proof that outer space is hot, regardless of our interpretation of the temperature gauge telling us what we wish to hear. One must look at outer space and judge outer space from the findings only considered in the terms which outer space insists upon. If helium remains a gas, it is hot. The removing of heat from the space that contained the heat makes the centre of the Earth cold. In our universe we see it as being terribly hot because the heat then forms a separate substance but remains a form of material (8) but that is because we see the heat and not the space derived from the separating of the heat. The only reason why the space can seem to be hot is because the space is cold and in such a cold environment the heat can gather in a much concentrated state and space can collect heat because the particles hold concentrated heat in the space separating the particles.

By removing such high concentration of heat from the space that used to be expanded heat, the space then must contradict the heat by being extremely cold. We look at the heat in the space, which by that time is another form of material and find the surrounding heat in the space hot while the space is extremely cold. The cold in the Earth centre causes the concentration of heat by space reducing, as all cold surfaces tend to do. The proton contributes to that reducing of space. If it was hot the space within the Earth would expand and explode but the space within the Earth where we think so much heat is concentrated is so much it does not expand therefore it must be cold. To gather and accumulate the space in a liquid means it became much colder when the space parted from what then is being a liquid. Finding the surroundings terribly cold will allow the heat to gather and not expand but when the surroundings are hot, it will not tolerate more concentration of heat and thus it will expand to rid the balance of excess heat within space. The concentration or release of space with heat or space from heat is a direct contribution of the singularity in control of the space-time. The regard of the singularity stipulates the conducing of heat in space or the release of heat to form space by means of bisecting the occupied space.

Look at the sun and see how the sun turned the hydrogen it holds captured in its atmosphere to a freezing cold liquid at 6500 K. Hydrogen is in a fluid state within the sun and yet it is still colder than the hydrogen we find in outer space that is in a gas form in outer space. The sun is without any doubt the coldest place in the solar system. That is when the protons oversupply the removing of space to produce the cold that is so apparent in the heat levels that the atom cannot absorb in normal growth and therefore do cannot find accommodation in the walls of the atom. By the reducing of space, it can concentrate heat to a fluid state. By producing the opposing cold that finally freezes the heat to a solid state, we find that is what matter is. The expanding of space is a way of duplicating space without reducing space and by duplicating in the form of expanding it becomes just the opposite to duplicating by motion therefore reducing space by halving space in time. That is what gravity does. By motion space duplicates and by space duplicating the material must be by dividing or bisecting - halving it removes heat in space as well as by dismissing space and in that concentrating heat. The density of the protons brings about space dense enough to harbour the heat in such quantities and visa versa applies in outer space.

The particles claim more space when heated to preserve the cold. The claim to more space produces more space and reduces more heat. Such expanding brings about cooling. When particles heat or cool motion applies in some form. Motion started at a point when the Universe was extremely hot and there was no space. By introducing motion space formed and the lack thereof produced friction that became heat that became space. It is natural, it is simple, and above all, it makes believable sense.

The application of gravity is that which condenses space by bringing about heat with the compressing of space. We apply the progress we have as a species in the way we go about with our skills to unveil ways we can tap into the energy that nature provides. Internal and external combustion engines all rely on this application for harvesting motion by driving power. Compress space even today with a piston in a cylinder and then pump the compressed air into a container and such confining of space will increase the heat by the piston effort to reduce the space brought about in the container. The heat coming about inside the cylinder has no relevance to particles colliding because all compressor cylinders cool down with time moving and not necessarily with the loss or release of particles. It is not only the discharging of air that will reduce the temperatures inside the container. The time flowing bringing motion about where the motion is not about particles escaping but heat escaping in the replacing of the heat density (not the density of the particles forming the material content within the container) but the space that compressed to heat will also bring about that the heat displaces through the container wall to the outside. This is bringing about equilibrium where heat will always flow from more dense areas to the lesser dense areas. This has no influence on the status of the particles on the inside of the cylinder but only concerns the density levels of the particles inside versus outside. After the pumping of air increased the heat in the cylinder which even can go to dangerous levels, will reduce back to room temperature when further pumping ceases and that stops further air movement into the cylinder and such surging of pumping air is what brings about heat stabilizing.

Mainstream physics ignored the clear connection completely, notwithstanding it being so very obvious. There is this far in their recognising of principles in natural physics not one single reference made to prove their appreciation of this matter. They are bent on particle colliding. When particles collide, such collision forms an atomic thermo release and that action we call an exploding atomic bomb. What principle this argument about particles colliding, ignores is that all atoms use negative charged electrons forming the atomic limit on the outside forming a definite border to the boundaries of all atoms and in both electrons from different atoms are being negative charged. In being negatively charged, it means both will come out and totally reject the other. The closer they come the more violent the rejecting will be and such rejecting is the production of heat that will turn to space. The electrons repel other negative charged sub atomic structures, which the electrons are that form the outer borders of all atoms. With all electrons highly negatively charged (being as negatively charged as any possibility will allow to match the utter extreme) such electrons could not touch.

It is about time scientists start looking with their minds and not their eyes at the Universe and see what is truly out there to see. All the difference we find is seated in the human mind. We humans set differences because we look at the cosmos by placing humans and the life we find on Earth in a pivotal centre in the cosmos instead of placing singularity in the centre and life where it belongs; only found on Earth. Einstein proved mathematically that in the presence of a strong gravity such a strong gravity slows time down. Surprisingly with that evidence being around this long, nobody in science since Einstein's discovery took those statements and made any further progress from that. It seems to have been left in some drawer to dry. Science still sticks to the opinion that time did not change, not even slightly, since the beginning of the time it held the same pace ever since the start of the Big Bang notwithstanding the implications this concept carries. Before the Earth took one year to circle around the sun and even before the sun was there a year was still the same duration of one year. How odd... don't you think ... that the only aspect in the entire Universe that is beyond change is the aspect of time? With the entire Universe including all the gravity now present and not excluding one Black Hole or dust speck pressed in such an area that was possibly the size of a lepton even then the gravity extending from that circumstances must have been beyond what words can ever describe.

When everything was that small when the Big Bang took charge, the gravity at the time was beyond light, because even today in the Black Hole the gravity is beyond the speed of light. If the gravity was that high and Einstein already proved that strong gravity slows time down, then there is one logical conclusion and that is that time was in fact at the time of the Big Bang standing still. Mathematically it is incorrect to allow gravity to compress the Universe into a spot smaller that an atom and exclude any other factors and relevancies to change.

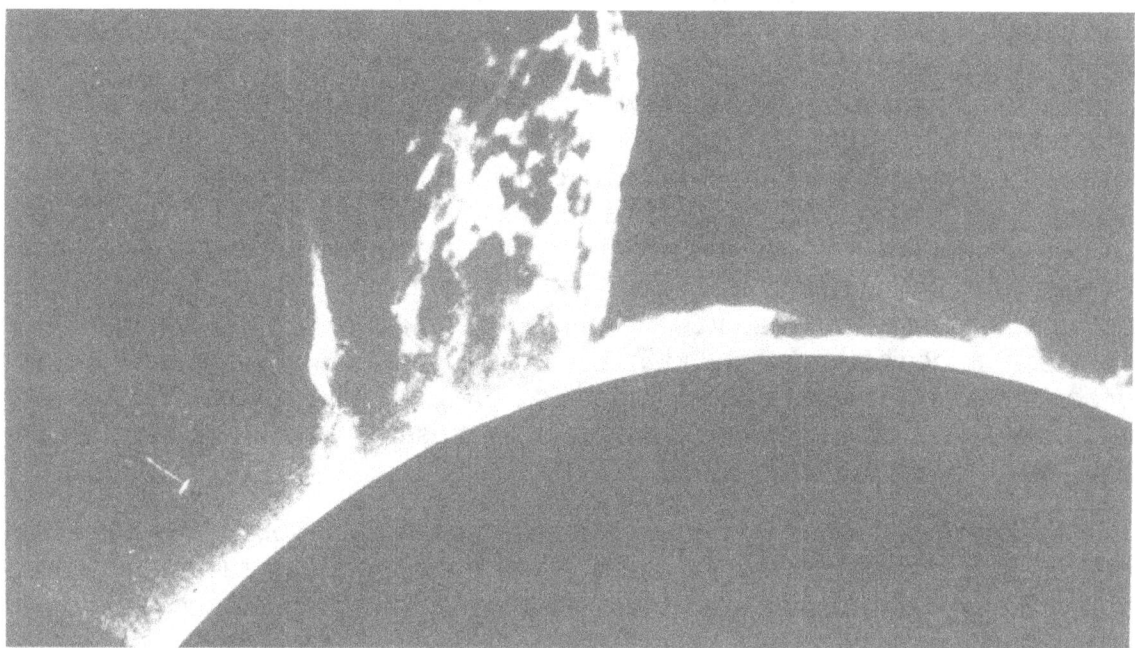

The fact that the prominence squirts out liquid in vast amounts is because there is a lot of space reducing going on in the Sun so there is less space inside the Sun. The fact that the prominence expands to outer space means that the prominence was expanding into a hotter area and does so as a liquid. However that fact that the prominence fell back into the Sun can only result from the prominence not being hot enough to return to a gas state and then through such density discrepancies had to return to a less hot area. That is science not magic.

By using Newton one cannot even begin to explain any one of or the combined efforts of the above cosmic phenomena that are all over the cosmos and forms all the laws in the cosmos. Newtonian definition cannot even recognise any of the principles but only Newtonian science are taught to students. No student can have the fortune to disagree with Newton and remain a student. If the student will dare to disagree with Newton it is the end of such a students academic career. By setting this firm condition Newtonian science becomes institutionalised mind conditioning of the concepts of thought forming in physics.

With my saying this I have not made one academic friend but neither have any one proved me wrong. Students are taught to accept Newton and to ignore Kepler and any student doing it the other way around will fail all examinations and other testing at Universities. Students accept Newton or they accept a ticket taking them home. Newton is an institution force fed to each following generation but saying that reserves only resentment towards me amongst Academics. According to Newtonian science space is simply nothing with no qualities but gravity separating space and space does not mingle, as one would expect if space was nothing because space does form borders.

Disasters of unprecedented magnitude arise from such borders. The Challenger disaster of February 2003 is pertinent testimony to those borders that was powerful enough to break the aircraft into pieces while the explaining contributed by Mainstream science is evidence of a shocking lack of understanding about what took place as cosmic laws were breached. I do not pretend to be of superior understanding and do not place myself on any pedestal. On the contrary the information is so simple and so easy to understand that the lack of any Academic understanding frustrates me almost witless. But academic taught culture demands all persons to miss the evidence, which is so clearly visible because academics demand researchers looking in other directions because students are forced to accept Newton's vision about Kepler's work. By the time they reach researchers status, they too have tunnel vision that can only acknowledge Newton and ignore Kepler. Our not understanding laws, provide a platform for future disasters occurring because it will lead to us ignoring more of applying principals that leads to space tragedies of magnitudes we have not thought of as yet. By not understanding the sound barrier, tragedies have and will again come about and will increase as misconceptions become more present in the future because the demand on space travel increases.

The book **an open letter To Selected Academics ISBN 0-9584410-9-X** is about that process adapted by the Big Bang, never ended and it is still bringing over, that which is in unoccupied space to material being in occupied space. Occupied space holds matter and unoccupied space is empty of solid materials. There is contraction, which we know by the name we gave as gravity. Then there is expansion, which we gave many names being the Big Bang and the Hubble Constant or better known as simply exploding or forming plasma with all the terminology accompanying that simple idea. This I show is antigravity. Apply heat and space and a balloon lift where such lifting is antigravity. There is a balance in the Universe where gravity contracts and reforms space and heat expands becoming space and produces space.

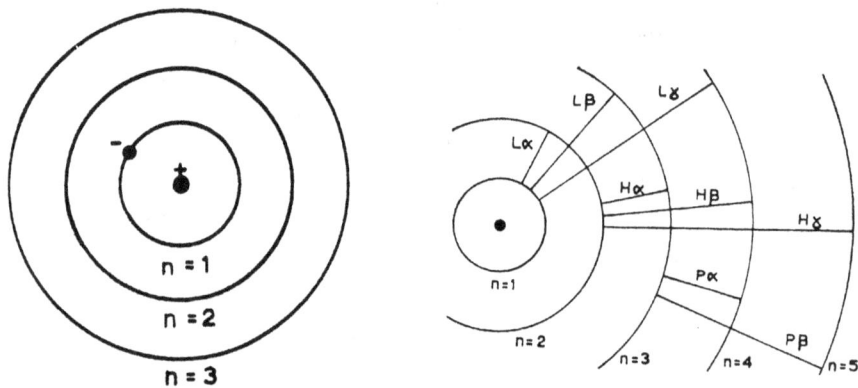

Whenever an electron jumps from a higher into the lower (innermost) orbit, the atom gives out radiation at a wavelength corresponding to a spectral line of the Lyman series. Jumps down into the second lowest level contribute to the Balmer series. The greater the jump, the closer the emitted radiation is to the limit of the series, which is reached when an electron enters from outside the atom. Outward jumps involve the absorption of energy and give rise to absorption lines. Even today the forming of heat increase becomes space and proves that material is time delay. This was the process that was started by the spot becoming the dot and the only difference is the increase of space to the demise of time.

This puts my theory in line with reality. The only way anything can get bigger is when heat is added to what there already is. The Bomb at Hiroshima and

Nagasaki showed how intense the heat in atoms are and how well packed the heat is which is contained in the atom. Bringing more heat to the atom brings about the atom having more of the same but in a higher measure.

Only heat can make material expand and with the Universe unable to expand because the Universe is what ever can be, it must be the material inside the Universe that is expanding. In the same measure we find where space reduces heat, is removed from that space. Gravity is about contracting and that is true. In the manner that Kepler put it material that is filling space (a^3) is equal and the same as the motion ($T^2 k$) of the Material moving. That means to have time then time must move and the only way time can move is to move the space in the time.

However to move is either to come closer and using Kepler we can see how Kepler would reduce space in $k^{-1} = T^2 / a^3$. That means the expanding subsides to a point where contraction is and contracting is about cooling or reducing the Heat that was expanding the space. Then there is expanding of material which is as Kepler would put it $k = a^3 / T^2$. That would be when the space becomes less and the only valid way for space to become less is when what there is becomes less than what there was. That amounts to cooling where the heat is being removes from the space.

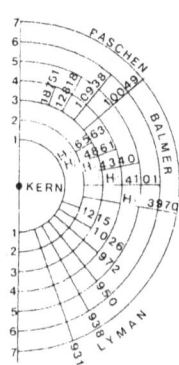

One should see the sound barrier as an electron with the Earth being the proton, the atmosphere being the neutron and the aircraft being the electron. The electron (aircraft) is attached to the Earth (proton) by a flexible membrane (atmosphere) where the Earth is in the Coanda principle of $a^3 = T^2 k$. The combination is the Earth a^3 being in place by the motion T^2 in the atmosphere that the aircraft as an individual structure k provides. The motion of the aircraft T^2 puts the distance k inn relation to the entire Earth package a^3. However as the case is with the Coanda principle as soon as T^2 increases so would k because a^3 is getting larger $a^3 = T^2 k$.

When the aircraft is stationary the aircraft holds the least energetic line being $a^3 = [T^2 = 7(3\Pi^2)] [k = \Pi^0]$. As soon as motion commences k increases because although the relevancy from the Earth perspective remains the same the relevancy from the aircraft changes drastically. From the view the aircraft holds the k that the Earth has becomes the T^2 that the aircraft has and the T^2 that the Earth holds becomes the k that the aircraft uses.

As soon as motion commences k increases because although the relevancy from the Earth perspective remains the same the relevancy from the aircraft changes drastically.
From the view the aircraft holds the k that the Earth has becomes the T^2 that the aircraft has and the T^2 that the Earth holds becomes the k that the aircraft uses.

The sound barrier is just another manner in the way the Universe brings about gravity. The aircraft has to produce excessive heat and by more heat delivers the space between the Earth and the aircraft increases. The motion becomes more and that stretches the space connecting the Earth and the aircraft applying heat to produce extended motion where the extended motion leads to extending of the space the aircraft covers in the same duration of time.

The heat (formed by the release of motion in the engines) allows more motion to apply than that which the Earth generates which puts the aircraft in a higher atomic bracket where the aircraft holds more space that the Earth normally would grant the aircraft. The aircraft has more gravity (granted that it is using the motion of the Earth which is the gravity of the Earth) than it would have being stationary. With an additional source of heat the aircraft can add to the earth gravity and the Earth gravity is unrestricted motion. It has no bearing on mass whatsoever. Gravity is about motion and mass is the restricting of such motion.

It is not the motion we must be after but what causes the motion in the first place (other than being Newton's pet force). We must find what produces motion and from that then we must think further than what we can recollect from thousands of years of culture that got us this far but is getting us no further. We must see why that which moves or tend to move as we do on Earth in our gravity. It is nothing to do with mass pulling everything about but is a flow of space-time.

When one applies heat to an object it expands. That is primary school science. This states that more heat applied leads to more space acquired by the heated object. In sharp contrast to this is the growth in space when heat levels rises but freezing brings about the opposite result. When I freeze an object that object reduces its occupied space as it shrinks. Removing heat reduces space. That comes directly as nature responds to heat and I can prove that easily.

By expanding it accumulates space to increase the improving of the size of the material. The accumulating of heat is for the sake of securing singularity, which accumulates the heat in the material whereas the freezing tarnishes the overheating symptoms by the removal of material in unoccupied space using external matter and setting motion to the material until it contracts into a form which we see as visible heat. The heat is in the form of dissolved singularity that became material as material used it as growth. That is why by freezing it will diminish the space as to accumulate the heat absorbing into the heat into the material to maintain the equilibrium needed in space.

The atom is the optimum proof of the statement. The atom is the absorber of heat as well as the release valve of heat. The atom regulates heat in relation to space acquired as well as space acquired. The atom is as much about heat as controlling heat and when the atom expands space it accumulates and store heat. When it cools it reduces and absorb space. The cosmos is the atom and the atom is what the cosmos use to regulate heat.

Taking this equation of nature to outer space we seem to confuse the natural law. With outer space as expanded as anything can get we regard outer space as incredibly cold. As heat sets in the normal flow will bring about expanding of heat into the form we think of as space that limits the heat overheating. Outer space is the very edge of expanding of space where heat cannot expand into space any more. Outer space is the limit, the epitome of expanding where heat meets space at the edge of all limits once more. Therefore being the representation of the very limit of expanding outer space has to be the hottest place there is. By applying heat to a kettle holding water, the adding of heat manifests as steam and steam is hot water that traded heat as it reviewed space. By allowing the receiving of the heat to continue the container will let loose steam in order to match the contributing of space.

The manner in which heat expresses itself when confronted by overheating is to provide additional space through expanding of space. Outer space is outer space because outer space has expanded all it can it is still expanding to the speed of Hubble's $1/H_0$ which inevitably does not only affect far-off places where we cannot be, but effects us on a daily basis. As outer space is stretched to its limit, its limit will continue to stretch but while it is stretching it has to having more than it had before in that outer space holds the limit of heats expanding possibilities. Singularity has been expanding since way back when but that means singularity is still releasing heat as space-time that turns out as space in the universal time of outer space. In outer space heat cannot expand more therefore except for the continual growth that benefits all singularity throughout on a continuous bases concerning all outer space.

Every element is in relation to the heat level it uses in forming the gravity it has. One can see how the forming of the numbers of elements available in the Universe stands related to the density of the elements total numbers. More pertinent to note of that the effect of gravity is not in the mass of the element but shows a much stronger relation with the density and the density is the relation the element has with the heat that marks a boiling point or a freezing point The density factor shows what we use to classify the element in relation to being a liquid, a gas or a solid. This factor is much more prudent than the mass factor and that I show later on as the book develops.

If singularity expands when heated and there is a limit to the point it can heat, and where that point forms the maximum expanding possible, then it has been reached in the area we think of as outer space. Outer space has expanded through the unleashing of heat, where overheating is turning liquid heat into space. Any explosion is a vivid reminder of this fact and the unleashing of space is so real it destroys the space holding solids by rearranging the construction of the solids. With that in mind we can declare with great confidence that outer space as the hottest place there is. Whatever expanding there possibly is, was done to secure the cooling and all cooling that can be introduced to bring about further cooling was performed in the area we think of as outer space. Forget schoolboy culture and the temperature scales and other Newtonian scientific defects I call Xepted mistakes. Think of reality and throw out culture teachings Use the mind and not the thinking power of the past. Any place that can expand no more is the hottest place there is just because of the shear implication that it can cool no further is as hot as it gets anywhere. If that is the case then it is safe to say that galactica is freezing cold notwithstanding our concepts of heat and space and heat in space given to us by our collective culture and not by our ability to reason.

The galactica is little frozen islands in a vast see of heat. That is the reason we can see the galactica because the galactica is space concentrated by a frozen space. The galactica is slowly heating and therefore it is expanding into outer space. Outer space on the other hand has expanded to the maximum that it can yet we think it is cold when it is the extreme there is in heat that introduced the maximum expanding. What I now am saying might be deemed by the most purist as the contradiction of the century and that much I do realise. At the inner core of a star all space shrinks into the oblivious but we consider the inner core area of a star to be the hottest spot in the solar system. That just cannot be because when material shrinks it becomes cold and by shrinking into the oblivious it has to freeze into a fusing element as newly formed units. Again that is the contradiction of the century. Why will that be? The space inside the star shrunk to the minimum there can be and that tells us the space has to be cold because of the shrinking took the space to a position where no space can shrink anymore.

That shrinking of no more space can only be inside the inner star and in that region is where we locate the strongest gravity. With outer space as expanded as nature may allow the space that grew could only grow in conditions of heat because heat produces expanding and expanding is the result of heat coming about. Space shrink because it is cold: that we know and taking this law to the star centre it means regardless of our interpretation of hot and cold, that area in the star centre is as cold as it can get notwithstanding what our nature may tell us. Then obviously the same must apply to outer space for precisely the same reasons because it is so hot there it can expand no more.

At this I have to redeem myself from being human. Only in the eyes of humans are there hot and cold, but as a reality in the cosmos we will find this nowhere. We look at the hotness of space and the coldness of space but it is the relevancy to the solidity that forms the actual heat and cold limits. It is so hot no expansion can produce more space in outer space, as the outer space seems to be the epitome of what can be cold while it is truly hot and quite the opposite reveals as the true scenario inside the star in the centre of a star structure. That means the number of protons in motion has a lot to do with the cold and hot scenarios because where the protons are most dense the cold is in extreme…well in most cases. Only in the absence of space can so much heat gather in excess and the opposite is true about outer space where the least denseness found brings about the space in heat found in outer space. Our human selecting of hot and of cold and what is hot and what is not prevents us the clear vision we would have when truly understanding the applying temperature. Temperature comes about from spin and the smaller the spin density is the colder the space becomes because the more duplication produces the most cold. We think of outer space as 0^0 Kelvin but in fact it is as hot as no other place can be in the Universe. The coldest is where material is freezing solid as material does when frozen solid and the hottest is when by boiling the material is going into a gas with liquid being the intermediate position where heat acquires the space to perform as a flexible substance.

When we look at particles in outer space we see the particles being frozen. It is because there is such a severe contrast between the particles and the environment surrounding the particles and not the particles that is so frozen. The particles are in a gas state because the particles do not form a part that is part of the space unit. Hydrogen clouds of hundred of light years in diameter are a common sight in outer space. The heat we find filling space is not part of the space but like the particles the heat is a separate issue. That heat filling the space is another form of material that could conduce by diverting from space or marry the union of space by becoming more space. If it were that cold which we think it is, it would not have expanded into such a massive cloud but would have contracted forming a cube of frozen hydrogen. But as we can see the cloud expanded the gas as far as the gas can expand.

That expanding is indicative of heat and has extremely little to do with gravity or is it just a matter what we think of as gravity. If you are of the opinion that those hydrogen clouds will contract one day into forming a star, well then think again as there is just no such a chance that that will ever happen because that is not the manner that form of gravity functions. Because outer space is completely overheating the condition it has in support of the particles makes the particles appear to be in a state of freezing but the particles is counteracting the heat limit it meets. However the particles do not contract, as the heat is immense. The space in outer space has absorbed all the heat by means of expanding and will appreciate still further as it will never depreciate. That is not because outer space is freezing the particles but it is because in contrast to the heat of outer space the particles seems to be frozen.

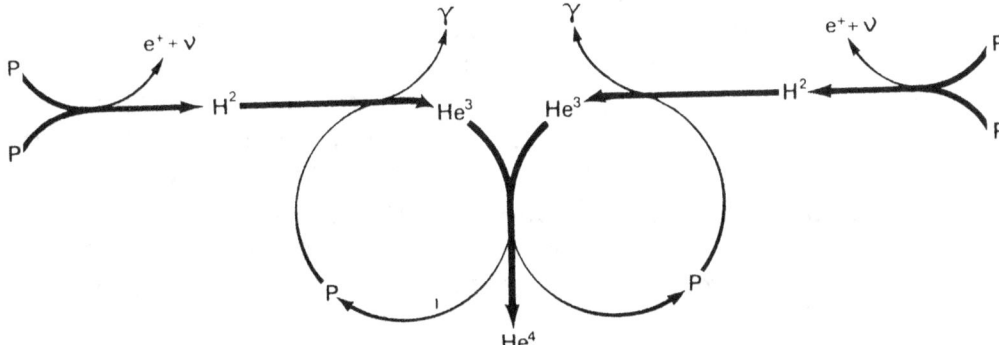

The atom must be the utmost coldest because the proton is even much colder than whet the electron can freeze. In fact the proton is 1836 times colder than that which the electron is able to freeze. We find that when cold escapes it turns to heat and the heat relieves by forming space, however it seems that that no one can understand. Motion brings about cooling. When the spin of the atom allow the cold of the atom to release the heat it had, which it had frozen the heat returns to space. This is what the atom shows in the electron bands or rings the atom holds. This must not be confused with uncontrolled release of heat. When the motion of the electron is interrupted such motion reducing results into the utmost expanding there possibly can be. When this heat releases from the containing form of the atom it brings about much more heat than the Human mind can cope with because no human mind can comprehend the total devastation a nuclear release of space may bring forth. In this I am not referring to the normal way material relates to heat. That is a totally different matter altogether.

One may not look at the material and judge the surroundings. The fact that hydrogen remains a gas and so does helium in outer space must serve as enough proof that outer space is hot, regardless of our interpretation of the temperature gauge telling us what we wish to hear. In the vent of outer space truly being the coldest we have, then hydrogen and helium should be frozen crystals clotted in balls of material. One must look at outer space and judge outer space from the findings only by considering outer space without the prejudgement of teachings about ideas when persons were still held in prison for being suspected werewolves. If helium remains a gas it is hot. However we can witness hydrogen being a liquid in the sun. That squirting from the sun is liquid heat that is frozen as a form of material in the hydrogen layers and holding the hydrogen in form in the hydrogen layer.

We might think it is hot in the centre of the Earth but that type of thinking is as Newtonian as thinking of big stars as mighty gravity pools. The removing of heat into a liquid makes the material in the centre of the Earth cold although we see it as being terribly hot. The only reason why it can seem to be hot is because it is cold and in such a cold environment the heat can gather and space can collect heat because the particles find the surroundings extremely cold. Then again we confuse heat and time altogether and completely but more about that later on...

The cold in the Earth centre causes the concentration of heat by space reducing, as all cold surfaces tend to do. When material reduces space, it parts the material from the heat within and places that heat within the electron bands on the outside of the electron bands. By removing the heat the atom contracts and by contracting the atom reduces space. That heat forming space has to go somewhere. If it was hot the space within the Earth would expand and the space within the Earth where we think so much heat is concentrated does not expand therefore it must be cold. To gather and accumulate the space in a liquid means it became much colder being a liquid. Finding the surroundings terribly cold will allow the heat to gather and not expand but when the surroundings are hot it will not tolerate more concentration of heat and thus will expand, to rid the balance of excess heat within space. That is the terms in which to think in when thinking in terms of cosmology.

Look at the Sun and see how the Sun turned the hydrogen to a freezing cold liquid at 6500 K. Hydrogen is in a fluid state within the Sun and is colder than the hydrogen that is in a gas form in outer space. The Sun is the coldest place in the solar system. That is when the protons oversupply the removing of space to produce the cold that is so apparent. By the reducing of space it can concentrate heat to a fluid state by producing the opposing cold that finally freezes the heat to a solid state. The expanding of space is a way of duplicating space without reducing space and by duplicating in the form of expanding it becomes just the opposite to duplicating by motion therefore reducing space by halving space in time. That is what gravity does. By motion, space duplicates and by space halving it removes heat in space as well as by dismissing space. In all the applying of gravity space bites the dust. The density of the protons brings about space dense enough to harbour the heat in such quantities and visa versa applies in outer space. However it is not purely the density of the protons that produce such cold but the exquisite motion forming a rapid duplication of material and such duplication brings the contraction by removing space. Removing space is also removing heat that is separating material.

We have to accept that the coldest place in the solar system is in the very centre of the Sun because there the most number of protons sharing the least amount of space producing the coldest area that can allow therefore the hottest density of heat within the cold environment. Later I will show why the star is so extremely cold it freezes material together and outer space is over boiling with heat expanding into more space. We have to see what forms space and why space can be the absolute basic container through which gravity can relay the influence it carries.

We have to realise that whatever forms space has to be that same ingredient which also is the basic component that forms the lot of everything in the entire Universe. It is than which becomes more making everything seem more and it is also by removing that which reduces every aspect of the Universe. That which becomes more is what the Universe is built with and it is that which the Universe uses to form its entirety. When particles heat up the particles expand the space the particles hold to limit which the rising heat demands in relation to the heat rising. The particles claim more space when heated to preserve the cold that the material is protecting. The claim to more space produces more space but that in turn reduces more heat exaggeration. Such expanding brings about cooling. When particles heat or cool motion applies in some form. Regarding this fact we can claim that motion started at a point when the Universe was extremely hot and there was no space. However I have indicated that hot and cold are only factors with little specific or formal value in the Universe. By introducing motion space formed and the lack thereof produced friction that became heat that became space. That must be the way the Universe then started.

The application of gravity is the same as the condensing of space and bringing about heat by the compressing of space we apply in the way we go about tapping into the energy that nature provide. Internal and external combustion engines all rely on this application for harvesting motion by driving power. Compress space even today with a piston in a cylinder and then pump the compressed air into a container and such confining of space will increase the heat by the piston effort to reduce the space brought about in the container. The heat coming about inside the cylinder when being compressed has no relevance to particles colliding because all compressor cylinders cool down or become colder when that cold escape through the walls of the cylinder. As soon as the pumping stops the heat releases from the inside space. There is an immediate stopping of the increase of heat as soon as the pumping stops. The material inside the container forms a secondary form of material that comes about since the space reduces and the forming of space is in a turnabout. The compressing of the space inside brings about a rise in the heat levels within the container but apparently that no one in Newtonian circles can understand. By compressing the spin of the atom increase and the motion of the material removes additional heat from the ranks of the inside of the atom. Thus, when the spin of the atom increases it allows the cold within the atom to release the heat the atom holds into uncontrolled space. This releasing of heat and unifying the released heat once again with space increases the levels of heat in the atmosphere of the containing cylinder. What that heat does is the heat that the material absorbed as material within the atom was captured as frozen heat because of the motion of spin to space that the atom holds remains in a frozen state under the guard of the spinning electron. But when this heat releases from the containing form that is the atom in being the biggest cosmic heat container the heat becomes in a frozen state through the motion within the atom. Forming a frozen substance by producing motion that is faster than the speed of light the heat is frozen by the spin of the electron. The spin of the electron brings motion and such motion reduces the heat to a frozen state which is the frozen state of heat we named material. Therefore one may not look at the material and judge the element state of form by its surrounding which is heat it surrounds its electron to the outer side of the containing spin.

Again we must look at the state of material in outer space and realize that the fact that hydrogen remains a gas and so does helium in outer space must serve as enough proof that outer space is hot, regardless of our interpretation of the temperature gauge telling us what we wish to hear. One must look at outer space and judge outer space from the findings only considering in the terms which outer space insists upon. If helium remains a gas it is hot. The removing of heat from the space that contained the heat makes the centre of the Earth cold. In our Universe we see it as being terribly hot because the heat then forms a separate substance but remains a form of material but that is because we see the heat and not the space derived from the separating of the heat.

The only reason why the space can seem to be hot is because the space is cold and in such a cold environment that rejects the heat within the atom. There the heat then must gather in a more concentrated state and space can collect heat because the particles hold concentrated heat in the space separating the particles. By removing such high concentration of heat from the space that use to be expanded heat, the space then must contradict the heat by being extremely cold. We look at the heat in the space, which by being in a liquid state should be by our standards considered as another form of material and find the surrounding heat in the space hot while the atomic material in space is extremely cold. The cold in the Earth centre causes the concentration of heat by space reducing, as all cold surfaces tend to do. But the numbers of protons contributes that reducing of space and the removing of heat captured by the material. If it was hot the space within the Earth would expand and explode but the space within the Earth where we think so much heat is concentrated is so much it does not expand therefore it must be cold within the solid parts. It is the motion of so many protons in such a little space that allow the heat to be contained as a liquid and the extravagant motion by the many protons in such a reduces area forms the ability to contain the heat as a liquid substance without allowing the expanding of the heat into gas or space. To gather and accumulate the space in a liquid means it became much colder when the space parted from what then is being a liquid. Finding the surroundings terribly cold will allow the heat to gather and not expand but when the surroundings is hot it will not tolerate more concentration of heat and thus it will expand to rid the balance of excess heat within space. The concentration or release of space with heat or space from heat is a direct contribution of the motion controlled by the space-time. The regard of the space-time providing the motion, which provides the cooling of the space, stipulates the conducting of heat in space or the release of heat to form space by means of seizing the occupied space.

Look at the Sun and see how the Sun turned the hydrogen it holds and which is captured in its atmosphere to a freezing cold liquid at 6500 K. Hydrogen is in a fluid state within the Sun and yet it is still colder than the hydrogen we find in outer space that is in a gas form in outer space. That must be because of the enormous motion of the particles within the confinement of the sun. The Sun is without any doubt the coldest place in the solar system and that is because as the ferocious motion within the sun. That is when the protons oversupply the removing of space to produce the cold that is so apparent in the heat levels that do not join outer space. By the reducing of space it can concentrate heat to a fluid state. By producing the opposing cold that finally freezes the heat to a solid state we find that it is what matter is. The expanding of space is a way of duplicating space without reducing space and by duplicating in the form of expanding it becomes just the opposite to duplicating by motion therefore reducing space by halving space in time. That is what gravity does. By motion space duplicates and by space duplicating the material must be by dividing or halving. Halving the material, which is heat, is at the same time doubling the space, which is bringing about cooling. By doubling the space as the duplicating of material removes half the heat from a single space and distribute that same quantity of heat over a double amount of space it removes heat in space as well as by dismissing space and in that concentrating heat. Again it is apparent that in all the applying of gravity it is space that bites the dust. The density of the protons brings about space dense enough to harbour the heat in such quantities and visa versa applies in outer space.

We have to accept that the coldest place in the solar system is in the very centre of the Sun because there the most number of protons sharing the least amount of space producing the coldest area that such intense motion can allow therefore the excessive motion brings about the hottest density of heat within the cold environment. It is the duty of scientist to look far beyond the ordinary and find why the inner star will be so cold and as to why outer space will be so hot while being seemingly so utterly cold or hot in humanly applied standards. It is the duty of the professionals to find matters as they are and not as they would seem to look from a human vantage point. Later I will show in much better detail why the star is so extremely cold and outer space is over boiling with heat expanding into more space. We have to see what forms space and why space can be the absolute basic container through which gravity can relay the influence that it carries. We must come to realise that whatever it takes to form space it has to contain something that is the same ingredient, which also is the basic component that forms the lot of everything else in the entire Universe. When particles heat up the particles expand the space the particles hold to limit the heat rising.

The particles claim more space when heated to preserve the cold. The claim to more space produces more space and reduces more heat. Such expanding brings about cooling. When particles heat or cool motion applies in some form. Motion started at a point when the Universe was extremely hot and there was no space. By introducing motion space formed and the lack thereof produced friction that became heat that became space. It is natural and it is simple and above all it makes believable sense.

The application of gravity is that which condenses space by bringing about heat with the compressing of space. Compress space even today with a piston cylinder wall in an engine cylinder and then from that action pump the compressed air into a container and such confining of space will increase the heat by the piston effort to reduce the space brought about in the container. The heat coming about inside the cylinder has no relevance to particles colliding because all compressor cylinders cool down with time moving and not necessarily with the loss or release of particles. It is not only the discharging of air that will reduce the temperatures inside the container but the time flowing bringing motion about where the motion is not about particles escaping but heat escaping in the replacing of the heat density (not the density of the particles forming the material content within the container) but the space that compressed to heat will also bring about that the heat displaces through the container wall to the outside. After the pumping of air increased the heat in the cylinder which even can go to dangerous levels, the heat will reduce back to room temperature when further pumping seizes and that stops further air movement into the cylinder and such surging of pumping air is what brings about heat stabilizing.

The atom is the Coanda principle that generates gravity.

k is the neutron

a^3 is the proton

T^2 is the electron

Mainstream physics ignored the clear connection completely, notwithstanding it being so very obvious. There is this far in their recognising of principles in natural physics not one single reference made to prove their appreciation of this matter. They are bent on particle colliding notwithstanding the much nonsense such an idea promotes. Atoms cannot touch simply because electrons are all negatively charges and will therefore repel one another long before there is any possibility of touching coming about. However in the case when particles do collide such collision forms an atomic thermo release and that action we call an exploding atomic bomb. What principle this argument about particles colliding ignores is that all atoms use negative charged electron forming the atomic limit on the outside forming a definite border to the boundaries of all atoms and in both electrons from different atoms are being negative charged. In being negatively charged it means both will come out and one totally reject the other as much repel the other or cast the other away. The closer they come the more violent the rejecting will be and such rejecting is the production of heat that will turn to space. However that rejecting will increase the motion and the increased motion will reduce the space occupied. The electrons repel other negative charged sub atomic structures, which the electrons are that form the outer borders of all atoms. With all electrons highly negatively charged (being as negatively charged as any possibility will allow to match the utter extreme) such electrons couldn't touch. When the pumping of the air container commences the balance at first favours the forming of heat from the space coming in and being reduced in the containing size they are squeezed into is reducing the space from what it was on the outside. The space distribution inside then changes considerably and reduces a great deal compared to conditions outside the cylinder wall and with the decrease of the space distribution inside compared to conditions outside that space then becomes reduced and charges with excess heat on the inside.

The electrons will disallow any contact directly between atoms. No force can be big enough to enforce such touching. It is because of that contact rejection electrons bring about that science has to use an overload of neutral neutrons putting them in the atom nucleus to fake a complying of charges that will eventually lead to atom touching each other but that is through enticing a neutral stance which is enticing a positive overload for a short while. When the touching of electrons does take place the event is called a thermo nuclear reaction where heat is released in unmatchable quantities and the atoms in reaction dissolves into a liquid heat. The increase of heat by the distribution of particles in the space that is forming the connecting space still keeps particles separate. The heat rising is a separate issue that has nothing to do with contained particles colliding because why does it stop when pumping is seized. This ratio of heat reduction is time connected as much as it is motion dependent. Motion reduces space by expansion as much as time contributes to space distribution by allowing the flow of heat. When the pumping stops the heat immediately starts the reducing thereof. Most important is the realising that every atom constitutes of two parts. In fact the entire Universe constitutes of the two parts, which I go about mentioning in this entire book. On the inside of the atom there is a circle formed by a rotating electron that contains the outer wall of the atom forming the sphere and holds material in contact with the protons. On the outside there is heat surrounding the inner material part within the sphere and distance the inner material from the space between it and the next atom. The electron forms the division between heat uncontained and heat contained. This is why the Roche factor is so very important. There can be friction between particles in reduced space under controlled circumstances where such particles are grouped together in a unit and as a unit elects a group singularity forming the centre of the chosen form of the unit.

The Universe separated heat from material by covering the exterior of material with heat that forms space. Some material became softer by uncontrolled overheating while others remained more solid by containing form through controlling the overheating. On the outside of all elements there are a layer that is the heat the element uses in relation to place relevancies between such an element and the rest of the cosmos. In the case where many atoms form a unit such as an aircraft coming in from outer space the space surrounding the craft becomes liquid heat as the space becomes more intense within the atoms combining as the structure in concentrated space that forms heat. In an aircraft coming in from outer space at altitudes that high there can be no particle in friction and even more so way up there in the atmosphere at the altitude where the cosmos meets the atmosphere just because the particles up there are so sparsely distributed in that part of the atmosphere. Above and beyond this lies the fact that all the so called air particles are very volatile and excitable by nature and they are known to turn the slightest heat into rapid motion thus establishing a scene where the particle that supposedly are in contact with the aircraft sheeting will move away from the hot incoming aircraft. The gasses will become more gasses when the heat levels surge. If then and not for any other reason why there can be no friction then it is because the particles are highly volatile and exceptionally sensitive to heat. Airborne particles are prone to motion just because it is the airborne element nature to change heat into motion and the motion comes about from their sensitivity to duplicate. No particle in the air being part of the space we call air which is in a free floating in that air can produce friction because of the volatile nature that those elements have. The craft's coming into the atmosphere produces a point where $a^3 = T^2k$ changes to $k^{-1} = T^2 / a^3$ (the explanation is forthcoming a little later on) The distance separating the incoming object from the Earth centre reduces rapidly therefore the object start to descend towards the centre of the Earth. We must also acknowledge the fact that there is one specific point of specific entry where this will occur more than before.

That point will rapidly increase the time factor where the incoming object crossed such a very visible border. By the reducing of distance k space a^3 will have to compromise in the relation of all the factors forming the equation since T^2 will very suddenly grow more acute. What happens is that the applying gravity reduces the space a^3 and the compromising factor comes about since the time factor T^2 moves back to a time where outer space was as dense back then as the density we now have within the atmosphere that then became the Earth atmosphere. It is outer space that remained denser than what the outer space currently is. I am now referring to a process that I introduce as this book unfolds which is by nature completely different to what is accepted by mainstream science (as you might have noticed in this short space of reading). That which I refer to as outer space back then was the same density as that which the Earth now supports but outer space in the meantime expanded while the motion that the material that forms the Earth structure provided, came about at a point just before the Earth established an atmosphere that grew through gravity and by the measure of the Earth gravity became separated from the atmosphere. While the gravity of the Earth contained the space surrounding the Earth in a much denser packed envelope the area not under the direct influence of the Earth governing gravity became more spacious.

The Earth contained its atmosphere and it relatively grew as much denser as the solar system developed into what it is today and outer space reduces its density. It is a matter of the kettle not being able to call the pot black. As the atmosphere released from what we think of as outer space that releasing from outer space made the atmosphere much denser in the space just above the Earth, which is using a reducing time factor. It is there that the applying gravity makes the Earth atmosphere more compact. That established the T^2 factor to be that more condensed when one compare in ratio the density with outer space. The density that was there at the time when the separation came about in outer space when such parting between the limits of the atmosphere and the limit of outer space separated and such separation allowed outer space as a separate object to move away. This parting brought a barrier that is in place between the Earth and the outer space and any object coming from outer space into the Earth's atmosphere will have to negotiate its entry by passing through that division. The incoming object then would have to reduce the measure of the space the craft holds as the containing singularity set new standards applying to the incoming object with which the craft then needs to confirms its form and its status within the contained space of the Earth. The reducing will then suddenly no longer use space as the compatible factor but the focus will shift to the time factor that dictates to the space what the space can be. Such reducing comes from the switch there is in space – time where it was in outer space performing as being $k = a^3 / T^2$ to what it has to be within the Earth atmosphere $k^{-1} = T^2/a^3$. When the atmosphere grew apart from the outer space there are two ways of looking at the event. One can think that outer space expanded by the implication of the Hubble constant or that gravity withdrew the atmospheric space of the Earth at the time that the parting of space came about. But however you look at it there was a time when both outer space and the Earth's atmosphere shared equal density as we find it still applies on the moon and on Pluto. Then the Earth became dynamic and now they do not share any density at all. Things were overall more compact back then than at the present time and that included all things in the Universe. The space component is reducing the time component by compacting space to alter the space – time ratio.

This is portrayed by Kepler's formula $a^3 = T^2k$ It shows space as the density of space decreases. The Earth still compact space by reducing the volumetric confinement of space $T^{-2} = k / a^3$. This we call the atmosphere, as the atmosphere becomes denser towards the Soil of the Earth. There is a change in the time component. Most evident of this is when studying the pendulum. Just as we can see in the pendulum swinging, we can see that the swing reduces. Such reduction is because as the space diminishes every time the arm rocks from side to side. With this there is proof that in the developing atmospheric space of the Earth the ratios change from outer space. This is proved by the pendulum arms that Galileo's experiment used to show that the swinging pendulum indicates $k^{-1} = T^2/a^3$. Further more it proves that Galileo was correct after all and unnoticed by science Kepler helped Galileo prove Galileo's point. In this the net outcome establish Kepler as being correct and the Newtonian argument of friction brought on by gasses fall apart which is at that altitude where such friction supposedly should take place, the material in friction is not even present in the atmosphere.

 Nevertheless science will stubbornly cling to the old theory with persistency that would warm any warring Field Commander's heart. In retrospect the following information is established in the past few pages: Every element stands in different regard to the heat surrounding the material, which makes us consider the material to be either a gas or a liquid or a solid. The material in every element there is as such is all three forms and not one of the forms in particular. It is the way under which the circumstances is presented that the element allows the heat to gather and accumulate as the surrounding heat occupying he surrounding space. Every particle is unique in the way it regards the heat to material ratio and how much heat it uses to form either the gas liquid or solid state. The fact of being a gas or liquid or solid is so much more complex but in time we will get to that explaining. If space a^3 declines then so must motion in relevance will have to compensate by reducing k and limiting T^2 because space a^3 must always be equal to motion T^2k

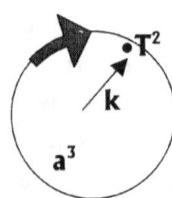

$a^3 = [T^2 = \mathbf{7(3\Pi^2)}] [k = \Pi^0]$.
The Earth holds a specific size in relation to the motion of the individual object being the Aircraft $\mathbf{T^2 = = 7(3\Pi^2)}$. Since the craft is stationary the distance between the craft and the object is $\mathbf{k} = \Pi^0$

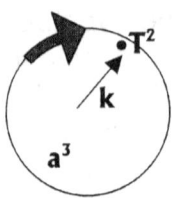

$a^3 = [T^2 = \mathbf{7(3\Pi^2)}] [k = \mathbf{2\Pi^0}]$.
As the motion of the aircraft accelerate the Earth still holds a specific size in relation to the motion of the individual object being the Aircraft $\mathbf{T^2 = 7(3\Pi^2)}$. However the connecting flexible link being the atmosphere which plays the role of the neutron has to extend in order to compromise to being $\mathbf{k} = \mathbf{2\Pi^0}$

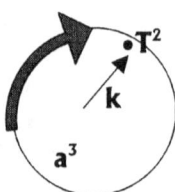

$a^3 = [T^2 = \mathbf{7(3\Pi^2)}] [k = \mathbf{2\Pi^0}]$.
The Earth holds a specific size in relation to the motion of the individual object being the Aircraft $\mathbf{T^2 = 7(3\Pi^2)}$. Since the craft is now in motion the first beacon to arrive at in relation to singularity Π^2 extends the distance between the craft and the object to $\mathbf{k} = \mathbf{2\Pi^0}$ then $\mathbf{k} = \mathbf{3\Pi^0}$ and so on.

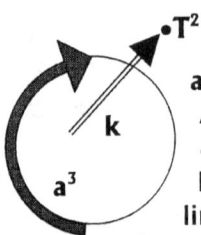

$a^3 = [T^2 = \mathbf{7(3\Pi^2)}] [k = \mathbf{2\Pi^0}]$.
As the motion of the aircraft further increases the Earth still holds a specific size in relation to the motion of the individual object being the aircraft $\mathbf{T^2 = 7(3\Pi^2)}$. However the connecting flexible link being the atmosphere now has to extend to being the most furthers that the neutron can possibly extend and such extending can compromise up to being $\mathbf{k} = \mathbf{5\Pi^0}$

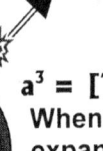

$a^3 = [T^2 = \mathbf{7(3\Pi^2)}] [k = \mathbf{2\Pi^0}]$.
When the motion of the aircraft further increases the Earth expands beyond what the limits will allow being the Roche factor of $\Pi^2 / \mathbf{2}$. This puts a cap on the specific size in relation to the motion of the individual object being the aircraft $\mathbf{T^2 = 7(3\Pi^2)}$. In this the neutron of the aircraft parts in a dimensional time value from the neutron the earth holds and the connecting flexible link being the atmosphere then cannot extend beyond what the neutron can possibly extend and such extending breaks down at $\mathbf{k} = \Pi^2 / \mathbf{2}$

We learn that heat always move from a hotter area to a colder area. Well there is another way of thinking about the issue, which is more accurate and much more scientific in explaining. When an object is heated the object expands and cooled it shrinks or that is what we are also been told. If we pump air into a compressor the air gets

> My theory on gravity is the flow of heat causing expanding and cooling bringing about reducing.

more inside the compressor. The air gets more. The size of the same while the air gets more inside the compressor is shrinking while the air is  compressor gets hot while the compressor remains the container and that means the remaining the same because the air cannot get more while the compressor remains unaffected.

In relation to the heat the heat gets more because the surface of the compressor remains the same while the size of the compressor within the relation shrinks. Because it is also the size of the compressor that shrinks the space outside the compressor has to accommodate the flow of heat because equilibrium has to be re installed. The size of the compressor therefore, will remain reducing up to a point where the compressor is just too small to accommodate the air. The air then will expand. However the air was always expanding from the pumping started because the compressor inside became smaller as the heat rise to expand the size of the inside of the compressor to match the outside of the air.

The same goes for material blown by wind to reduce heat. The object has an initial size to start with. Then we put heat to the object and the heat makes the object increase in size. That is hardly the increase worth noting because more important it is the relevancy of heat in the air to the heat in the heating object that goes array. The heat has to increase the size of the object in relation to the match it has to find in the space it is within. With the heat coming into the object the relation what the object has with the heat or air outside makes the object that many times bigger because the ratio in the heat balance is disturbed. If we blow air over the object we increase the size of the object by allowing the surface of the object to make contact with much more air in the same period of time, which will bring the size of the object back to the normal ratio because in relation and considering the contact with air the object expanded by the motion that increases the amount of air being in contact with the object.

In the normal flow of time the object has a heat to space relation set by the time the Earth dictates. Then we go and increase the heat on the object and in that event we actually increase the size the body has in relation to the heat in the air. By blowing air over the body we increase the air and therefore we increase the size of the body during the same period of time. There is now a dispensation of many times more where the body is carrying more heat and contacting many times the heat or air, which bring the equilibrium back to normal. There was a body size ratio and by applying heat the balance shifted to the reducing of the body size in

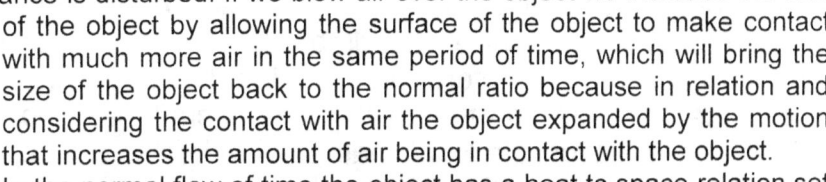

Same time frame

relation to the heat. The body then had to heat because the body was too small to incorporate the large heat. Then by lowing the air over the body it increases the size of the body and heating the body decreases the size of the body in comparison with the air it comes in contact with. The body is either expanding or the body is reducing and the balance in heat places the body in relation to either gravity cooling by contraction or expanding by overheating. The very same principle applies in the sound barrier.

Space shifts as heat releases space and converts the Universe in one direction bringing about expanding into more space but less dense space. Remember how the heat came down from 10^{34} to 0 K at present? The density of heat in space surely diminished considerably since then to now. Gravity on the other hand is exchanging heat through the concentration by removing space bringing about space loss with increased density of particles and therefore heat concentration. In the centre of all spheres, which all stars are it is hot. In fact the heat in the centre of the star is the product of the space it concentrates to form heat and in that we can read the gravity the star can produce. The ability to secure heat by reducing space becomes the measure of the star. Momentum is the second form of gravity symbolised by Kepler, as **k.** The Big Bang is the result of heat expanding into the forming of space. Gravity, on the other hand is about concentrating space back to heat, and take recouped heat through to material, acting out a balance of expanding while contracting. This way gravity is applying the onset of the Big Crunch by destroying space while space is converting heat to material occupying space. The Big Crunch is coming about because the Universe is expanding where the two processes are one principle.

The relevancy there is between the aircraft and the Earth is precisely the relevancy we find between the proton and the electron in the atom. When heat released provides more space between the aircraft and the Earth the distance between the aircraft gravity relevancy and that which the Earth allocated to the aircraft by only providing Earth gravity without the adding of heat by the aircraft allows the aircraft to respond exactly as the electron does in the case of the Atom. The aircraft falls into the role of the electron, the atmosphere takes up the role the neutron has and the Earth retains the aircraft therefore the Earth has the role of the proton. When the aircraft has more heat than that which the Earth provide through the atmosphere the neutron position has to expand in order to facilitates the new dimensions which the additional heat that drives the aircraft provides. The ratio that the Earth initially holds becomes stretched as the aircraft suddenly finds more heat and therefore more motion that becomes more space between the allocated position and the position the aircraft claims by individual motion in addition to the motion the Earth has provided. It is all about heat released that generates motion and motion provides space differentiation.

Throughout the entire cosmos is leaning on the four pillars which is the phenomena and the four culminates in one which accommodates all the others an it is the Coanda principle that establish space which provide the gravity which allocates the motion a position within the space that forms. This very principle of electron / proton is in gravity. Gravity I shall prove is motion and not mass inspired. In fact of mass being a factor corrupts gravity by restricting motion. Gravity is anti mass and mass is anti gravity because the neutron is all motion with no mass. The gravity of motion is heat driven because it is heat that drives gravity. When an atom is in outer space it is surrounded by an atmosphere of 0 K. That puts a limit on the atom as far as structural differentiation goes.

The outside of the atom is zero Kelvin and that is because zero Kelvin making the temperature on the outside of the atom that surrounds it, in relation to the inside zero. The zero therefore should also apply since the structure is zero Kelvin.

When standing still the body holds a relevancy of $7(3\Pi^2)$. That is the contact in gravity motion

In motion producing more of the relevant contact with time in relation to the Earth increase from Π^0 to $5\Pi^0$. The relevancy increases by $7(3\Pi^2)\Pi^0$ to $5\Pi^0$. It is because more material contacts more space or time in time duration as motion produce more space over time in time.

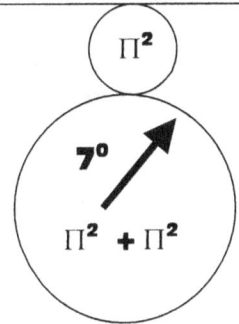

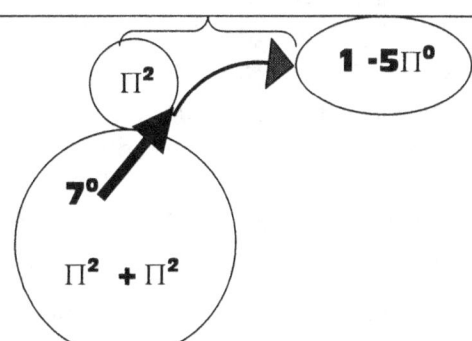

From the space we find the space extends by the layer the liquid increases the space holding the solid as well as the liquid connected to the space. That is when applying Kepler's formula putting everything in the right context of solid controlling liquid $k = a^3 / T^2$ as the liquid flowing controls the boundary space has.

From the liquids position we find quite another and opposing perspective as the liquid attaches to the space by the layer the liquid confirms to the space holding onto the solid while providing the gravity k through the providing of the motion as the liquid connects to the space. That is when applying Kepler's formula putting everything in the right context of solid controlling liquid $k^{-1} = T^2 / a^3$ as the liquid flowing controls the boundary space has.

The inside as well as the outside must be zero Kelvin because outer space has no other scale that being zero Kelvin. When the atom is on the Earth the relevancy goes that the atom is 40^0C, which is 313 K.

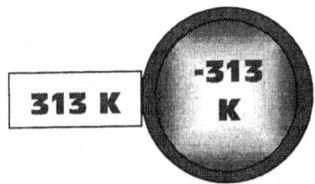

The outside of the atom has a relation to what applies on the inside of the atom. The normal atmosphere temperature on my farm is 40^0C, which makes the temperature on the outside of the atom also 40^0C. This is what we humans realize because that is what we humans feel and experience.

If the outside is 40^0 of the atom is hot then the inside of the atom must be 40^0 cold. The heat on the outside must generate a condition on the inside, which opposes the condition on the outside. The inside is in relevancy or in division of the outside because there is a mass differentiation of 1836 times.

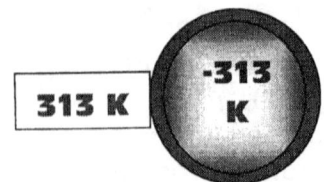

What applies on the outside is a response to what applies on the inside because what applies on the outside will be met with a response on the inside. When the outside is getting hotter, the response must be that the inside is also getting colder.

It is true that when concerning the Earth and outer space where there is little to choose from when comparing what changes is occurring in the atmosphere of a star. The Earth is as close to outer space as common civilized decency will allow.

If the outside of the atom is zero Kelvin there has to be a temperature on the inside that responds to the temperature rising or falling on the outside of the atom. The atom is an autonomous and independent moving object that is maintaining a structure.

When an atom finds a location in a minor star such as the Sun we are filled with surprise. It seems to us that Sun is very hot and with the Sun that hot the atom has little validity to stay intact. The atom should explode being in such a hot environment, and yet it is there and very it is much undeterred. Any atom we would heat to a temperature of 6500 ^{0} C as the sun temperature is, will destruct with an enormous bang. Well it is good and well to say gravity keeps it from destroying but when saying that we should use that as a clue and not as an answer. It puts what is in the sun in another class of structure and confinement.

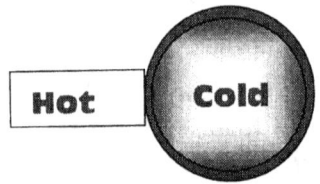

The outside of the atom calls for a direct response on the inside off the atom since the outside can change little if the inside will not bring just as much change on the inside. In the relevancy there are always three factors performing as gravity and in that is the Coanda effect in charge of committing gravity.

If the temperature on the outside of the atom rises to 6500 ^{0} C, then the temperature on the inside should respond to what applies on the outside.

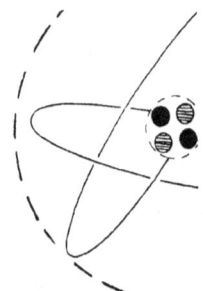

There is a definitive relevancy between the electron and the proton and the neutron fills that. When the atom is hotter the electron jumps a band but that is not altogether the whole truth.
The proton shrinks as much as the electron jumps because the neutron becomes more which is what is between the electron and the proton.
One cannot gauge the electron by ignoring the proton because in the process it is the neutron that suffers

It is quite true that when temperatures rise the electron jumps a band. The electron moves apart from the proton as the circle widens. It is not the amount that the circle widens that should be of any interest to us but the total response. On Earth the electron ring would enlarge but at the same time the proton should equally respond by reducing. Place the atom in the circumstances we find in the Sun where the atom heats to 6500^0 C. On earth the atom would explode but in the Sun the atom remains well formed and very intact.

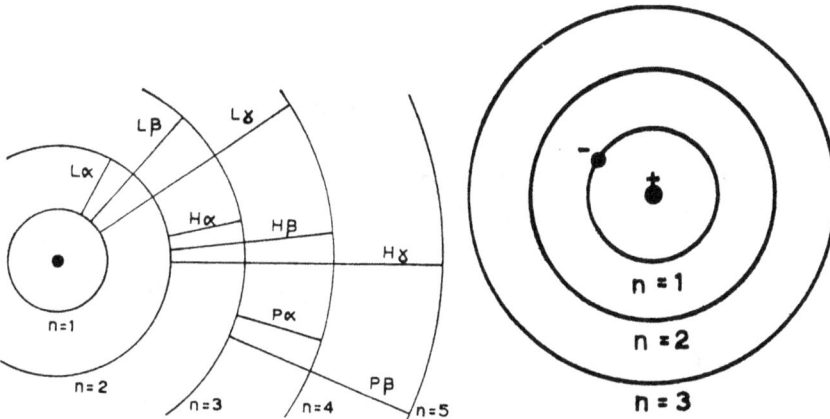

The atom does not explode because the atom does not get bigger and extends to outside its proportions. In such an event where the heat rose enormously and the atom remained as it is in outer space it would mean that the atom therefore must have gotten smaller because the enormous atmosphere kept the atom in tact. Yet with such temperature rising there has to be a change to the form the atom has and that means the atom shrunk in size. The proton became smaller when the temperature rose because the atom had to respond in some way to the rising of the temperature. Putting all this down to gravity is tiresomely attributed to laziness on the part of the human thinking capability because it proves how far we will go to restrain our ability to think. If gravity controls size by heat contribution then gravity has more to do with temperature than it has to do with what mass contributes.

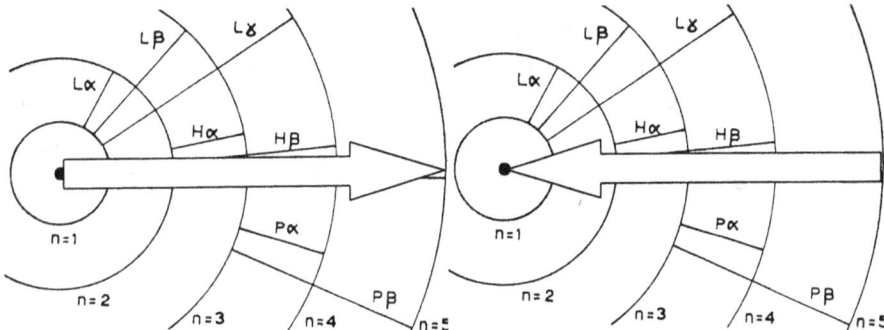

What we see as heat is relevancies because as the relevancy within the sun changes the atom adapts to the changes. The atmosphere of the sun becomes denser, which we see as being hotter and the containing becomes stronger. The atom has to reinvent it by adapting to the changes or different surroundings. In this manner the motion that the star provide which is so much more than what is the motion is we find in outer space that the hot / cold dynamic changes all together.

When any object is in outer space, the object encounters through motion and size a specific relation with space. The space puts a value to the material as the material moves through the space and is therefore in contact in ratio with a certain volume of space.

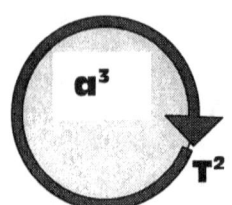

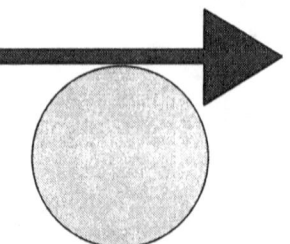

The revolving through air is a ratio of material that makes contact with a ratio of space and between the ratios; there is a heat ratio that is coming about. This forms a circle that we can put in a straight line.

We call this dynamic speed or velocity, which is just another name for a motion in ratio with time. There is a volume (meters moving) in time Seconds flowing and that ratio produces the size of the object in relation to the time the object allow the ratio to be in contact with the time the object moves a distance.

We also know by blowing over a body the "air" cools the body. That means the more "air" that the body is in contact with, the colder the body will get. To this argument there is a lot more and later in this book I return to the matter. However It is a ratio that is coming about.

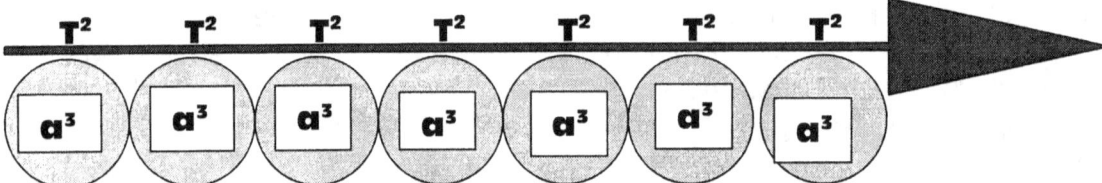

If the motion of the material is more it is more in contact with "air" which is not "air" or even "space" but it is time, the time (or space or air) effects more of the material since the same volume of material moves through more time. The time is a constant and therefore the material cannot increase the time but the ratio can produce more material (in contact with time) than moving at a slower speed. The material reduces in relation to the time it moves through and therefore the material shrinks allowing heat to flow from the material to the time aspect. This is the same, which we find when compressing air into a container used for storing compressed air.

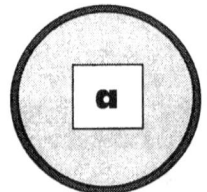

On Earth which is little different from outer pace there is a certain ratio of heat to space that allow the material, a certain size as the material move about in the space making contact with the space. The material is hot because of the large size the material is allowed to be holding the ratio of space to time.

That means the relevance between "space" which the material moves through or is in contact with, reduces the size of the material in ratio to the space it encounters. But we know that this effects the heat balance more than the size because the material moves through more time therefore more heat is transferred from the material to the space surrounding the material It is for this reason that we blow or radiators with fans. By the excessive motion of the massive Sun the material reduces allowing the material to become so cold that the space outside the atom, become 6500^0 C in a normal day on the Sun.

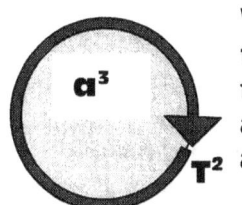

With the enormous motion going on in the Sun the material in ratio shrinks to freeze and this freezing/ shrinking allows the material to accommodate massive heat on the outside while all the shrinking is going on, on the inside.

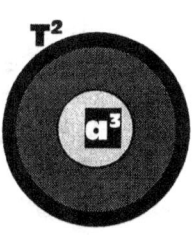

It is a case of the motion cooling the material and the cooling is shrinking the material while it is excelling the heat in response to the material cooling. That way gravity is all about motion and heat that contracts material as motion cools material in relation to the outside of the atom that has to rise because of the lowering of the coldness and size of the material.

Gravity has very little to do with mass and has so much to do with motion and it is gravity in motion cooling down material that shrinks material to accommodate more dense heat on the outside of material which becomes prevalent within stars.

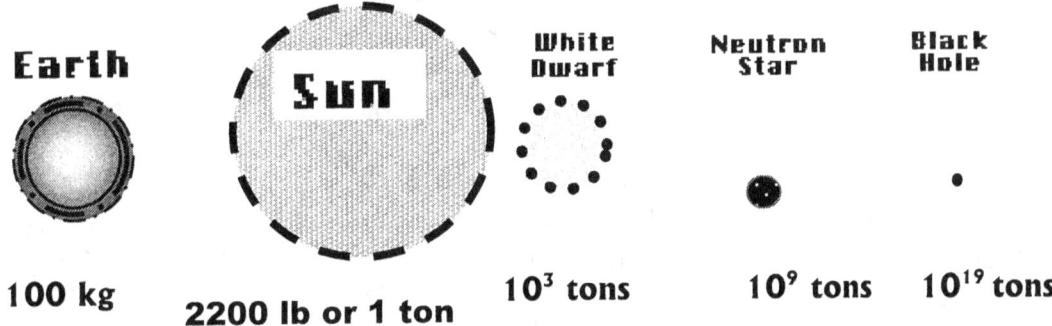

Earth	Sun	White Dwarf	Neutron Star	Black Hole
100 kg	2200 lb or 1 ton	10^3 tons	10^9 tons	10^{19} tons

That is the reason why so much "mass" would fit into so littler "space" by such a lot of material, and even moreover is the fact that the smaller the star gets the more material can fit into so much less space while the temperatures get so much higher. Gravity is motion that reduces heat which brings on cold which reduces material size which compacts space into more dense heat that multiply the gravity or motion of the relation there is to "space" or one actually should call it time and space which we actually call matter.

The gravity motion of the Earth is 7 ($3\Pi^2$) which is the distance of space in relation to the time it takes to displace that space and any motion above that is an extension of the atmosphere where the atmosphere accepts the role it has as being the neutron of the Earth. The extending can go from Π^0 to $5\Pi^0$, which will then be the moving, object extending its neutron part while still attaching to the Earth atmosphere. Above that Lagrangian limit of 5 the Roche limit in sharing neutron status sets in at $\Pi^2/ 2$ and the attachment there is between the atmosphere linking the aircraft and the Earth is severed.

Some years ago long before I dreamed of becoming the author of _**an Open letter to Selected Academics**_ or any of the books this letter is about a certain remark that Einstein once made on a realisation or a conclusion that Einstein came to in his younger days while still being a clerk at the patent office. Apparently the idea Einstein came to was concerning the subject of gravity. This happened while Einstein was still being a patent clerk in his younger days. Apparently Einstein was looking out a window of the multi story patent office, when Einstein suddenly realised that had he, Einstein fall out of the window from the roof to the ground of the patent office where he was working at the time, then he (Einstein) would feel as if he was weightless during the time of his fall. Not only that but also so would all the articles in his office that surrounded him at the time being his office chair, his desk and a pen. By falling with him those articles would feel equally weightless should they accompany his fall down as being part of the falling process in his imagination. As the objects were travelling alongside Einstein down the building to the ground the lot would travel at the same speed from the top to the bottom of the building. That is what Galileo concluded about five hundred years ago. Then I went one step further by supposing the Einstein group's falling was real and no imaginary thoughts were set in the fall, then what was the imaginary factor then? Let's pretend Einstein did fall with his pen, his chair and his desk and Einstein was not imagining his fall. Einstein as a human being can imagine but his falling companions can't. Then during a true fall Einstein may have had an imagination that could tell him about his feeling and in particular about the condition of his weightlessness, but the pen, the chair and the desk had no such imagination and they were travelling at the same speed as he did downwards and therefore had the same weightlessness as he (Einstein) had while they all were being in a downwards fall. If Einstein was imagining his weightlessness, it might be psychological, but in the case of the other travelling companions it was not possible to imagine anything. The falling companions had no such a luxury as having an imagination, however they too had to be weightless as they travelled next to Einstein all the way. There is an immense difference in size between the falling companions and that notwithstanding they travelled the same speed while descending. If they travelled the same speed as Galileo proved and they all hit the Earth the same time, which then indicated that their weight and mass, that which gravity used to drive and what propelled them downwards and that which was causing the drawing of what the mass was instigating to allow the motion of fall to commence, was equal. Size changed nothing to the equality there was in speed. Einstein should only have thought a little further than he did at the time because that would have made him realise what gravity exactly was and what Kepler found gravity to be. Kepler found space a^3 being equal to the motion thereof T^2 in relevancy to a centre point **k.** Kepler found space had to move.

When reading this that evening so many years ago, I came to realise that Einstein could only feel weightless if it was true that he (Einstein) was weightless. He could not feel as if when the as if was part of his imagination because he was truly falling, and in truly falling the falling was then without his imagination doing the pretending. Einstein had to feel his weightlessness as a cosmic fact in the true sense because if he was truly falling, then the part, which was the falling experience, was what he was experiencing in reality by three dimensions with one dimension in time. Then he (Einstein) was feeling weightless through falling and that feeling came as a result of what was happening to him as a cosmic interpretation of reality. He was not pretending to fall whereby he then would feel as if...he was really falling and with that there is no "as ifs". What he then would have experienced came by means of what he was experiencing in reality because of his cosmic state in relation to his relevancy with gravity. If Einstein was experiencing weightless ness, it would be because he was weightless while falling, then Einstein would not imagine the weightless ness because Einstein was truly falling, thus carrying out his cosmic state he was in. His body being in motion ($a^3 = T^2k$) was at that moment truly weightless while experiencing unrestricted gravitational motion. Einstein, the pen, and the chair had the same weight since they were all weighing the same in falling. If there were any mass differences there had to be speed differentiation for the force of the one would generate more motion than the force of the other onto the different mass components but since there is not mass discrepancy amongst the falling while falling the lot is having the same state of weightless ness, they adopt the same speed in the fall. After all it supposedly is the mass that is doing the pulling and more mass does more pulling…except if the mass is not doing the pulling in the first place. With more force applying to different masses there had to be more speed involved and an increase in mass in some participants has to generate more force. All four items including Einstein, would be equally weightless during the falling…that was what Galileo found because objects of different size and different mass travel at an equal pace (distance over time or space moving divided by time flowing while the object changes position in relation to the Earth ($a^3 = T^2k$)) while descending. The bigger objects do not fall quicker than a smaller object and that can only be attributed to one fact; it can only be true if the four weighed the same while falling and no one weighed anything while falling. That means the gravity applied while time flow in relation to the space that was applying the motion, which was what gravity is $k = a^3/T^2$ according to Kepler. The single line falling is represented by the factor k being the relevance of space a^3 that was relocating its cosmic position while all that was happening in relation to the motion of the Earth T^2, which was in relation to the Earth spinning around the sun and that rotation gives us our time T^2. While in motion the four different objects weighed the same since they travelled at equal speed downwards. However, when they stopped moving and came to a standstill, they then weighed different, which then indicated a difference in mass factors amongst them. By standing still the objects had mass differences and when they were in motion they weighed the same. When the motion became frustrated by being blocked by another space that was also filled with material and that was holding the spot too where the motion was directed, they then had different weight. The two had different levels of frustration with the larger party being more frustrated in the inability to move. The pushing resulted from the bodies striving to remain independent. It is the independence of the two bodies and the desire the bodies have to remain independent and not to share space that bring about the mass or weight. The two objects were in a fight to claim the position each desired, and that was to fill the centre of the Universe. Being ($a^3 = T^2k$) was being in the centre of the Universe because the centre of the Universe was $k^0 = a^3/T^2k$. $a^3 = T^2k$ $k^0 = a^3/T^2k$

From this one can deduct that gravity is motion or the intent to commit motion and mass is when the motion of gravity is frustrated by some solid structure blocking or preventing the continuing of the motion. Then one may conclude that gravity is motion of space and mass is the restricting of the motion of space. Having mass does not bring about gravity but it does restrict gravity's motion, which is what brings about the mass and weight. Gravity produces mass but mass does not produce gravity or in fact mass produce weight but mass is not responsible for the intended motion. Gravity on the other hand is the intention that the body has to move the very instant the blocking is removed. The intent on moving while being blocked by another object is frustrating the motion of gravity in both cases and the higher the frustration on motion is the more mass there is co0ming the way of the bigger object who then has the greater desire to move. The reason why it has the desire to move and why space is equal to the moving in time of the space in relevance to the centre of the Universe (which at that point might be the Earth or be the sun) is what I am trying to explain. Mass is the restraining of motion and gravity is material moving about by committing gravity. Mass only comes into the application thereof when two objects filled with space moves into a position where both want to claim the very position in space the other occupy. It is the motion and the independence they show to hold onto their individuality as independent cosmic structures that prevent them the sharing of space which in turn prevent further motion that causes mass. Gravity is in essence where mass is present, still in a tendency to commit motion but is then in the frustration of motion and gravity at such a point is the commitment to move once the blocking of space is relinquished. Because the one object that has more "mass" would put in a more assertive effort to move in relation to a smaller object and the effort to move will constitute to a greater resisting effort by the blocking object in a fight not to relinquish its position on the space both object claim that the tendency to move and the tendency to block the movement will bring the effect of greater or smaller mass being present during the effort and in line of resisting the effort. However while any space is in motion, the gravity of motion is equal to all and puts everything on an equal basis. Therefore there is no big and small and the big sun does not pull the small Earth closer. The big sun allows the small Earth to glide past in a circle year after year without interfering because the two does not claim the space each other has. Mass is when the motion is prevented that a differentiation in motion effort becomes part of the picture.

Do not be fooled by the seemingly innocent explanation that space is the motion thereof which is what gravity produces because of all things the cosmos creates, motion of space through time is the utmost complex manoeuvre and without bringing a restraining of mathematics into science, it is so complex there is no viable explaining in physics about how the cosmos produce the act of motion of space in time. To get every atom to spin as every atom follow the lead of the atom in front and gives direction to follow to the atom just behind while giving coherency to the structure the lot of atoms are holding as an individual unit times the units there are going around in the entire Universe is beyond what the human mind can absorb. While the atom in front is vacating space to fill the space of the atom in front that is vacating at that instant, the atom behind is filling the space that the atom in front has vacated in order to vacate and relinquish the previous position in favour of the following position to honour the direction gravity is insisting upon. Times that with every atom there is in the Universe and one may grasp the significance of the calculation. The coordinating of moving one atom from one point to a next point requires the skills that the human mind may never conquer. We may see the moving of an object through space being as simple as merely accepting it as a given fact, as science has done in the past, or we may reason about the complexity as civil person's should do, and come to realise that the complexity of motion of matter is beyond the scope of human understanding.

Removing material from space by filling material into a position of new space sounds simple because the complexity has never been realised. This was all a result of understanding the dynamics of Einstein's arguing about gravity and mass. Then with this information I further realised gravity is motion differentiation between objects. It is the independent motion providing a different speed while sharing a common centre of attracting that allows a discrepancy to establish mass under specific conditions applying between the two in relevancy. While falling the gravity applies as moving of space that is putting time in relation to the distance travelled. That means there is a speed relevancy between particles in motion and synchronised motion would bring about equal orbit around a shared centre. That is the result of gravity functioning. While the object falls the motion confirms gravity. When motion ends mass sets in and becomes the constraining of the object preventing further motion. The motion is still there but now it is reduced to a tendency to move thus establishing the object mass as the limiting of further motion. Preventing the motion by implementing mass is the resting of objects against each other by resisting the motion to continue, which then is where the mass takes the place of the motion. Where a confronting of objects restricts gravity the action then implements an introducing of the mass as a substituting factor for motion that then replaces motion as substitute for the motion that would be and the mass is providing the tendency of gravity being the motion of space. However mass then restricts motion and becomes motion in a tendency to apply motion. While falling gravity applies and motion neutralizes size, mass or weight. Mass counters motion being when the Earth restrains further motion of the falling object and the moving object is stopped from further movement where mass is then preventing or hindering gravity. This is the result of objects claiming an individual and personal claim to space occupied in a dual or in fighting for their individuality and independence of each other while wanting to be in the **centre of the Universe**. While falling or moving there is no opposition to the body being independent. When the motion seizes the falling object remains individual and still tends to move while Earth individuality resists further movement of the falling body's movement. Further movement is disallowed as other material fill space that falling body wants to lay claim to. The only manner to remain independent by the falling object will be to relinquish to motion in the securing of mass as a substitute to motion where it then finally comes to rest. Mass then sets in not causing the motion but substituting the motion and from that motion restriction becomes resistance that becomes mass. While falling the object is experiencing gravity because the object is in gravity but when on the soil the object experience mass which is the restricting of gravity or motion by other space filled with material. It is a fight of objects to secure and retain the position they have of being in the **centre of the Universe**.

Moreover, the author of **an Open letter to Selected Academics** came to another conclusion of equal importance. When any person is standing on any place anywhere, while viewing the Universe, that person is filling the **centre of the Universe**. Let's get more personal. When you, the person that is reading this, are standing at night and are looking at the Universe you are seeing the Universe from the position that one only can have if that person is filling the specific spot in the **centre of the Universe**. All the light, every single beam that ever left any destiny at any time acknowledges this fact. You are the most important person in the Universe because you are holding the most important position in the Universe.

All the light that come across and travelled all of the vacant space from any and all possible positions in space runs directly towards your position using a straight line towards you where you are filling the **centre of the Universe**. Not excluding the effort of one photon, all light is heading to meet you where you are in that centre spot and not one photon will pass you by. Not one photon dare miss you because if they do they miss the effort that all light has to accomplish and that is to locate you as the person filling the **centre of the Universe**.

Should you decide to shift your position to any other place in the Universe, you will shift the **centre of the Universe** to that location as well. If you install a camera on Mars, the light is obliged to acknowledge your relocating the **centre of the Universe** at your will to reposition you're being that **centre of the Universe**. All the light that ever left its destination crossing the vast spaces of the Universe, excluding no particular light, travelled all the way just to find you filling the **centre of the Universe**, right where you are. By you're standing anywhere, you fill the **centre of the Universe**, and the entire Universe admits to that because all the light comes to meet you there. If you shift from the North Pole to the South Pole you will shift the **centre of the Universe** because all the light travelling throughout the Universe will find you where you then moved the **centre of the Universe**.

The light left its destination billion years ago as it travelled through space at the speed of light anxious to acknowledge you're being in the very **centre of the Universe**. No photon will be able to pass you by where you are in the **centre of the Universe** because all light is heading your way from their starting positions. No wonder every person born has the idea they were born to fill **centre of the Universe**, which we do fill. The Universe is spinning around you or I, which is filling a centre where all motion is connected. That is the Coanda effect on the utter-most grandest scale imaginable; nevertheless it is only a manifestation of the Coanda effect. It implicates gravity as wide as can be... Some things mathematics is able to explain but other explaining goes beyond mathematics. Try to explain mathematically the colour of the sky being blue in a clear sunny day and changing to black when nighttime falls. Do the explaining in mathematics to a blind person that had no vision since birth in such perfect mathematical detail that would allow the person afterwards be able to explain the difference between blue and black to other blind persons by using only mathematics. Some aspects of the Universe go beyond mathematics and some even go beyond words. It is our task to find space, to find time and moreover it is our optimal task to find the Universe. We have to see what is solid, what is liquid and what causes gravity. It is therefore very important to see what is a solid and what is a liquid. Again we must put culture in the background and value the cosmos by using cosmic standards. Everything that moves, do so in relation to another that is relevant stationary is a liquid notwithstanding that it may or may not contain material. It is a liquid nevertheless. Everything that is relatively stationary is a solid in relation to the liquid that moves about the solid that anchors the liquid by gravity. Gravity **is to move or apply the intension to move** space a^3 **at the** distance or relevancy of **k** while T^2 is the time it is going to take to **apply gravity** or move the space filled with material space a^3 at the distance of **k** in the time period of T^2. That confirms Kepler's attribution to gravity where according to Kepler space a^3 is equal to the movement T^2 (time it takes to move) at the distance **k** from the centre specific.

I then subsequently reviewed my vision I received from the vision Einstein received and applied such a vision on the findings Kepler received from the Cosmos. It puts all aspects of gravity in the Universe in new dimensions. But the visions formed the beginning because the visions unleashed many new questions. If gravity is motion, what causes motion? What stops motion? That answer is in the Black Hole. In truth the explaining of the Black Hole is as complicated as the Universe may represent and as simple as the cosmos truly is. If a star is about fusing atoms and with such fusing of atoms is thereby growing, what happen when all the atoms fused into one all collective atom in one already all—atom-accumulated star? What is the gravity if the star has melted all atoms it had into one all-inclusive atom and this all-inclusive atom is providing all the gravity that the star had when the star still had massive volumetric space? If all that space that once filled an entire giant star fused into one specific space less centre holding singularity 1^0 then the enormous gravity is applying to the centre of such a non existing space-less atom and that entire enormous force has been secured in the space less than that which one atom holds. In that case the atom would then show a force that would pull the surrounding Universe flat. The purpose of fusion is to reduce space and magnify space less ness inside the sphere. Where does the gravity of the star end when all the atoms in the star became one giant atom by fusing all atoms into one nucleus? Gravity is smallest where space is least. Where space of an entire massive star is left in the size of one atom the gravity coming from that will pull the Universe flat at that point. However fusing means freezing together because only by reducing the heat can the removing space be accomplished and by reducing space to the point of freezing material permanently together is getting material frozen permanently. That means the Sun and all stars are as cold as they can get and not hot!

The motion is a product of heat and the motion produces a cold that sets in as the motion comes about. When the object moves it moved because the heat became excessive but by moving it is doubling the area it holds by halving the area as it divides the space between where it goes and from where it came. As soon as the motion halves the space used to move the halving of the space halves the heat which produces the cold which brings about the containing or the reducing of the space. Then with the motion completed the contraction retains heat where the heat increases to bring more heat rising that leads to more expanding coming about and the cycle once more repeats.

I am not getting into that argument now, but because of the size the Sun has and the size moving through such distance the Sun in its very centre is the coldest place in the solar system and outer space is so hot it is over boiling. That is why outer space is expanding. It is because the heat rises as much as the stars reduce the heat by containing the heat as material inside the atoms. Nevertheless, gravity is motion and motion comes from overheating which the motion then produce the cooling that contains the overheating by accumulating the contained heat inside the atom. To do that the Coanda principal is employed and the Coanda principle sets the Titius Bode law in operation and the Titius Bode law produces gravity. By spinning liquid heat around solid space a relation between 10 / 7 and 7 / 10 produces a relevancy that contacts heat.

It is about. Gravity is all in Kepler's $a^3 = T^2k$ and $k^0 = a^3/T^2k$ where then relevancy $k = a^3/T^2$ and in response to keep equilibrium applying $k^{-1} = T^2/a^3$

Matter in relation (part of) to the total dimension of space.

(10 / 7) \ **(7 10)** = 2.04

1.4285 / **0.7** = 2.04 Taking from both orbiting influences

SPACE DIVIDED INTO TIME

(7/10) / (10/7) = 0.49

.7 / 1.4285 = 0.49 Taking from both orbiting influences

SPACE MULTIPLIED WITH TIME

7/10 / 7/10 = 1 and   10 / 7 X **7/10** =1 Therefore not influencing change

THE PROCESS PARTED USING THE ROCHE PRINCIPLE

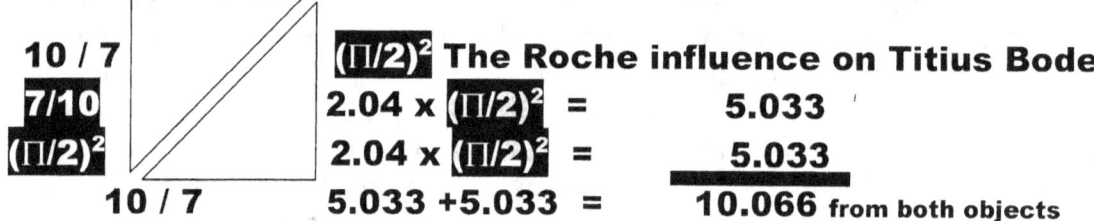

10 / 7   **(Π/2)²** The Roche influence on Titius Bode

7/10   2.04 x **(Π/2)²** =   5.033

(Π/2)²   2.04 x **(Π/2)²** =   5.033

        10 / 7   5.033 +5.033 =   **10.066** from both objects

SPACE DIVIDE INTO TIME

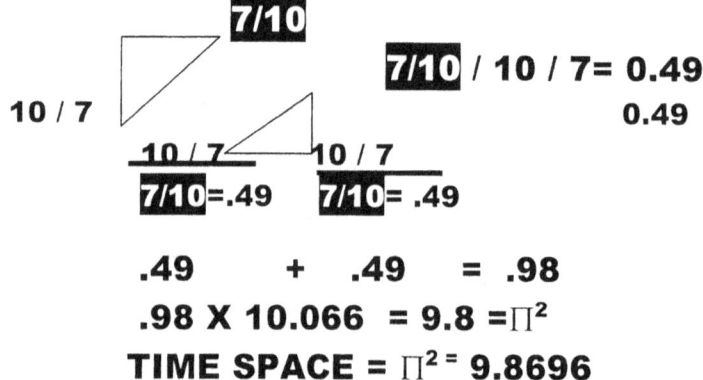

7/10 / 10 / 7= **0.49**

0.49

7/10=.49   **7/10**= .49

.49   +   .49   = .98

.98 X 10.066 = 9.8 =Π²

TIME SPACE = Π² = 9.8696

TIME SPACE =Π²=9.8696= Space and time in a dimensional implication.

This proves the reality of the Titius Bode law, which too I have to add, the Newtonians put down to a coincidence. This proves that the Titius Bode law is part of the chain that brings about gravity. Most of all this disqualifies mass as having any importance what so ever in the producing of or the conducting of gravity. This proves that gravity is the motion where space interacts with time to give singularity the significant control it has as not being part of the Universe and yet being responsible for all action in motion taking place in the entire Universe. This is what keeps the top erect while spinning and it keeps the Earth in gravity as much as it built the solar system to a mould that built the entire Universe. The fact that motion brings about gravity in line with singularity must be proof to all Newtonians that their perception on mass has no grounds. Even where those Newtonians are unable to show what brings about gravity even after so many centuries of investigative research while trying and getting no results does this simple arithmetic proves more than all the multitude calculations on their part that proves zero about the manner they promote gravity as being a pulling force.

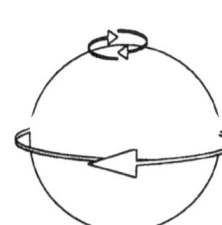

On the inside, there are the seven markers of which singularity is the focus point in the centre of the centre. The markers are representing one aspect of space, which for argument's sake let us call it cold. Then there are three more markers on either side being part of the space but not captured in the space. It is space in motion by the influence of the motion of the Earth.

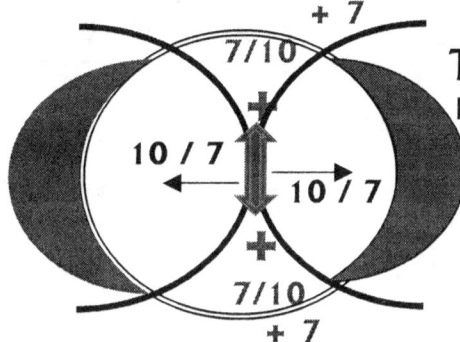

TITIUS BODE LAYER CONNECTING
Inside the cosmic sphere

$7/10 + 7/10 = 1.4$
Singularity in the square of matter
$10 / 7 = 1.42$
Singularity in the square of space
$1.4 / 1.42 \times 10 = \Pi^2$

MATTER HOLDING THE SECOND PROTON COUPLING THAT TO THE NEUTRON TO COMPLETE THE NEUTRON. Due to the influence of the matter dimension on the space dimension, the curvature of space-time comes into affect by dominating outer space.

The Titius Bode Principle is equal to gravity @ $= \Pi^2 = 9.8696$

Proving that the Titius Bode Principle is a product flowing Directly from the growth of singularity forming space-time The Titius Bode principle directly valuating TIME to SPACE $= \Pi^2 = 9.8696 =$ MATTER HOLDING THE SECOND PROTON COUPLING

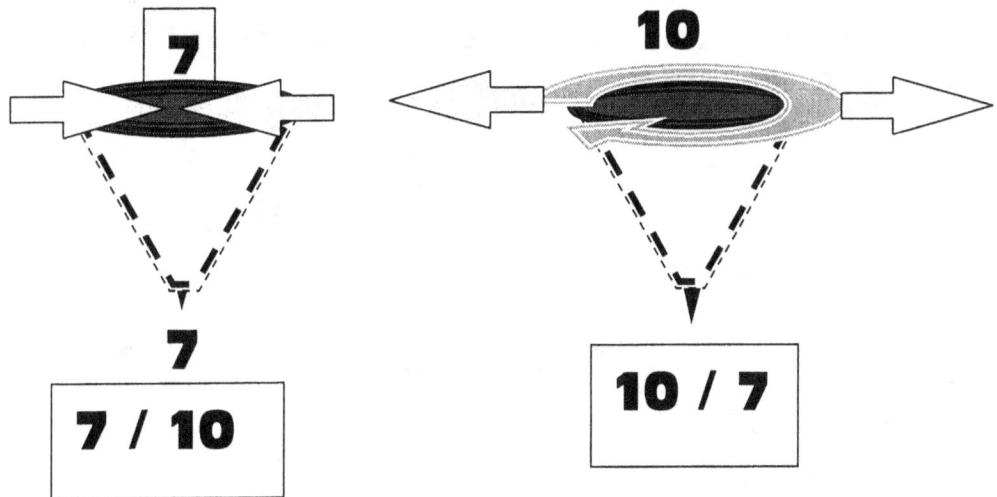

In this maintaining of cross referencing of singularity located in individual atoms providing spin to the governing singularity that maintain structural form in solids, many factors of singularity all form a close knit network and being inseparable as one unit, by the same margin it also is strictly individual to a point of destructing. From the inner or governing singularity outward all is concerned as space-heat. The relevancy in the material sector always includes the governing singularity and the very next one to the inside. All the others do not form any part of such a relevancy to the object forming the relevancy. On the time issue it is only the relevancy forming a connection with the one in question and the governing singularity. All other objects in the line are merely space-time with no value to the object that holds the relation. The fifth object will link in the material sector to the fourth and then directly to the governing singularity skipping or excluding all others from one to there. In the case of say three, it will connect to two and skip one while all points holding singularity to the outside is no consequence to the rotating object.

SINGULARITY BY DIVIDING SPACE INTO MATTER AND MATTER INTO SPACE, ANG ALL OF THIS ACCORDING TO THE TITIUS BODE LAW OF 10 / 7 AND 7 / 10 IN CONJUNCTION WITH THE ROCHE PRINCIPLE OF $(\Pi/2)^2$

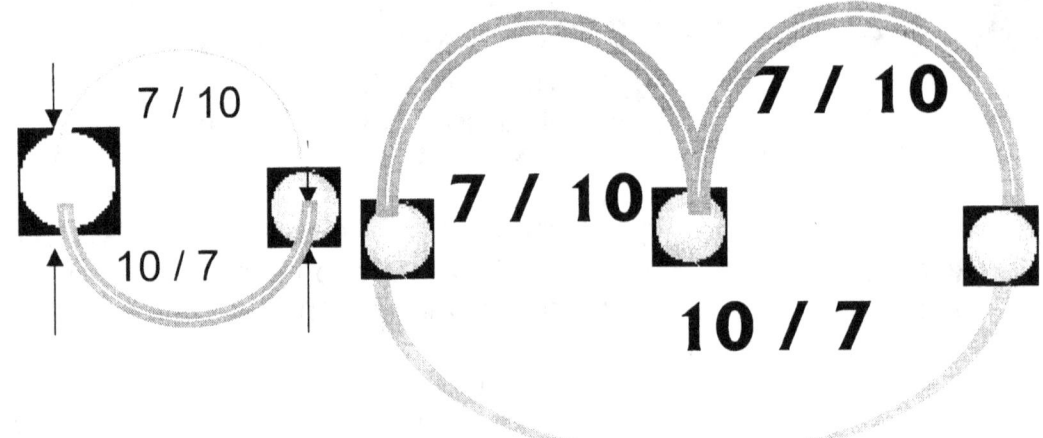

Time started at zero, eternity, whatever you wish to say, as long as you say time did not move at all.   Then the command came and time overheated for the first Π^2 in time. That brought space into play.

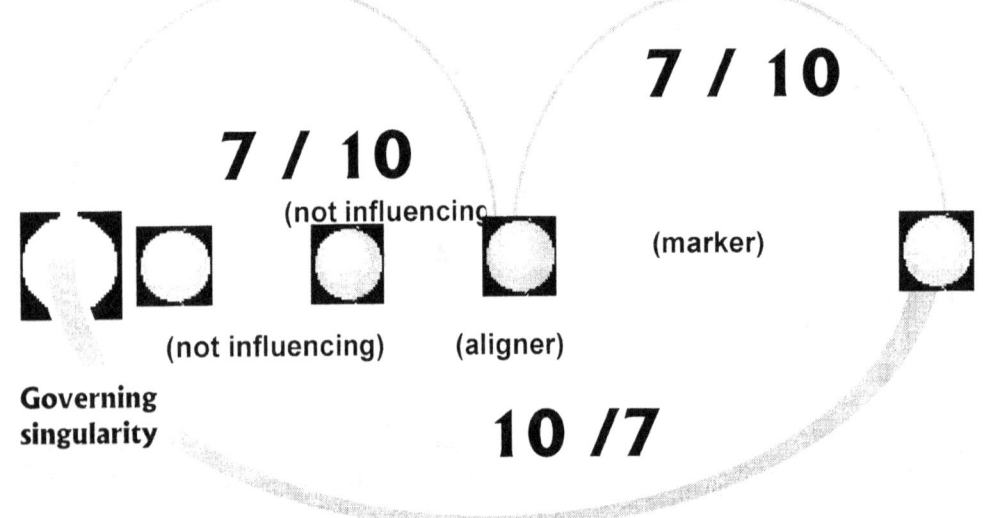

The spherical positioning layout forming the Titius Bode Principle

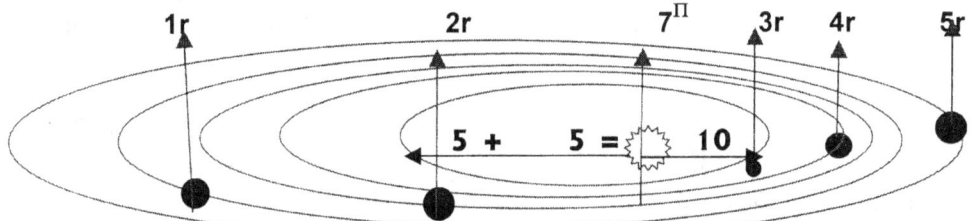

From the matter-to-matter relation in the Titius Bode configuration there are 7 / 10 + 7 / 10 = .7 + .7 = 1.4

From the space-to-matter relation in the Titius Bode configuration there is 10 / 7 = 1.42

(7 + 7) / 10 = 1.4

(7 + 7) / 10 = 1.4

(7 + 7) / 10 = 1.4

7 + 7 + 7 +   7   +

10 / 7 = 1.42

10 / 7 = 1.42

10 / 7 = 1.42

= .7 /|\ 1.42

The 5 + 5 = 10 is a position of dimensions as space loses value to singularity. The 7 that matter diverts in points from singularity may seem as coincidental but is valid. Still in accordance to our perception valuing the number in degrees, it seems coincidental but if it is coincidental, it is nevertheless a figure of diverting proven as accountable in all other calculations and plays a most dynamic role.

The Lagrangian 5 point system results as much from the Curvature of space-time as does the form the Black Hole holds. The Galactica is the opposing equivalent of the Black Hole and has identical but opposing similarities being the five points positioned to singularity. The galactica is generating space and the Black hole is degenerating space.

= 1.4 /|\ 1.42 Because the space-to-matter is in the square at 10 placing the matter-to-matter at a square of .7 + .7 = 1.4 the space-to-matter forces the matter-to-matter to double the distance by number as structures are place father from the mainΠ^0 maintaining singularity.

1 3 6 12   24   48

Reasons why this does not fully apply to the solar system I give in book # 7.

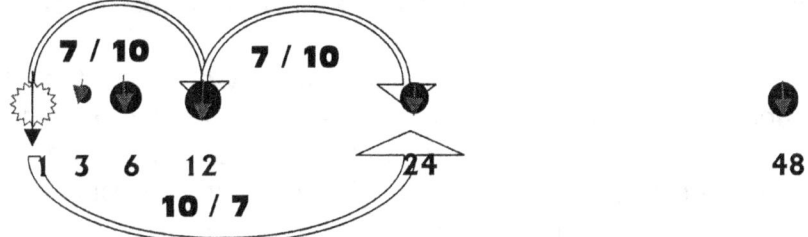

7 / 10   7 / 10

1 3 6 12   24   48

10 / 7

From the dividing singularity only one reference holds a matter value forming the position next to the governing singularity and therefore 7+7 becomes a factor and not all the dividing singularity between the point of reference and the governing singularity. That way the star to the outside takes a position doubling the distance every time. In balance everything in space to the outside of the governing singularity is space be it space or matter that makes no difference therefore that is 10.

The extension of Π is well received as a dimensional implication to matter holding seven positions from singularity and space having four quarters through out the rotation of singularity forming the centre to the five dimensions (one side lost to the cube's six sides connecting to the five remaining sides) making the total sides facing space from the point holding singularity at any given instant at a value of twenty (4 X 5 = 20). Then adding the singularity cross of Π being (1+1) = 2 the relation becomes 22/7. This is crude because in more precise calculations it becomes .91 + 1 = 21.91/7 = Π

Planet	Mercury	Venus	Earth	Mars	Ceres	Jupiter	Saturn	Uranus
Bode's law distance	4	7	10	16	28	52	100	196
Actual distance	3.9	7.2	10	15.2	28	52	95	192

The sectors provide individual singularity as a means in sustaining governing singularity by which provision comes through maintaining governing singularity the required spin in maintaining cooling. If this process did not apply, there would be no connecting individual singularity to major singularity. The sectors provide individual singularity a means in sustaining governing singularity by which provision comes through maintaining governing singularity the required spin in maintaining cooling. If this process did not apply, there would be no connecting individual singularity to major singularity
SINGULARITY BY DIVIDING SPACE INTO MATTER AND MATTER INTO SPACE, ANG ALL OF THIS ACCORDING TO THE TITIUS BODE LAW OF 10 / 7 AND 7 / 10 IN CONJUNCTION WITH THE ROCHE PRINCIPLE OF $(\Pi/2)^2$

This ratio applies between the governing singularity and the marker (inner next-door inside planet and planet marking a position according to the Titius Bode. There are reasons why some diverting stems from this but in other books I explain that with better detail.

From the orbiting structure (planet) aligning singularity only one structure the very inside singularity applies as a position of reference and that is reference to the distance applied between the points in governing singularity. From the Sun (governing singularity) the matter marker is 7/10 = 0.7 with the only one other forming a marker 7/10 = 0.7. The two form 1.4. From the Sun(governing singularity) the outer planet forming the marker in search of position holds space in the square 10 / 7 = 1.42 in aligning with the 7 forming material of the sun. Therefore there are two sevens relating to ten forming the material positioning of the structure in orbit and from the governing singularity all outside the Sun is the square of space (ten) aligning with one particle (seven) and not one of the other structure to the inside or the outside holds any value. Because .7 + .7 = 1.4 and 10 / 7 = 1.42 the distance doubles every time there is an aligning of three orbiting object. In this there is definite proof of influences coming about between particles sharing gravity. But then again the entire Universe shares gravity and as such then all will influence everything.

Mercury	Venus	Earth	Ceres	Mars
4 – 4 = 0 14 – 4 = 10	0 + 7 = 7 20 – 4 = 16	7 × 2 = 14 32 – 4 = 28	10 × 2 = 20	16 × 2 = 32

Jupiter	Saturn	Uranus
28 × 2 = 56 56 – 4 = 52	52 × 2 = 104 104 – 4 = 100	100 × 2 = 200 200 – 4 = 196

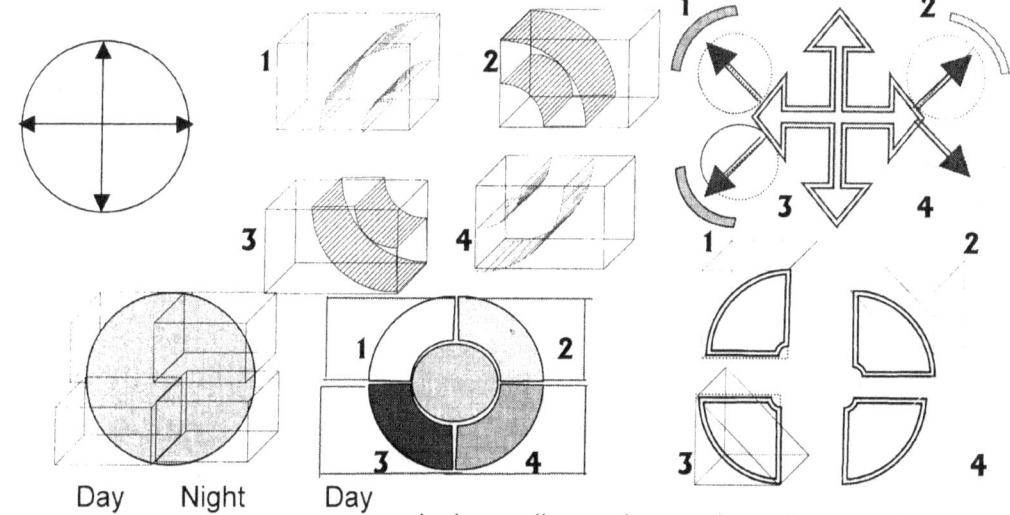

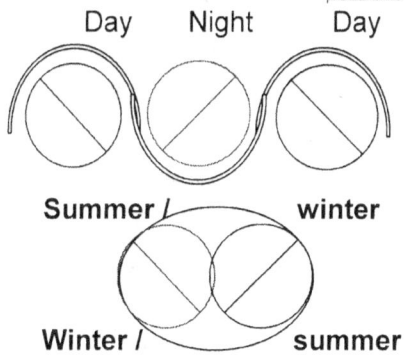

Day   Night   Day

Summer /   winter

Winter /   summer

I do realise science do not recognise a relevancy between the rotation or orbit of the earth and the position of the sun as Newton claimed and I shall come to that in a brief time. On monitoring the rotation of the earth to a graphic display one find that the earth movement displaying in accordance to change in positional location does indicate a relevancy that imitates the flow of current to an almost exact. Seasonal change has all to do with the graphs influence derived from the cosmos and little to do with the position of the sun and the Earth.

It is the position singularity holds in relation to the Universe and the Milky Way forming currents and seasons moreover than the Sun shining brighter or not. The Sun in size over dominating the Earths in comparison disqualifies any positional influence that can alter the Earths heat standings. Through shear size the Sun can shine at the top and the bottom of the Earth simultaneously without effort from all normal possible angles. I show a relation between singularity in different positions maintaining seasons and north/south polarity, not only as far as concerning the Earth but also outside influencing polarization. This has to do with the second position singularity holds in accordance to matter and space and is an "*electromagnetic*" (used for the lack of a better word) sustained positional opposing derived precisely from the graph in the manner when calculating electricity.

In this it is clear why the Titius Bode ([10 + 10 + 1 + .991] / 7) and the Lagrangian 5 \\ 7 systems part their ways when applying the different processes they hold. With all the differentiating, the observer must also consider the dual massage that light uses in travelling through the vastness of universal space. The thought of nothing is just what it is, a thought of nothing and although it is in the human mind common nature to present nothing as a value in the recalling of something, nothing is a presentation of the figment in the human mind. There can be no number such as nothing and that was (possibly) Newton's biggest error. Nothing represent non-existing and that is just what nothing is, it is non-existing.

The Titius Bode influence in a manner that on the one side holds the matter-to-matter relation of 7+7/10 whilst on the other side during the same time holds the space-to-matter relation of 10/7 forming equal and opposing values. From this the orbits of cosmic structures are always oval favouring the singularity dynamics of the one structure at one point and switching the favouring to the other structure on the opposing side. Because the structures can never be equal in size (singularity will not permit that where the Roche principle will intervene) the shape is always "off centre" as well.

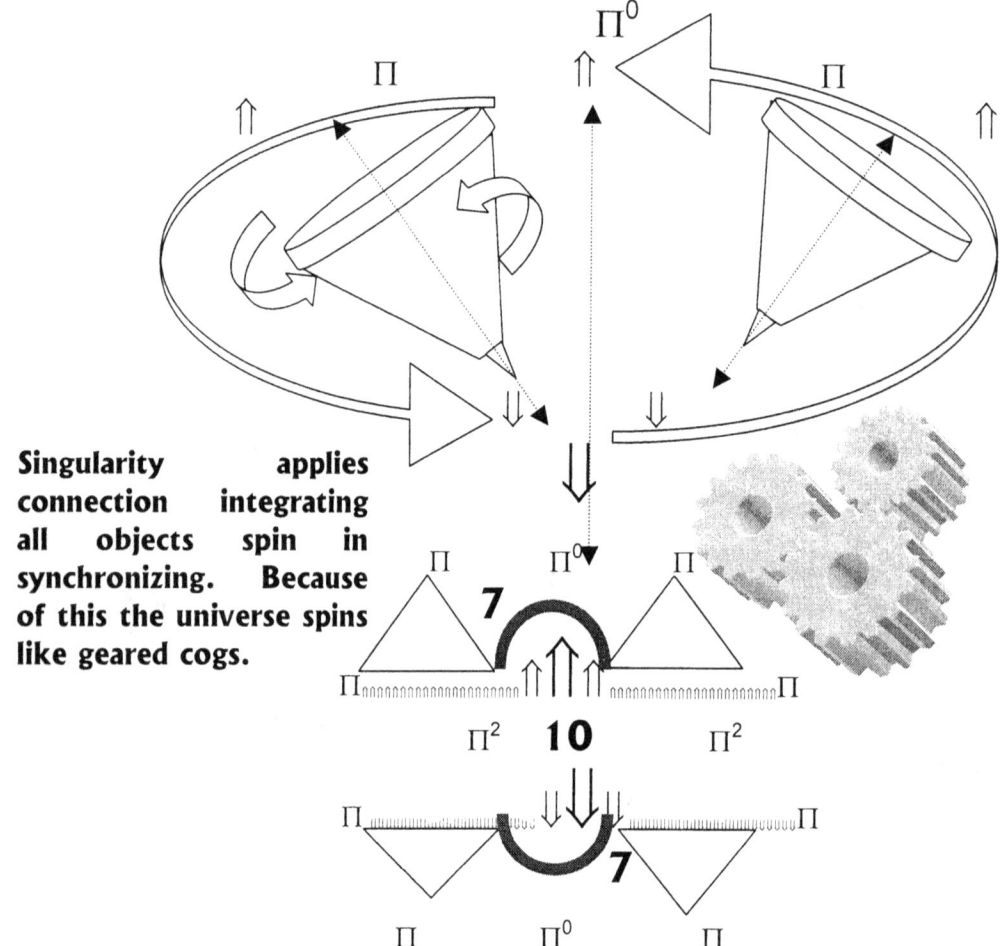

Singularity applies connection integrating all objects spin in synchronizing. Because of this the universe spins like geared cogs.

The ten dimensions I named the atomic relevancy is also showing the double value of singularity as singularity extends into as well as beyond space. The atomic relevancy is $(\Pi^2+\Pi^2)(\Pi^2 \text{ X } \Pi \text{ X } 3) = 1836$ that is the mass relation between the electron (3) and the proton. Proton = $(\Pi^2+\Pi^2)$ Neutron $=\Pi^2 \Pi.$ The atomic relevancy holds the dynamics of singularity control. In the ratio and dimensions we find in the atom, all space-time derives from the atom, whatever the atom is.

It started with a dot, because that is the only form, size and dimension mathematical logic will allow our brain to accept. From the one dot had to come a second dot and a third dot. The dynamics of such a dot is smaller than we can understand because such a dot is in negative relation to what we see Π to be, and the deeper we delve in finding the smallest fragment where space started, in the spot where time is still eternal as much as we can accept eternity to be. This we find in the aligning of planets. Where the one dot from which the aligner stem becomes the reference too the distance applied between the aligner and the original dot, or governing singularity or structure in charge of holding position to all orbits following.

The reason why we should first locate the spot is because we can only work from that point forward. By working forward we have to work backwards to locate where we are heading. The cosmos started at a point and where such a point is, we will find the Universe. Every one knows where the Universe is, because we can see where the Universe is, but if we can see where the Universe is, then we should find the centre of the Universe in that spot. Einstein theoretically positioned the point of beginning at a place he indicated where singularity should be.

With the cosmos the size it is and space so large compared to our smallness we have no chance in finding the centre of the Universe. The Universe started where singularity is and singularity is the sure indicator of the Universe. With all spinning objects holding singularity we then have located singularity in as much as finding the centre of the Universe. The Universe started with a dot forming. That answer arise from taking mathematics back to a point of being the smallest possible position, far smaller than we may be able to calculate form.

My approach might seem unconventional but through the abandoning of the accepted, it enabled me in locating the precise location of a universal singularity forming a connecting basis of the Universe (this I say with some degree of confidence). The smallest figure there can be must be a dot. The dot is the only form that leaves all the options open to extend in any and in all directions should the opportunity arise. The only mathematically sensible option about extending a line from the dot will be non-bias progress in all directions equally in order to give a meaningful flow of mathematical equilibrium.

The Pythagoras mathematical principle is the proof and that I explain. The obtaining of singularity is in my rejecting of nothing by replacing it with something being the dot. With the clepsydra or "water thief" Empedocles deducted that air was composed of innumerable fine particles, braking the thought that what we now know is air, was also believed to contain nothing being altogether a space filled with nothing until proven to be wrong so many years ago. Never did science take the lesson learnt back then to the future and out onto outer space. If there is space, there cannot be "nothing" as space is something. The claim becomes obvious when observing the connection between the half circle, the straight line and the triangle, which could also promote all the qualities lurking behind the pyramid. Consider the connection between 180^0 sharing and then one may realise much of the pyramid mystique becomes less spectacular in considering the very basic in mathematics being the Law of Pythagoras on which all mathematics are based. Once the water thief was eliminated by some human intelligence the matter was left at that. Nothing shifted out to an area we think of as outer space. In outer space we now find nothing. There is nothing but an atom here and there and even the atom is covered in nothing.

I wonder why the nothing landed there? Could it be that the reverse came about and because there was no visible "water thief" the very limit of man's suspicions came into practice. Man has always been extremely good in flying from one outer edge to another and if the water thief proved something was present, then the mere absence of a water thief must therefore prove that nothing must be in outer space. But what is space as such. What can space be, because with explosions we can clearly witness space created from heat. Our culture prevents us from admitting our vision, but the release of heat produces a *"shock wave"*. That *"shock wave"* is nothing less than space created from heat released. We have to brake free from culture of the past and a rigged mind set narrowing our vision.

Einstein's Critical Density lacks the accepted matching facts we need in proving the critical mass factor. But our inability in securing such required evidence defies the most basic logic. It seems all new evidence we receive from outer space is disputing all Newton laws findings that disprove Einstein's Critical Density as the answer. The Universe will not reach a point of contracting, not withstanding whatever dark matter astronomers try to locate in the vast space.

Why would the expansion turn around and do a reverse by going back to where it came from. Consider the momentum alternation such a change will bring about.

The Sun is not a gas-filled sphere holding hydrogen in its "natural gas" form, but it is all fluid and is in a liquid form where singularity is liquid- freezing hydrogen at 6500^0 C while outer space is boiling over at -276^0 C. This book explains the Roche limit in the practical sense... when applying cosmic laws instead of improvising cosmic laws uncovers that reality then becomes awesome. It becomes clear the Universe is as much expanding as it is contracting and contracting by expanding. As there is no hot or cold, no big or small, no grand opposing but relevancies in ratio to one another. If you do not believe me, then believe your eyes when looking at the picture. What ever the Sun is it is fluid falling into fluid.

Consider the time it took from 10^{-43} to 10^{-5} seconds to create a cosmos the size of a neutron. Compare that to what is happening now and see how many events took place by the creation of every lepton and every hadron and it is true that that period took longer to complete than it took the Universe to create the solar system. The flow of light through the density that space produces heat gives the speed of light the relevancy of time in space. The thicker the "soup" of heat is that space forms, the longer it will take light to cover a distance. It is very important to note that the speed of light is a relevancy between time (seconds) and space (kilometres). The speed relies completely on the value **k** holds on space –time.

The speed of light is forever a constant but the constant is part of the relevancy of space-time

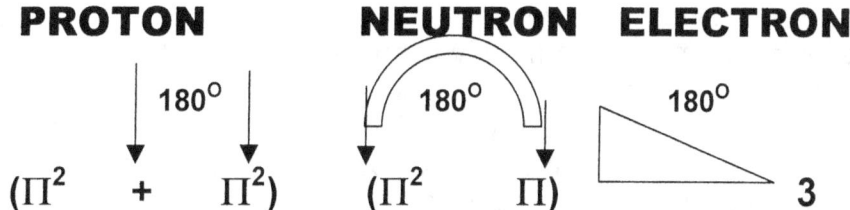

PROTON   NEUTRON ELECTRON

180°   180°   180°

$(\Pi^2 + \Pi^2)$ $(\Pi^2 \quad \Pi)$ 3

If one looks at the transmission of sound, it too depends on the relocation of matter, but to a very small degree, and in this process lies the transmitting of sound. To make the error of judgment in confusing the process with the breaking of the Doppler rings are quite understandable.

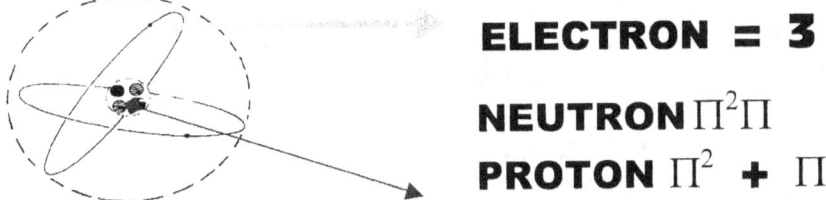

ELECTRON = 3

NEUTRON $\Pi^2\Pi$

PROTON $\Pi^2 + \Pi^2$

It is about confirming space conforming space and converting space.

ELECTRON is about confirming space

NEUTRON conforming space

PROTON converting space

The Universe is an atom and every system in use is using the format of the atom. The role of the electron or the neutron and the proton is very commonly used and the mould is accepted in every case to be similar to that of the role that each sub atomic particle plays. The atoms forming galactica and stars are combining as groups in units that are just as much an atom as the atom is and group together as just more cosmic atoms playing their part in the very same way as does the sub atomic particles.

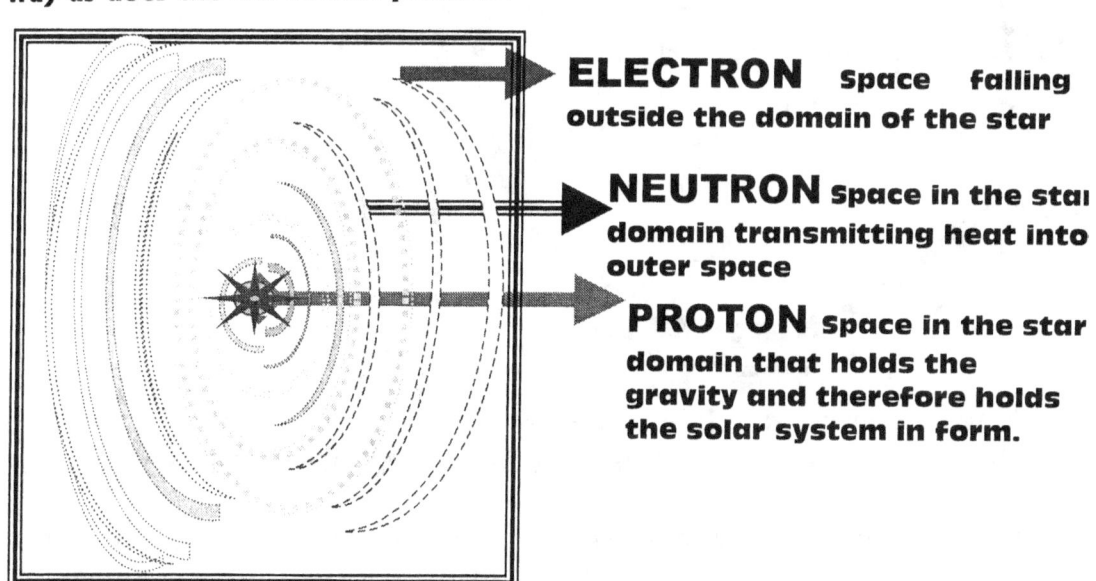

ELECTRON Space falling outside the domain of the star

NEUTRON Space in the star domain transmitting heat into outer space

PROTON Space in the star domain that holds the gravity and therefore holds the solar system in form.

The Universe connects in a way Kepler established through his relevancy theory. Those not convinced answer this: where would the Planets be if not for the Sun securing planet positions. The relation proves the ratio of one in all cases to be valid. It proves much more than merely connections at liberty of holding positions where ever the randomly opportunity placed the structure. The structure does not come closer by a pulling and tugging. Kepler's figure must still be around and by repeating the task but this time made much easier with the help of computers and telescopes of magnificence compared to those which excited the likes of one Tycho Brahe in his time. Science should become serious about science and not about self-protection and self-preservation. I found on all and every campus I went that any remark about Jesus Christ supposedly making a mistake generated immediate interest with even the most adhering Christians coming to hear the argument. Making a remark about Newton making an error gets you marched off the campus by security. Why not test Newton's $F = G (M.m)/r^2$ from figures Kepler left us and see how far did planets shift closer. I guess this will again make this book as successful as the others with my openly criticising Newton and Newtonians but Universities are not about knowledge but t about protectionism. Universities protect their own without any willingness to test that which it protects. It should be clear in confirming that the basis on which the entire world science union is founding all their policies and beliefs are correct and not only that, how far did the structures move closer. From that we then can see what we are waiting for and how long before the big solar clashing will begin. The absence in they're just mentioning such possibility confirm to me they know as well as I do there is no tugging and the Universe is in synchrony more than any person may ever be able to prove.

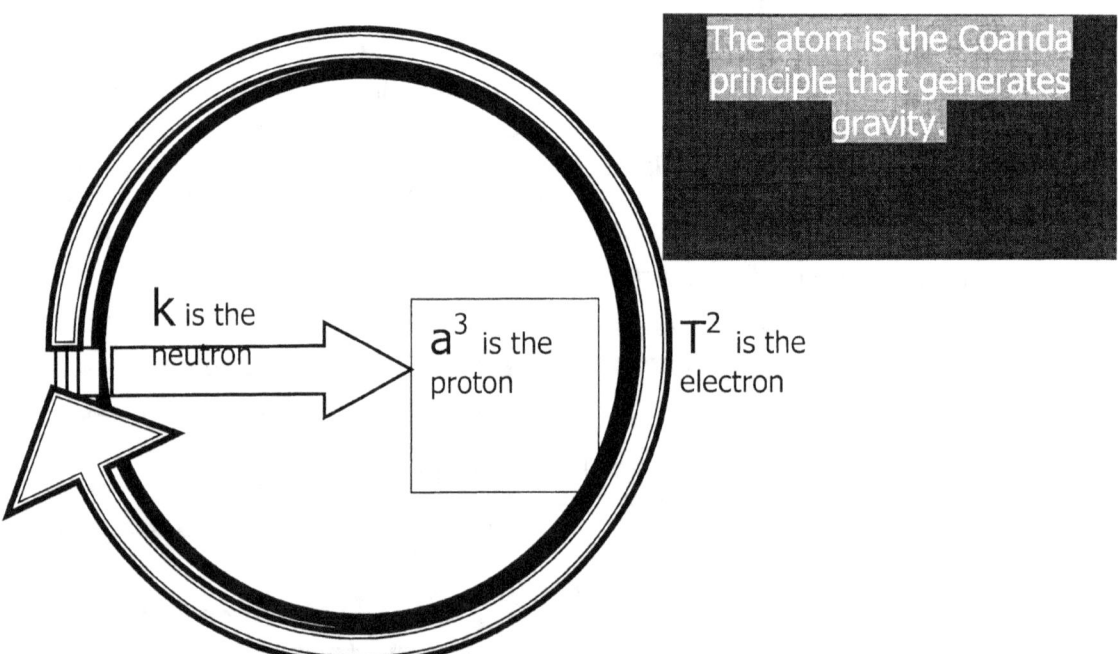

Everything in the cosmos is moving, either by own individual accord, or under the influence of some other singularity dominance. In explaining we return to the top.

When the top is in a state of motionlessness on own accord it is everything but motionless. The motion it adapts are synchronised with the Earth in harmony with the solar system and according to the greater picture of the cosmos. When an energy source not related to the cosmos called life intervenes and energises the top's motion, the singularity in that top suddenly jumps to life. By adopting a rotation energised to an unnatural state of energising because of life's intervention, the singularity of the top is not in charge but as it applies more and more energy, it will begin to find a means whereby it can escape and apply individual singularity as the top starts to separate from the singularity the Earth holds. The singularity holding the Earth would then allow the singularity of the top to rotate within a specific band where that a specific band of being active before the Earth's singularity will start to destroy the singularity in rebellion. The top on the other hand will try its outmost, when the singularity it holds gets by individual spin is too strong to remain in domination of the Earth's singularity. The motion of the top is an attempt to begin applying an individual singularity space-time defying and standing apart from the Earth's gravity. That action we see as the top starts rotating in a manner where the top does not align with the Earth's singularity, but est6ablished a driving singularity independent from the Earth's gravity. With the adding of spin, the time the top holds becomes unrelated to the time the Earth holds and the top will start a campaign too escape from the singularity domination the Earth has on the top. When the time or spin of the top exceeds the limits the Earth places on the top, the top would emerge by trying to escape from constrains placed by the Earth. The view I represent at this point is known to science for almost as long as science knows mathematics.

Not long after the law of Pythagoras was understood where Pythagoras introduced mathematics Eratosthenes of Syene made as big a discovery as Pythagoras did. But in the one instance the world took notice because the world could see and understand and the other instance the world disregarded the findings because the world did not see what the implications was. The same apply to aircraft flying and when the aircraft wishes to escape the earth's singularity hold it has to comply with the laws laid down by the earth. The seven becomes as big a part of the concept as does Π as it all interacts

If we wish to find the future we should locate the past. If the cosmos is contracting, where to is it contracting? The direction of contracting must be in the opposing direction the direction of expanding. If we wish to locate the past from where the cosmos came and through that in what direction the cosmos came, we must take an effort to backtrack the direction it came from. Should the argument come about that all came from nothing, then everything either still has to be at nothing, or our understanding of nothing leaves much to desire. Nothing means not existing, not being, never found and unable to produce any multiplication of any growth.

The above questions, but mostly the fact of what is more nothing and what is less nothing draws me to the realisation that there can be no such a quantity in space as nothing because even space has to be something. Heat expands as the levels rises and clearly it is for any one to see that the releasing of uncontrolled heat creates space, which is no more evident than by releasing a nuclear explosion.

The wind is shock waves, but what is the shock wave other than new space coming into prominence. In that way it is clear that releasing heat brings about the expanding of r as part of the sphere forming space. Hubble proved the Universe is expanding. Then by backtracking we have to set about reducing the sphere constituting the expanding Universe. If r in the circle is growing we have to reduce r to backtrack.

When the circle reduces, the value located to r will become implicated because r determines specific size. Not so in the case of Π, because Π in the true sense only indicates that the circle is a square without corners and therefore Π dictates form and not size. By reducing size only r comes into contest and will point to such reduction. By reducing the circle radius r by half continuously will lead to an infinite small circle but Π will remain because the circle as a form remains even being infinitely small. In the past, and even in some quarters today, science is on the search for the 100% efficiency machine. That theory runs on the surmise that a machine can drive as an output delivery without receiving input of energy. A few hundred years ago many Kings were fooled by such notion and some scientists truly spent a life in honest search of just such a device. Mostly the accomplishment came from cheats that very well new their machines were not up to the task, but in fooling a rich investor, brought about wealth to the inventor. As science progressed the no input giving all output machine became less and lesser a feature of the honest inventor. But the idea does not exclusively come from crooks finding a way to cheat the world.

The practise of receiving without giving comes from science in the form of physics. It is physics taking the world on a wild goose chase in the way physics presents the cosmic motion. Physics propagates that the cosmos is all about running without input driving energy. The cosmos is all about wasting matter to a supply of motion. This idea prevails even after the world of science saw clearly in the past that there could be no such machine anywhere. Even the cosmos must be a machine driven by an input and an output. It is the input / output driving energy that must be located and the driving ability we have to locate. Science holds the mass drawing power to prominence, but what if it is not the drawing power of mass that holds prominence, but it is the reducing or contracting of space that is the driving motor behind the cosmos. All energy we humans at present use to accomplish matter motion, holds some form of heat redistribution. Even electricity is a form of pure heat. I say that in mind of what applies when the energy of electricity becomes over abundant and the machine overheats. By overheating it means that the motion the machine creates comes about from heat control and precisely planned heat distribution.

When I realised that it is not me that is drawn towards the Earth, it is the space in which I find myself that reduces, and that produces the effort bringing me closer to the Earth. The formula $F = G (M_1.m_2)/ r^2$ suggests driving, moving in a direction and contracting. It suggests the reducing of space and not merely drawing or moving closer. When looking at any machine in practice, the machine draws power from space reducing whereby heat increases. Not releasing the heat to form space will lead to the destruction of the composition forming the machine. There is no form of matter, or element strong enough to resist matter deformation brought about by overheating. Having this in mind that matter does not resist heat, it is of importance to recognise that it is heat that is allowing space to give matter form. Looking at the manner in which energy is utilised it is space and heat forming matter allowing motion that allows work to achieve value.

At this moment science is all about a body falling where the two bodies are producing a force whereby the bodies draw one another closer. The bigger the mass, the bigger the drawing that comes about from the force unleashed by the mass of matter. The idea about this practise was phenomenal in 1602, it was impressive in 1802, but it is really ridiculous in 2002. Why would Boron form a solid having 5 protons weighing 10.811 g / mol and Argon a gas having 18 protons weighing 39.9 g / mol, but the "heavy" element with the biggest drawing power is a gas and the lightest element is a solid. That denounces the contracting force theory. The way we compile and use energy must be in a similar manner to the way the cosmos uses energy distribution. **We humans can create nothing, but nothing is all that we humans can create**. The rest of our achievements are by duplicating whatever nature provides. To establish what drives the Universe except for blaming some medieval magical force coming from nowhere going nowhere we have to find what drives us. The energy we use in all forms is producing heat in space by either converting space to heat or heat to space. Explosions are about converting heat to space. Compressing is about reducing space to heat. That is all energy composing work and is the only method of producing energy notwithstanding the immeasurable many names we use to express the same function in different forms.

Arriving at the question about locating the space and time forming the centre of the Universe one has to realise the centre of the Universe are in every singularity forming matter be it is big or small, size carries no significance. It is the impartiality of singularity that is claiming the value and not the differentiation of matter. One must realise there are no big / small or hot /cold or near / far. It is all relevancies between matter claiming space and space is heat in a turnabout manner. Every aspect in the cosmos is locked-in Universes, sealed off from other Universes and inclusive or exclusive depending on singularity holding relevancies relating to one another. The relevancies rely on inter dependence and inter linking, but there are no differences according to human sizes or standards. Accepting that principle unlocks the "so called mysteries" of the Universe and brings about clear understanding. It is all about accepting, acknowledging and interpreting the role singularity maintains on matter.

One should not try to focus on an image of such a spot or dot because there is no image. The line dividing the cosmos and that runs through every particle, no matter how large or small is beyond our vision. Such a small line, so small it is not even noticeable is large enough to part the cosmos into sectors. It splits the biggest there is into particles and we are not even able to notice the precise location of such a split. In truth there is no top or bottom that we living in 3D can see. We shall have to use a general conception brought about by intelligence. Your intellect tells you about such a spot, but that is all because that spot is on the other side of the Universe (quite literally). From the centre of the dot there is a top and a bottom spot. From those points there is connection with four quarters. That produces six connecting points that are all aligning to the centre. Because it serves big and small, hot and cold equal and alike, and it is the smallest cutting the biggest into equality, size is of no issue. Size is what man makes of it. In the

Universe there is no size in hot and cold, large and small. For the smallest there is, it is serving the largest there is equally.

Our instincts, our logic and our calculating process all indicate that the sphere holds a centre point from where six evenly positioned point's position matter to be. Using The formula $F=G\ (M_1.m_2)/\ r^2$ it indicates to a force pulling objects closer, where each force is coming from each centre point the body in question has. The contraction must commit the two bodies towards a point in each case being spot on in the middle, not withstanding what direction the force is applying, the body will draw to the centre. If the Universe spins around a centre point holding singularity, and singularity confirms the centre of the Universe, then every particle holds the centre of the Universe making the number of universal centres immeasurable many, and every atom and sub atom particle presented outside the atom in smaller bits, are all not pieces of the Universe but they are a Universe surrounded by many Universes. If every atomic particle no matter how small is holding the centre of the Universe, then the gravity is coming about from that point because that is where the gravity applying in the Universe is applying contraction.

It then is the atom in the most centre part where space and time meets singularity, that Einstein found a Universe collapsing to a single dimension, and every atom at a point post of the proton where gravity initiates in according with the proton dimensional colas of $(\Pi^2+\Pi^2)(\Pi^2 \text{ X } \Pi \text{ X } 3) = 1836$

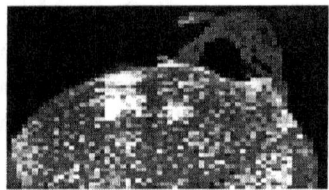

See the fluid push out of a bowl of liquid, spilling both sides as it falls into liquid. The inside of the sun is not gas but it is fluid.
In all of nature there is no NATURAL GAS as much as there is no NATURAL SOLID.
No element is either a gas or is a fluid or is a solid.
We arrange the elements in such a manner, but that is only applying to the situation the earth grants the elements.

When an element freezes it is solid notwithstanding...

When an element melts it becomes a liquid

When an element boils it is a gas again notwithstanding..

Hydrogen is as much a liquid as iron is a gas and neon is a solid. It depends on the element relating to the space/heat in the circumstances surrounding the substance at that very precise instant in time. We have to stop telling the cosmos to show us what we wish to find and start accepting what the cosmos is telling us to find. The culture that I am referring to is all about **nothing.** At present we find that there is something we think of as nothing in outer space. Because nothing is what we wish to find and nothing is precisely what we are getting because we think of outer space as nothing. If you accept the cosmos to be nothing, then please define nothing to yourself and find the definition in the cosmos.

The liquid the Sun has is the driving force that creates the duplication in motion. Without such liquid heat the Sun would become stationary and only depend on contraction while the contraction then passes the motion onto the heat in outer space. While the Star is in liquid all motion comes from the accumulating spin effort of the combined motion all elements together accomplish. In the case where the star is still in liquid the heat is stored in the atoms and as the star develops the heat transfers to the governing singularity, which makes the star immobile as the governing singularity takes charge of the entire star.

The formula of $F = G (M_1 \times m_2) / r^2$ only apply in a very specific range, and at a very determinable point the formula does not effect objects in the air. After such a point one will find satellites able to orbit, be it art a definite pace that matches the rotation of the earth. Still...below such a point (B) orbiting objects will come crushing down to the earth.

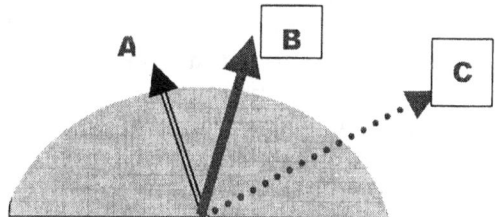

From point (B) to the earth Newton's formula apply and from point (B) upward Kepler's formula apply, but my pointing this out brings about all sorts of annoyance concerning academics. It must be clear to all persons that there are a big difference between the applying of Newton's $F = G (M_1 \times m_2) / r^2$ and Kepler's $a^3 = T^2 k$. When the objects reach some point they will drop to the earth and when that happens, mass do not play a part in the speeds they come to reach.

When examining the case where two balls drop vertically, gravity, as a force does not apply and therefore gravity does not come into effect because there is no difference in speed or duration.

With out any apparent reason the formula is substituted with the following formula:

$g = G(M . m) /r^2$ where:

G = the gravitational constant,

M = the mass of the body,

M = the mass of the lesser body

r^2 **= the radius between the two bodies.**

Let us take this formula back to the accepting of the Big Bang and find sensibility amongst a lot of confusion that I can see.

There was a beginning that saw a radius between objects so small that the size will never again be repeated. The diameter of the particles were also next to nothing but that should not be a contributing factor surely...the main focus point is that particles were as cramped as it shall never again be repeated.

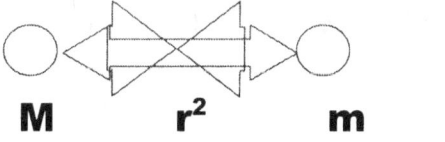
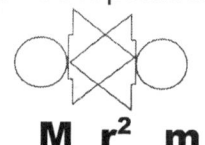

M r^2 m   M r^2 m

With the radius in the square dividing the shared and combined mass of the particles the relevant mass of the particles rises by the square as the radius reduces. If the radius becomes infinite, the relevant mass that the particles will produce goes up eternal. No force in the world would keep particles apart drawing on each other with an applying force but such a force is divided by an infinitely small separating radius. This is a recipe for joining and not dividing. Still according to the Universe I am able to witness that the dividing became enormous and the joining practically irrelevant. The gravity was more than words can describe, the heat was able to melt it all in one structure, but that did not happen. It split into billions of individual atoms.

Another point I question about the Official Policy is that they as I am are in agreement that the heat melted particles onto particles and in those joining better combinations of particles came about. How it happened is another bone of contention but more about that a little later on. There was heat on the outside and there was matter on the inside. The heat was liquid because the sun and other stars still indicate masses of liquid fluid inside. I can only imagine that that liquid inside the sun holding temperatures as low as 6500^0 K and up to 1.8×10^6 K the heat already is in a molten form. What about the heat then when the frozen outer space was 10^{34} K and such temperatures were the general order of the day back then. If the sun is liquid now then those temperatures raging back then must put the heat in form available in outer space at the time as thick as mud.

From the outside drawn onto the particle inside the blanket of heat came a flow of soup that became matter. That much I do understand. This carried on until...when? When did this stop? When did the Universe run out of heat? When could one consider outer space as the coldest all around? Where to did the Universe dismiss the heat that was once there but now is empty? How did the process stop of bringing from space intense heat and from that particles grew stop? When did it stop affecting the growth of a particle, the growth of space? In fact the growth of everything that grew came from this first growth. What you see or do not see grew since it was part of the Big Bang and everything in the cosmos at present was part of the cosmos during the Big Bang. I say this process of collecting heat from outer space never stopped but was an on going process we now give a nice name calling it gravity. Outer space never became empty and void but relevancies changed concepts where centres formed that should not be as it then interferes with concepts about relevancies Gravity is not and never was about particles pulling each other closer. If it was, no Big Bang was possible. Gravity is about turning space, which is released heat back to heat and concentrate the heat where gravity is the strongest and heat is the least. Space is the transverse form of heat and visa versa is also true. Should any one not believe me, try a bicycle pump by compressing the plunger while blocking the valve bit. The heat will burn your finger to blisters if the force on the plunger is strong enough, the plunger seals enough and your ability to withstand pain can last that long. Then answer your own question about where the heat came from because sure as hell is hot, it did not come from friction with air particles such as oxygen and nitrogen escaping through the valve bit. Heat is unleashed space and space is concentrated heat. Reducing space to heat is gravity and antigravity is expanding from overheating blowing into space accumulation.

When looking at a sphere the inside has always (in a cosmic relevancy) the location with strongest heat also always has the strongest gravity in any given cosmic sphere. The centre of the sphere clusters the combination of particles forming the sphere into unity. By holding a specific centre the sphere becomes the strongest form any object can be. The sphere is without any doubt the favourite choice in forming gravity. Where gravity has the last say without other influences changing possibilities as collisions leaving debris in space or natural out burst like Super Nova explosions, gravity will enforce the sphere to be the form taken by the particle. But there is no evidence of particles of similar size joining in matrimony through gravity being the shotgun at the wedding. In cases where there is a mismatch of size outside any proportions of equality then there is a contracting of the lesser by the greater. In such cases the lesser is not qualifying as material (and that I prove later on) but the greater considers all the lesser to be heat. It is humans bringing distinction to matter in form.

The two objects should have their own value of gravity and _gravitons_ and in comparison with the _gravitons_ of the Earth; their value is insignificant. However, these two balls are in their own individual deuce to see who reaches the Earth first, and the iron ball's _gravitons_ should give it a superior advantage. This comes about because the two objects are in a position where they compare in relation to one another and share a common second factor, which is the Earth. In relation to the Earth, the gravity - motions of the two balls do not come into consideration, but this does not play a part since the Earth is a common factor. The balls, however, is put in a situation where they stand in relation to each other. When compared to one another, the _gravitons_ should give the heavier ball a sizable advantage. The sensible example one can show to prove that where some structures matching in size come into conflict about occupied space sharing. In such an event one of the structures are turned to heat as it is liquefied flowing in the space dominated by the other and larger structure. If the structure proves too large the superior structure turns the lesser compatriot into heat. Then being heat it will apply gravity and admit such heat into the ranks of its atmosphere, but not before it turned it into fragments good enough to be heat.

The Official Policy Protectors never tries to explain the relation between Newton's laws as mentioned above, and the binary star system forming the principle we know as the Roche limit. The binary stars are systems where two stars spin around each other and never collide. These stars are many times over the size of our sun. When one applies the same Newtonian formula as given above, these massive giants must crash into each other, destroying themselves in the process. The enormous mystery is not in the apparent misbehaviour of these giants, but the fact that this is known to science since the previous century. Relate the binary once again to the comet/ Sun relation and there is a distinct similarity.

With the comet, the Newtonians regard a force to attach to the Sun in some way where this force pulls the comet towards the sun. At the same time another force joins in that pulls the Sun closer to the comet, but such is the mass difference between the Sun and the comet, the force the comet applies never realizes. In view of this, only the force the Sun applies, comes into effect. The comet proves this force by speeding up its movement as it comes closer to the sun. If the force did not become greater, why would the comet gain momentum?

With the arrival of the comet in the sun's domain, the Newtonians leave the argument to be. The Sun applying the force should remain applying the force and the force should increase all the time, accelerating the comet to the point of splash down. We must all argue that gravity is a force, which pulls an object to the centre of the larger object wherever that centre may be. The very same force that pulls the Chinese down, is pulling the Americans, and if not for the surface of the Earth's intervention, the next world war would be between the Chinese and Americans for King and country, honour and glory and to find who has the most powerful gravity force that will provide space to live in. If not for the Earth stopping matter falling right through the Earth because of conflicting forces on both sides of the Earth the Chinese and Americans will then have to establish border checkpoints in the centre of the Earth. The checkpoints will indicate where the Chinese gravity meets the American gravity and by allowing the force of gravity to find borders, we will finally have world peace. The only problem is to find the position where the Chinese gravity meets the American gravity and the two forces nullify each other. Just think if the forces of gravity, and not man, will intervene to set border standards: that must be the answer we were always finding a question for. This is a study far too complex to bother the United Nations, so we can find a more suitable group to investigate this fact to bring about world peace.

I am personally part of Africa, born and bred in Africa as an Afrikaner. I know the African solution to such a problem. In Africa, appoint a committee to investigate and then wait for everyone to forget about the problem in investigation. Therefore, such a problem is far better solved in Africa, because those in government aim to receive maximum western aid but never aim to solve problems, you make it go away by postponing the solution to the unsolvable. The African way is to ask the west for aid in order to create another useless committee to become over paid and under worked, quite capable of dealing with any non-existing issue of any magnitude that will never find an answer. Then sit around at leisure and wait each month for pay day to come for many years while the west is paying the committee to be bored until their pension dates arrive. By then no one would know the name of the committee and much less remembers the problem investigated. On the other hand, the Newtonians are doing quite fine by their method on their own using a technique they apply for three and a half centuries. To solve such problem, the Newtonians will apply a very different solution: Blame gravity's boundaries on a non-existing force, brainwash all future students in accepting it to be a force by telling them they will accept the force and forget the problem or fail the examinations and be chucked from campus, because that solved the problem so far. By the time, the student reaches a senior position he (or she) will no longer bother their mighty brainpower with the little aspects. They will advance to a point where they can move Black Holes around, travel at the speed of light, and divert time back to the past while others calculate all the mass seen and unseen in the Universe and any other ridiculous notion they may find to test their personal brilliance. If you for one second think everything about this last paragraph was silly, the silly ness started with GRAVITY ON BOTH SIDES OF THE WORLD, opposing each other, and that idea is not mine!

This is where the century, old trick of the Newtonians work best; do not think any further and no further problem will arise. Leave it at a force because with a force and thoughtlessness applying even-handedly, the problem never surfaced yet and that continued for the past four hundred years or so. So why bother with a problem that bothers no one. When a fellow like Hubble proves quite the opposite to Newton's claim of attraction, get a man who has a bigger ego than a brain and tell him to measure the Universe. It will keep every one involved occupied with something senseless while the problem vanishes through the many centuries to come. It is a force, and the way of all forces is mysterious, but never admits in believing magic. Those that do not accept forces to be of a mysterious nature should just contact astrologists and come to their senses about forces being not understandable. With everyone in agreement about forces, their nature and unpredictability, who then needs more real problems to solve?

No big-brain should bother about little issues like comets when there are so many galactica to conquer. Apparently the comet-problem just will not disappear. Something broke the force, something interrupted gravity. Let us see what happens. A force means it acts the same way as tying a rope on one object and start hauling the tied object in. The longer the rope is, the less control will be on the lesser star. As the rope shortens, the better the control will become. By implying that gravity is the force, our Newtonians tell us that we have to regard gravity in the same way the rope is hauling in a comet. It is something like fishing where the comet pulls, and the Sun pulls and eventually the angler gets his fish. One may argue that the rope is not the force because the force is actually the hauling, or shortening of the rope. I have had Newtonians trying to avert the problem they refuse to see by bringing in this argument. This manner of reasoning has the same value as introducing the African committee of investigation that will never uncover an answer. The rope is the extension of the force in a way being the sole representative of the force and the instigation acting out the force. The rope therefore is the force, extended somewhat, but still acting out the application of the force.

Weather the rope eventually broke, or hauling stopped, the effect as far as gravity applying its force, the process came to discontinuing … and we know that gravity is a force that pulls something to the centre of the body in control of the gravity. What made the force act in defiance of its nature? Why did gravity change its mind? What stopped the Sun applying its ferocious onslaught of the body holding the poor defenceless comet? Non – Newtonians will blame me for exaggerating, but I know that there is no Newtonian that can understand my argument. In that light, I ask non-Newtonians to show patience, because there may be a few Newtonians that will also read the book and to them everything said this far does not make sense.

This is where tutoring comes in best. Should a student bring up such non-academic and spiteful thinking about the mysteries of a force, then the lecturer sets a date on testing all the students' reaction about how much they accept the force. When any student shows signs of defying the force the lecturer can fail him on the spot and have a good reason to drive the silly youngster from campus.

By ignoring the problem as to why this comet brakes free from the gravity of the sun, and continue in its freedom until gravity is at a point where it is most weak, may not bring answers, but it surely avoids nutty questions! Questions are not there to interrupt Newton's laws! Ask any Newtonian High Priest and he will either tell you that in a very roundabout way or he will simply ignore you by telling you to your face that you are incapable of understanding Newton. The best way to get out of the answer of course is to tell the sod with all the questions he has not the qualifications or the mental capacity to understand Newton. That will make the pest retract to some ditch he should be in, in the first place without bothering the greater minds with some stupid minor issue. How do I know this you may ask? I have been down that alley many times and treated with that precise treatment on occasions more than I care to remember.

Still, the comet defies the force of gravity and my questions remain unanswered.

Dear reader, if you wish to read the funnies, jokes and laughs -a- minute, treat yourself to some real good clean jokes. Read the Newtonians explanation about how comets came about; how they get to the Sun and where they came from. It is going from the ridiculous to the thoughtless and ending in the realm of the mindless. However, be warned! Only do this on occasions where you feel very depressed. The jokes will otherwise drive you in a state of laughing hysteria. Poor old Newton was considered a very dry humourless chap in his day. To think what silly ideas can come from his forces.

Hauling in and releasing something caught on a line is called playing with fish. We might say that Newtonians love fishing and confuse planets, comets and fishes when they regard the interaction of comets with the sun. There is only one small problem with that argument and that is that fishermen and fishes form part of a second natural force named life. Life stands apart from the cosmos. Life and the cosmos only share time in space, not a joining of forces. Beside that, comets were part of the cosmos long before life had any role to play, so blaming it on someway life interacts with life does not cover the solution.

Why would the comet brake free from the sun's gravity? That is defying the law of gravity. Far worse than that still, is the fact that the comet's actions have the nerve to defy Newton. No one alive can defy Newton and remain alive. Does the comet not realize his actions contradict the all- important Newton and the gospel of the Newtonian - Priesthood. The best way the Priesthood of Newtonian gospel can deal with such defiance is to ignore it and no one will notice the actions of the comet. That is the scientific approach. Ignore and forget the problem. It is as simple as that.

With that let us conclude comets and really enter the world of forces at work! Let us now apply our attention to the forces of planets.

The next formula is very simple to understand. It is the fight of understanding the applying that becomes not applying it that is troublesome. If you understand the applying of the working and never spotted it not working, then forget it. You are a brainwashed Newtonian and if not..., well there is still hope that you have a clear mind left. The resentment you carry with you from childhood about the formula is in, not understanding it, but accepting the outcome of the formula you never could understand. Newton said that the force between two objects depend on the mass of both objects multiplied with each other and with the gravitational constant and the derived product you divide by the radial distance square that separate the objects. I shall put this in a mathematical language for your enjoyment that will explain the life-long not understanding to better effect.

$F = G\,(M_1 \times M_2)\,/\,r^2$. What does this say?

The greatness of the force depends on the masses of the two orbiting objects, aligning that product with the contribution of the gravitational constant. This then, you divide by the square of the distance between them at any given point.

Please, in all fairness to you, the reader, I have to warn you that quite a number of professors in physics told me that by reasoning in the manner I do, I only prove that I know nothing about Newton and understood even less about his work. Considering such allegations, I shall explain to you what I understand in as much as telling you what I know.

The Newtonian's formula states that the force between the planet and the Sun will improve as the mass of the planet increases (becomes bigger) and by multiplying that with the universal gravity constant you will get a value that will become lesser, the larger the distance are between the Sun and the revolving planet. With the reducing of the distance the mass on either side must therefore be on the increase because it holds an inverted relevancy. This means the Sun is pulling according to its mass. The planet is pulling according to its mass. The gravitational constant is influencing the pull evenly at both ends and the distance between the objects will reduce to the square value of the force's total application. I could never see what part I do not know and what I did not understand. No professor ever explained to me what it was that I did not understand either. That left me in a place where I did not understand what I did not understand and I never could see what I never could see. I shall try and make sense of my not understanding my not understanding as follows:

This is like having two balls attached by a rope on a floor that holds the same drag on both balls. When I reduce the length of the string, the bigger ball will show a greater resistance than the smaller ball, therefore the larger ball will apply a larger tug than that of the smaller ball. The rope will reduce (become shorter) at the end where the larger ball is than at the point of attachment where the smaller ball is. What is wrong with my argument? When the two balls are so miss-matched in mass as is the case with the Sun and the comet the one ball will do all the moving, leaving the larger ball stationary.

Surely the tugging at the larger end must bring the smaller object closer. By comparing the mass differences, you will find there is no comparison. The smaller object just has to come closer with the application of such a force as gravity. We know that gravity can really pull. By standing on a tall building you will find proof of this. Drop a tennis ball down from the building's roof and see for yourself how it falls. The distance between the Earth and the ball reduces by some speed. With that being obvious, the distance between the Sun and any planet have to reduce as the planet orbits the Sun each year. Even if it is small, there has to be a visible reduction after four and a half billion years of pulling and tugging! Today after wrestling this problem for the duration of twenty-five years I can say (with a clear mind) I finally know how it works. It does not work!

I was always looking for mistakes on my part. At first, I thought there are a fifth force that I am unaware of because of my slender education, a force the academics can obviously see, but I cannot through obvious lack of education. I thought that my personal ill literacy gave me a blind spot that every non-educated have and was born with. The blind spot cleared only thorough education as education removes it in the way only education in science brings knowledge. I thought the removing process similar to the way washing removes stains and spots from whites; education can remove blind spots through the process of intensive tutoring. All I wished for was some academic to help me remove my blind spot about comets and their behaviour. The comet's behaviour, I could see, was an exaggeration of orbiting patterns applied from our planets orbiting around the sun; in the way, we observe galactica in the sky.

Then finally I came to the point of accepting defeat. It was not I, with the blind spot; it was all the academics brainwashed into a state of having such a blind spot. Science insists on repeatedly ignoring mathematical principles, because Newton had his claim to fame with one single calculation, THAT HE, IN FACT, DISCARDED, BY THROWING IT AWAY.

He made a brief calculation as a young man that saw an apple fall from a tree. Seeing this he jotted down a formula and the chucked it away. His piers and elders picked up the trashed paper with the calculation, and got all excited by the logic implication it had. $F = r^2 / (M_1 M_2)$. The mass of the two objects destroys the radius between the objects. Everyone went ballistic, proclaiming him as an instant genius, the one the world was waiting for after the crucifixion event.

I do not, for one second, deny or dispute the revelation. What I do encourage is place the event into its correct context. It was merely, and simply an apple that fell from its branch to its roots. The apple did not pretend to be a meteorite that fell from the heavens. If it were a meteorite, I am sure, with the man's genius, science would be somewhat different at this stage. However, as a young man, being very impressionable, as all young men are, and with the attention this brought about in the world of science, the matter overshadowed the fact.

I am not disputing Newton; I am disputing the relevance of Newton's scientific breakthrough. It was not two objects of cosmic proportions, colliding in a show of spectacular. It was, after all, only an apple falling from a tree. With this miracle revealed Newton found he was competent to improve on the work of Kepler and if I may dare say this, there must have been some political agenda behind this act and the accepting of it for Kepler was a German and what German can ever teach any Brit. The very same politics are still the order of the day forming international rivalry on all fronts.

Newton, and science, made one enormous blunder, from this stance. They took the radius of a wheel not to have any influence on the wheel. In doing that, they removed the very fact that keeps the universal attachment together. They put two objects in an attaching relevancy and then announced no relevancy. Doing that is breaking the most fundamental mathematical principle.

$$\frac{dJ}{dt} = 0$$

This disputes mathematics. DJ / dt can have any number from eternity to infinity, only excluding one; it cannot be 0. By placing the one in division of the other, you bring in relevance. You cannot then say there is no relevance. By doing such, you proclaim that one of the factors is non-existent.

$$\frac{dJ}{0} = dt \text{ or } \frac{0}{dt} = dJ$$

In both cases, one of the factors then does not exist. Such a claim is incoherent, because you proclaim that a circle has no radius, or a radius has no circle. When calculating a circle, you multiply either the square of the radius by Π, or the quarter of the diameter at a square by Π.

$$\frac{dJ}{dt} = 0 \text{ constitutes a circle and is also therefore } \Pi \times r^2 = \text{CIRCLE}$$

If you remove r it then is $\Pi \times r^2 / r^2 = \text{CIRCLE}$.

You cannot then say $r^2/r^2 = 0$ and therefore $\Pi \times 0 = 0$. That is nonsense. $\Pi r^2/r^2$ will always be $\Pi \times 1$, and that is the eternal circle.

When looking at any rotating object, there has to be a point of no rotation and no rotation means "no rotation", not no existence. No rotation means a factor of 1, not zero. That then is singularity. The eternal Π, the Π that may not have significance but still it is a Π of value.

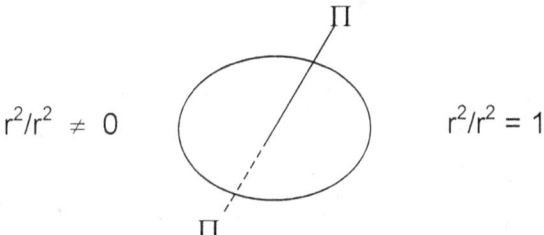

$r^2/r^2 \neq 0$ $r^2/r^2 = 1$

The relativity remains one, eternally zero. Therefore, dJ/dt cannot be zero dJ/dt can be eternal or infinitive or (dJ/dt = 1 but dJ/dt ≠ 0

When explaining this to any child, they can immediately see that. Explain this to any Newtonian High Priest and he may have you removed forcefully from campus. I cannot find one Newtonian, large or small to accept that. By not having a wheel rotate the rotation seize, not the wheel. When the wheel begins to rotate, you cannot state that all things remained as it was. With the wheel in non-rotation the rotation still exists forming the infinite possibility of rotation. Then afterwards the wheel starts to rotate and by the start of rotation the circumstances surrounding the wheel changes. A wheel in rotation is very different from a wheel not rotating and therefore cannot be the same thing. By establishing non-rotation, the wheel becomes the factor of one, and the rotating action becomes zero. The wheel does not disappear. But in the same manner does a wheel in rotation not remain still.

In the cosmos, everything is rotating because nothing ever stands still. Therefore the mean equilibrium, the common factor there is to share, has to be one, eternity, the eternal Π, because all rotating objects has Π in singularity, and sharing singularity, gives every object in space a relation with all other objects in space.

After trying for many years to bring our Brainy Bunch the candle, I concluded that Newtonians are incapable of realizing that mathematical principle as a reality. They maintain they know mathematical principles far better than an ill literate such as I and yet....

The comet rotates the sun, and the Sun by itself has a point of singularity where Π remains without r. The comet, holding the orbit, also has a point of singularity, but since there is space separating the two objects, they cannot share a mean point of singularity, the very point of existing. Since singularity means just that, being single, there cannot be two. The comet and the Sun have a mean point of singularity but the space they occupy divides their common singularity. That is why they orbit in an oval path, a path where the one structure holds on to more space from its point of singularity towards the space it claims. Since they do not claim equal space, BY THE DENSITY they hold, the space will not be in proportion. They do share in the common fact of singularity and singularity cannot   be   two,   because   then   it   will   be "dualarity" or (in case there is no such a word) duplicity where both find the space they occupy, with the space they hold, will be their individual eccentricity from singularity. The two objects are holding eccentric space around their individual but common singularity forming a point of mutual singularity in accordance with the individual singularity both claim space from. That point of singularity is Π the circle without the radius because the singularity removes all forms or values of r, leaving Π to be singularity.

That is why Newton is bullshit, and his $F = G(M_1M_2)/r^2$ is utter nonsense. The moment you say Newton or any of Newton's laws, the Newtonian brain stun. For all the life in me, I could not once find one single Newtonian to see this. If you say Newton is wrong, they spiral down to frenzy, and just mention gravity and they all fall on their knees, cover their eyes in the ground, start praying and you cannot make them say anything other than Newton is correct. Dare say there is no such a thing as gravity and Newton is wrong, they have you in an armed escort patrol, straight to the department of mental disabilities and psycho diseases in preventing you committing acts of extremely dangerous life threatening behaviour to yourself and others.

What is it the Newtonians fail to see? They fail to see the relevance applying. They fail to see that the Universe not only holds the atom, not only comprises of an accumulation of the atom but that the cosmos indeed functions with gravity and all as one massive atom. The Universe is exactly the atom that the Universe formed which then forms the Universe. If an electron is orbiting around an atom, the inside of the atom must be a circle. If the atom was not a circle, it then had to be a cube. The electron cannot rotate around a cube; therefore, the inside of the atom is a circle.

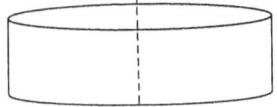 **In a circle, there is a radius that initiates the circle. The calculation of such a circle is $\Pi \times r^2$.**

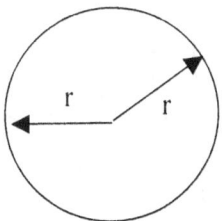

The radius r runs from the circle outwards, from a circle centre point towards Π, the value of the circle. In the centre of the circle, there is a point where the radius starts. It runs outwards from that point in all directions towards the circle Π. Technically, there then has to be a point where r is infinite and not zero, an absolute infinite. However, the circle therefore remains Π. The circle does not disappear; it remains there for all to see. It is only the radius that almost disappears into the infinite, but it does never become zero!

$$\frac{\Pi r^2}{r^2} = \Pi$$

If one removes the radius from the circle, the circle remains, only holding the value of Π. By removing the value of r, Π becomes singularity with no place to be. Singularity is the place where there is no space to be in place. However, Π remains because once r receives the slightest of space Π will find space. Then the circle will grow to Πr^2 and r would determine the space. Without space, there is no r but there is a circle with the value of Π. Singularity is in every single rotating object, be it the proton or the combining effort of all particles in the Universe. That is what light and the photon is. It is concentrated heat that the Sun (or any other generator of electricity) connects heat to singularity where the heat receives either temporary connection to singularity or a small piece of individual singularity.

At first you as the reader may think I am trying to create a mountain from an ant  heap, but in scientific terms the human race is preparing for the start of the cosmic journey. By completion of this book you will realize how _Xepted_ science believe they built science on a solid foundation, and, boy are everybody in for a rude awakening. Compared to the leaning tower of Pizza, science is about to start with the next section of a much bigger building adding many levels and already the view at the bottom where I am looks far worst than the leaning tower does.

If I contacted and argued with one Physics Lector or Professor about Newton, I have been in correspondence with at least a couple of hundred. What prompts me was the comet's orbit. The commit truly fascinated me from my childhood days in the way it defies all the laws of gravity. Since my very young days, I was in search of what I at first believed to be a fifth force. I have raised the argument with just as many people not schooled in the art of physics and received a very different response. The most amazing aspect was the fact that the two groups were that far polarized. The non-physics group reacted astonished, amazed, disbelieving and reserved about my view about comets at first, but with their distrust not withstanding, everyone saw my point. The non-Educated responded in the same manner that I did at first. They argued that I was missing something of vital importance because "why do the wise not see it", was their argument. They always were of the opinion that I was too little educated to understand, while the educated was of the opinion that I was too little educated to understand. Neither party had the same view about my not understanding. The non-Educated understood my argument, but dismissed it on the fact that it was so obvious I missed the rest of the knowledge behind the facts that makes my arguments too difficult to understand while the educated dismissed my argument that I could not see anything they could not see. Education brings the ability, which then made me unable.

In short, they thought I was to stupid in order knowing the rest of the story. Polarized to the non-Academic view was Official Policy Protectors where not one academic could understand my argument. The academic response was as much defending the Newtonian view as it was drawing a blank about my questions. They all seemed as if their ability understanding my view was completely locked behind some wall. The non-Educated, of which I am a member, at least understood what I was saying, but dismissed the simplicity about the argument. In the corner of the Official Policy Protectors, was no response of any kind, but to feverishly defend Newton by raising the dumbest arguments I have ever heard. The arguments, even the most highly educated brought about, seemed motorized and non-responsive. When it seemed their accepting the points I raised with my questions would demise their senses, in defence they put up a block. There is a peculiar sense of numbness in the way they could not understand what I did not understand. The academics showed no signs to indicate that they could even argue my point of view, by responding that I have an argument, and from that launched a responding argument to explain how or where I made my mistake. Their abilities in even understanding always seemed to hide behind a wall of not understanding that someone may not approve of Newton's arguments.

Newton says two pieces of rock will draw each other closer by reducing the distance keeping them apart. That we all can see by merely jumping in the air. No sooner have you lift off than you are back on the ground. That is what Newton said about three hundred and fifty years ago. Even Trying to tell the Official Policy Protectors that Galileo said mass of an object has nothing to do with the falling, seemed to pass the Official Policy Protector's sense's of comprehension by miles. I was told on so many occasions that I did not understand Newton, but there it stopped. No one could explain to me what it was I did not understand about the comet missing the Sun by miles, where it was supposedly to hit the Sun with a dazzling impact. To this point, I cannot get through to them as much as they cannot get through to me. Our understanding is so far apart, we do not share the same planet, and yet after all my arguments and investigation no one, and I repeat: not one could once clearly tell me what it is that I do not understand.

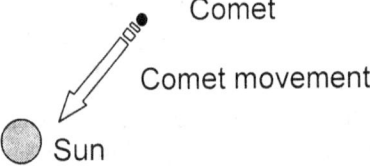
Comet

Comet movement

Sun

You have the Sun and you have a tiny piece of rock covered by water also better known as the comet. There are thousands of them flying around, but never aimless. At first Newton's formula makes pretty much sense. The Sun draws the comet towards the sun, as Newton said it does. The comet responds by speeding towards the sun, also as Newton predicted.

Anyone can see a collision coming ten miles away. The Sun applied gravity, the comet applied gravity, the Sun is far too massive to fly to the comet, so the comet with much less mass does the flying on behalf of both objects. Every person with even the least of knowledge about science knows how the gravity application works.

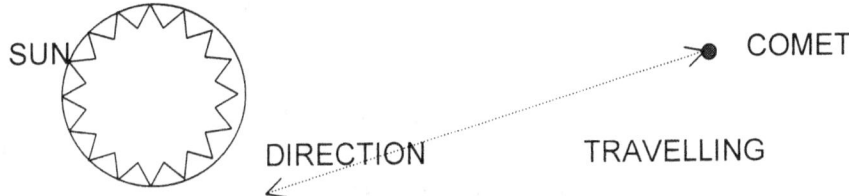

The gravity of the Sun collected the comet from no-one knows where, pulled it through billions of kilometres to the area where the Sun produces the gravity with which it pulls the comet where the comet is to find its last resting place. The mass of the Sun is obviously so large, it could produce gravity that can locate any comet hiding anywhere and collect it as a souvenir. What is there to understand?

The gravity of an object always points directly towards the centre of the object, the very, very middle point. Concluding from the fact that the comet is heading towards the centre of the sun, just as much as the Sun is heading towards the centre of the comet, would not be out of line. The two centre points are heading for a direct collision, the collision becomes more and more unavoidable as the radius reduces by the value of the gravity that is produced the mass in accordance with the gravitation constant. The comet is heading towards the sun, and by not even moving, the Sun is moving towards the comet by attaching the movement the Sun were suppose to have, on the comet. Newton's law proves to be exceptionally correct.

As the sun/comet, radius reduces, the radius separating the mass of the Sun and comet effectively increases the relativity of the mass influence on each other in the form of gravity. The mass of the Sun and the comet increases by the factor of reduction of the radius separating the two objects. That will produce a growing gravity force as the comet / Sun radius becomes smaller. By the time, the radius becomes one the mass will grow on either side by a relevancy of 100, and when the radius becomes infinitely small, the relevance to the mass of both structures will raise a force with eternal power.

At a point, where the comet / Sun apply a force of immeasurable strength, the comet brakes this immeasurable force. Remember the direction of gravity always point to the centre of the object, and that is where the collision is heading. As the objects draw closer, the distance reduces, but in accordance to the relevance the objects also become that much bigger in drawing power. It depends how one considers the relevancy to grow by the approaching nearness diminishing the distance between the objects.

$$\frac{M_s \times M_c}{100} = 1 \times F \quad (r^2 = 100)$$

$$\frac{M_s \times M_c}{50} = 2 \times F \quad (r^2 = 50)$$

$$\frac{M_s \times M_c}{25} = 4 \times F \quad (r^2 = 25)$$

$$\frac{M_s \times M_c}{5} = 20 \times F \quad (r^2 = 5)$$

Then out of the blue, the comet finds the ability to eliminate the eternal powerful force of gravity, and keep at a safe distance around the sun. At this point, Newton goes sour. Nothing Newton predicted is happening. The comet and Sun not only stabilized the force, the force begins to decrease as the radius between the comet and the Sun is on the increase AT THE POINT WHERE THE FORCE IS THE STRONGEST, THE COMET BRAKES FREE AND SLIPS AROUND THE SUN, UNSCATHED.

UP TO THIS POINT I STILL SEE WHAT THE BRAINY BUNCH AND NEWTON SEE, BUT IT IS FROM THIS POINT ONWARDS THAT THERE COMES THE POINT THAT OUR MUTUAL POINT OF CONCENT DIVERTS POINTING OUR MUTUAL POINT ABOUT THE POINT OF AGREEMENT TO THAT OF OPPOSING POINTS WHERE OUR VIEW SEPARATE BOTH HEADING IN OPPOSING DIRECTION ON AN ETERNAL DIVERTING PATH.

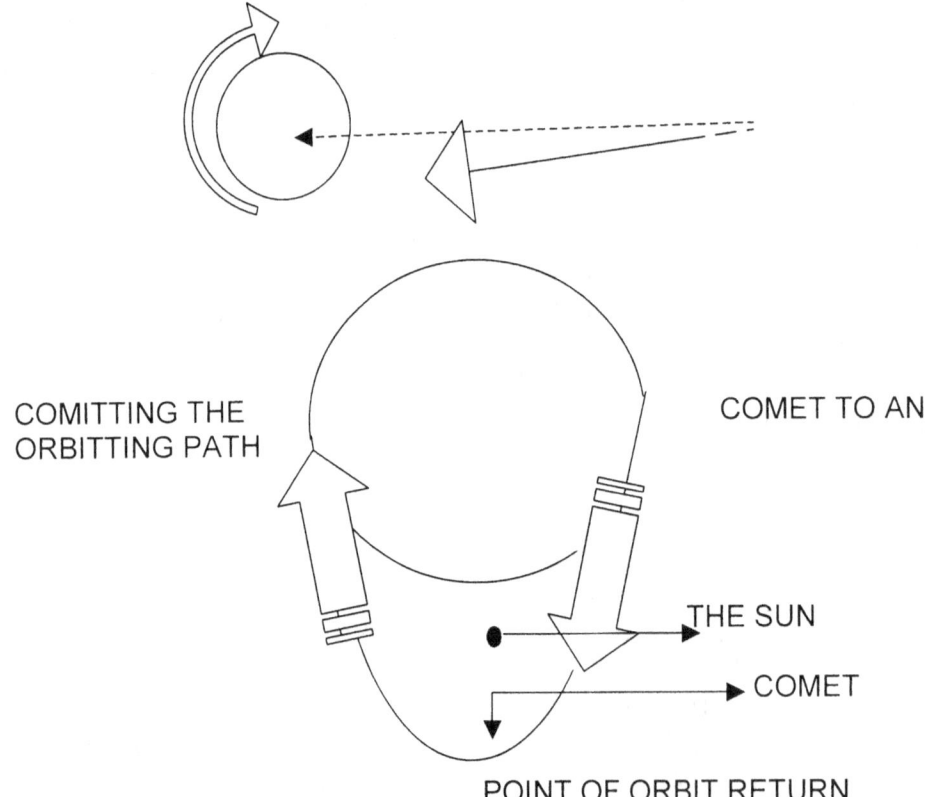

COMITTING THE
ORBITTING PATH

COMET TO AN

THE SUN

COMET

POINT OF ORBIT RETURN

Then, in complete defiance of the Newton Law on gravity, quite the opposite applies. At the point where the radius that is separating the two cosmic objects is at its strongest, it will also bring about that the gravity force is at its weakest. At the point of almost no ability the gravity force suddenly releases enough strength to break resulting in the parting of the two structures. The force now curbs the rebel comet on its way escaping the Sun's gravity for the very last time.

At the point where the force was the greatest, the comet overcame the force, but where the force was the weakest, the force overcame the comet's rebellion.
The correctness of my argument is no longer the issue. It was twenty-five years ago, when I still held the impression that I was missing some point here. I do not state this phenomenon any longer in the hope of bringing across some flaw in my understanding. The flaw in my argument is not there because the flaw is science as a whole.

I could never understand the reason why "the ordinary", like me and others with my development level, can see what I can see, yet academics that has more brainpower in their heads, than I have life in my body, were unable to see such an obvious conclusion. You; those Official Policy Protectors are my superiors in every sense a human can have, with the brainpower to break a wall, and yet you cannot see how far the tower of Pizza is leaning over.

I make the point to help you, the reader, to judge yourself. If you are able to see the validity in my argument, you are not brain dead. Education has not yet bashed your thinking ability out of your scull. However, if a cloak of not understanding roll over your brain, and a numbness sets in on your ability to reason about this phenomenon, beware, you are a Newtonian. Newtonians should read this book very slowly because the effort you are about to launch, may be the most painful you shall ever experience throughout your academic career. You are going to suffer from reconditioning and Newtonian withdrawal, not that dissimilar to that of an addict in rehabilitation. You are going to reject me, hate me, despise me, loath me as you never felt about anybody else. If you think I am sarcastic, I am not. You will reach a point where you will abandon the reading of the book. You have my sincere sympathy and with all the soothing it may bring, know that you are not the first I saw getting such painful Newtonian withdrawal in rejection of Newtonian doping.

Once more, this phenomenon should not occur with Newton's presumptions about gravity. These bodies will and must collide and destruct, without a doubt. When the formula $F = \dfrac{M_1 M_2}{r^2} G$ applies, there should not be any force which is able to keep them apart especially when r reduces to almost infinity compared to what it is at maximum. However, they do exist and what is more, they maintain a certain distance apart.

With the "force" of "gravity" "pulling" the stars closer using the accumulative mass of the stars and multiplying that value with both objects by the mass component, this will reduce the radius r^2 progressively until r^2 reduces to zero. Seen from this view, it is little wonder that the significance of this was lost in the notion that this is yet another "mystery" of the Universe. The scientists of the day (and the past) lost the importance, which this holds for us as Earthly dwellers.

A most surprising aspect of this is that it is not that an unfamiliar or rare phenomenon. However, any answer to this would clash with Newton's presumptions, and before the scientists allow that to happen, they would much rather ignore what is obvious. However, what is the obvious?

Space is equal to the motion thereof $a^3 = T^2k$

The cosmos work in relevancies to singularity. The earth has the value of $4\Pi^2$.

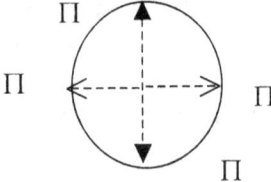

From the earth's perspective matter is 7 and time is 4 therefore
$4 \times (\Pi = 7) = 28$.

The moon holds its singularity in relation to the Earth in singularity. Because the moon is in a Roche limit (near enough) the proton value the moon accepts is $\Pi^2/4$. The neutron position the moon holds, relating to the Earth is 10 and the moon has an own point of singularity, forming the electron edge of the Earth.

The Earth is $4\Pi^2 = 28$ days to rotate to one moon cycle of 1. The relevancy of the Earth, taken from its point of singularity is 7 (matter in relation to space will always be 7 to the space value of 10).

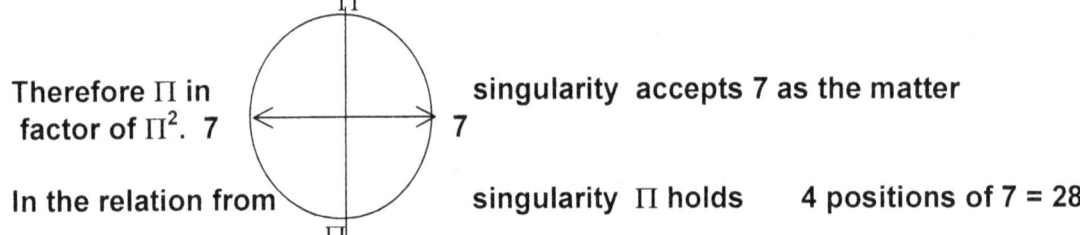

Therefore Π in factor of Π^2. 7   singularity accepts 7 as the matter

In the relation from   singularity Π holds   4 positions of 7 = 28

OBVIOUSLY ALL RELEVANCIES HOLD TWO SIDES TO EQUAL PROMINENCE. THE RELEVENCY THE MOON HOLDS TO THE EARTH WILL BE DIFFERENT TO THAT WHICH THE EARTH PLACES ON THE MOON.

The Earth holds singularity and the Earth's singularity holds the moon's position in singularity. To the moon's relativity, it holds value to space from the Earth's singularity of $1 \times (\Pi/2)^2$ (the relation between the Earth and the moon) x 10 (the fact that the moon is within the space of the Earth, as the moon has an individual point holding singularity (Π). Therefore the moon holds $(1 \times (\Pi/2)^2 \times 10) + \Pi =$ 27,8 days as one and the Earth holds to moons single day of individual singularity at one to 28 days.

Stars will never collide because stars can never collide.

The only absence in the cosmos is zero and without zero there cannot be an end to eternity but only an everlasting cycle that breaks to start one more eternity now and then. With the cosmos created minute by minute from no space within the cosmic centre, the cosmos is ruled from a position with every thing but nothing is but where we know God must be. By accepting singularity and the rule there of brings into the cosmos things physics are unable to explain, mathematics are unable to calculate and man is unable to dismiss. If you accept physics you have to accept God because you cannot except one proving singularity without the other coming through singularity.

If it is that simple then why is it complicated.

BEST WISHES,

PETRUS. (PEET) S. J. SCHUTTE

There is more about this in other books with the titles:

This was the prologue letter announcing
MATTER'S TIME IN SPACE: THE THESIS
ISBN   0-9584410-8-1

FROM THE ORIGINAL AFRIKAANS: "MATERIE SE TYD IN RUIMTE" I. S. B. N.   0-620-27041-1
WRITTEN BY PEET SCHUTTE
© KOSMOLOGIESE EN ASTRONOMIESE TEGNIKA

THIS WAS

An open letter
TO SELECTED ACADEMICS
3

ISBN 0-9584410-9-X

NOW TO FOLLW THIS THEN READ

An open letter
TO SELECTED ACADEMICS
4